LITHIC TECHNOLOGY: MEASURES OF PRODUCTION, USE, AND CURATION

The life history of stone tools is intimately linked to tool production, use, and maintenance. These are important processes in the organization of lithic technology, or the manner in which lithic technology is embedded within human organizational strategies of land use and subsistence practices. This volume brings together essays that measure the life history of stone tools relative to retouch values, raw material constraints, and evolutionary processes. Collectively, they explore the association of technological organization with facets of tool form such as reduction sequences, tool production effort, artifact curation processes, and retouch measurement. Data sets cover a broad geographic and temporal span, including examples from France during the Paleolithic, the Near East during the Neolithic, and other regions such as Mongolia, Australia, and Italy. North American examples are derived from Paleoindian times to historic period aboriginal populations throughout the United States and Canada.

William Andrefsky, Jr., is a professor of anthropology at Washington State University. He is the author of more than 100 articles and books, including *Lithics: Macroscopic Approaches to Analysis*.

LITHIC TECHNOLOGY: MEASURES OF PRODUCTION, USE, AND CURATION

Edited by
WILLIAM ANDREFSKY, JR.
Washington State University

CAMBRIDGE UNIVERSITY PRESS
Cambridge, New York, Melbourne, Madrid, Cape Town, Singapore, São Paulo, Delhi

Cambridge University Press
32 Avenue of the Americas, New York, NY 10013-2473, USA

www.cambridge.org
Information on this title: www.cambridge.org/9780521888271

© Cambridge University Press 2008

This publication is in copyright. Subject to statutory exception
and to the provisions of relevant collective licensing agreements,
no reproduction of any part may take place without
the written permission of Cambridge University Press.

First published 2008

Printed in the United States of America

A catalog record for this publication is available from the British Library.

Library of Congress Cataloging in Publication Data

Lithic technology: Measures of production, use, and curation / edited by William Andrefsky, Jr.
    p. cm.
Includes bibliographical references and index.
ISBN 978-0-521-88827-1 (hardback)
1. Stone implements – History.   2. Tools, Prehistoric.
3. Stone implements – Analysis.   4. Tools, Prehistoric – Analysis.
5. Technology – Social aspects – History.   6. Land use – History.
7. Subsistence economy – History.   8. Social archaeology.
I. Andrefsky, William, 1955–   II. Title.
CC79.5.S76L58   2008
930.1028–dc22   2008001390

ISBN 978-0-521-88827-1 hardback

Cambridge University Press has no responsibility for
the persistence or accuracy of URLs for external or
third-party Internet Web sites referred to in this publication
and does not guarantee that any content on such
Web sites is, or will remain, accurate or appropriate.

In memory of Yukiko Akama Andrefsky

CONTENTS

Preface and Acknowledgments   *page* xi

List of Contributors   xiii

PART I: INTRODUCTION, BACKGROUND, AND REVIEW

1 An Introduction to Stone Tool Life History and Technological Organization   3
   WILLIAM ANDREFSKY, JR.

2 Lithic Reduction, Its Measurement, and Implications: Comments on the Volume   23
   MICHAEL J. SHOTT AND MARGARET C. NELSON

PART II: PRODUCTION, REDUCTION, AND RETOUCH

3 Comparing and Synthesizing Unifacial Stone Tool Reduction Indices   49
   METIN I. EREN AND MARY E. PRENDERGAST

4 Exploring Retouch on Bifaces: Unpacking Production, Resharpening, and Hammer Type   86
   JENNIFER WILSON AND WILLIAM ANDREFSKY, JR.

5 The Construction of Morphological Diversity: A Study of Mousterian Implement Retouching at Combe Grenal   106
   PETER HISCOCK AND CHRIS CLARKSON

6 Reduction and Retouch as Independent Measures
   of Intensity   136
   BROOKE BLADES

7 Perforation with Stone Tools and Retouch
   Intensity: A Neolithic Case Study   150
   COLIN PATRICK QUINN, WILLIAM ANDREFSKY, JR., IAN
   KUIJT, AND BILL FINLAYSON

8 Exploring the Dart and Arrow Dilemma:
   Retouch Indices as Functional Determinants   175
   CHERYL HARPER AND WILLIAM ANDREFSKY, JR.

PART III: NEW PERSPECTIVES ON LITHIC RAW MATERIAL
AND TECHNOLOGY

9 Projectile Point Provisioning Strategies and
   Human Land Use   195
   WILLIAM ANDREFSKY, JR.

10 The Role of Lithic Raw Material Availability and
   Quality in Determining Tool Kit Size, Tool
   Function, and Degree of Retouch: A Case Study
   from Skink Rockshelter (46NI445), West Virginia   216
   DOUGLAS H. MACDONALD

11 Raw Material and Retouched Flakes   233
   ANDREW P. BRADBURY, PHILIP J. CARR, AND
   D. RANDALL COOPER

PART IV: EVOLUTIONARY APPROACHES TO LITHIC
TECHNOLOGIES

12 Lithic Technological Organization in an
   Evolutionary Framework: Examples from North
   America's Pacific Northwest Region   257
   ANNA MARIE PRENTISS AND DAVID S. CLARKE

13 Changing Reduction Intensity, Settlement, and
   Subsistence in Wardaman Country, Northern
   Australia   286
   CHRIS CLARKSON

14 Lithic Core Reduction Techniques: Modeling
  Expected Diversity   317
  NATHAN B. GOODALE, IAN KUIJT, SHANE J. MACFARLAN,
  CURTIS OSTERHOUDT, AND BILL FINLAYSON

Index   337

PREFACE AND ACKNOWLEDGMENTS

In 1968 George Frison introduced the notion of artifact transformations as a result of use and resharpening. This "Frison Effect," as it has come to be called, on stone tools can be viewed as the life histories of individual tools. Such life histories are intimately linked to tool production, use, and maintenance. This collection of chapters grew from presentations at a symposium entitled "Artifact Life-Cycle and the Organization of Lithic Technologies" that took place at the 71st Annual Meeting of the Society for American Archaeology in 2006. The focus of that symposium and this volume is upon the relationship between the manner in which humans organize their lithic technology and the life history of lithic tools.

Researchers interested in lithic technological organization realize the importance of artifact life histories in understanding the intricacies of tool form and shape as they relate to production strategies for those tools. In an effort to better understand those relationships, lithic analysts (including contributors to this volume) have explored lithic reduction sequences, chaîne opératoire, tool curation, tool production effects, retouch measurements, and the role of lithic raw material as these relate to lithic technological organization and stone tool life history. A great deal of imaginative and compelling research has occurred since the Frison Effect was first recognized, and this collection of papers provides a fresh new look at all of these topics from both a methodological and a theoretical perspective.

I would like to thank all of the participants of the original symposium for their participation. For various reasons, not all symposium participant chapters are included in this volume. Also, as chapters were reviewed, revised, and adjusted, some chapters gained authors and some authors contributed different written research. This blend of chapters captures opinions and ideas about lithic technology from some of the most respected scholars in the field today, but it also includes research from many young new researchers who will one day guide the field of lithic technology. It was a joy to bring this group together under a single cover. My best wishes go to all volume contributors and symposium participants.

I must also thank the team of editors and production staff from Cambridge University Press and their associated collaborators. In particular I thank Publishing Director Beatrice Rehl and her editorial assistant Tracy Steel for managing this book project. The production manager for Aptara, Inc., Maggie Meitzler, helped me navigate through the technical challenges of today's high-tech world of publishing. William Stoddard was a fabulous copy editor. Lastly, I thank the Cambridge University Press peer reviewers for making important comments on the original draft.

LIST OF CONTRIBUTORS

WILLIAM ANDREFSKY, JR., is professor and chair of anthropology at Washington State University. His interests include hunter–gatherer land use and technological strategies, archaeological ethics, and the Neolithic transition. He is currently the president of the Register of Professional Archaeologists. Some of his research has been published by *Journal of Archaeological Science, American Antiquity, Geoarchaeology, Journal of Archaeological Research, Lithic Technology, Journal of Middle Atlantic Archaeology, North American Archaeologist, Journal of Field Archaeology*, and Cambridge University Press, University of Utah Press, Elsevier, Blackwell Publishing, AltaMira Press, Westview Press, and Cambridge Scholars Publishing.

BROOKE BLADES received his doctorate in anthropology with a specialization in the European Upper Paleolithic from New York University in 1997. He has undertaken archaeological excavations or research in France, the Czech Republic, Luxembourg, Northern Ireland, and eastern North America. His research interests focus on lithic material procurement and reduction in the European Paleolithic and in hunter–gatherer societies in eastern North America. He was the author of *Aurignacian Lithic Economy: Ecological Perspectives from Southwestern France* (Kluwer Academic/Plenum 2001) and is the co-editor, with Brian Adams, of *Lithic Materials and Paleolithic Societies* (Blackwell forthcoming).

ANDREW P. BRADBURY received his M.A. from the University of Tennessee, Knoxville, in 1994. He is currently a principal investigator and lithic analyst with Cultural Resource Analysts, Inc., in Lexington, Kentucky. His research interests include southeastern U.S. prehistory, lithic technology, and evolutionary theory.

PHILIP J. CARR is Associate Director of the Center for Archaeological Studies and specializes in the prehistory of the southeastern United States and lithic analysis. Since joining the University of South Alabama faculty, he has engaged in the study of the rich cultural heritage of the Gulf Coast. Recent publications include the co-edited volume with Dr. Jon Gibson, *Signs of Power: The Rise of Cultural Complexity in the Southeast*, and he edited the volume *The Organization of North American Chipped Stone Tool Technologies*.

DAVID S. CLARKE earned his B.A. in anthropology from Mercyhurst College and his M.A. from the University of Montana. Currently he is an archaeologist for the Delaware Department of Transportation. His research interests include lithic technology, evolutionary theory, Pacific Northwest hunter–gatherers, and the peopling of the Americas. He has contributed to articles in *American Antiquity* and *Journal of World Prehistory*.

CHRIS CLARKSON received his Ph.D. from the Australian National University in 2004 on long-term technological change in the Northern Territory of Australia. He has since held positions at the University of Cambridge and the Australian National University. He is currently employed in the School of Social Science at the University of Queensland, where he continues research into the lithic technology of Australia, India, France, and Africa. He is the author of *Lithics in the Land of the Lightning Brothers* (ANU E-Press) and the editor of *Lithics Down Under* (BAR).

D. RANDALL COOPER is a staff archaeologist at Cultural Resource Analysts, Inc., in Lexington, Kentucky. He has a B.A. in anthropology from the University of Tennessee, where he developed an interest in lithic technology beginning in 1983. He has since analyzed lithic assemblages from North Dakota, coastal Alaska, the Tahoe National Forest, the Mojave Desert, and several regions of Kentucky. Other interests include relational database design and Appalachian folk culture.

METIN I. EREN holds degrees in anthropology from Harvard University (A.B.) and Southern Methodist University (M.A.) and is currently pursuing an M.A. degree in experimental archaeology from the University of Exeter. He has participated in fieldwork in Ohio, Colorado, Turkey, the Georgian Republic, Tanzania, and China. His recent awards include a National Science Foundation Graduate Research Fellowship and the 2006 Society for American Archaeology Student Paper Award (with Mary Prendergast).

BILL FINLAYSON is the Director of the Council for British Research in the Levant and a visiting professor at the Department of Archaeology, Reading University. His undergraduate and postgraduate degrees were in prehistoric archaeology at the University of Edinburgh. He is a director of two major PPNA projects in Jordan, at Dhra' and in Wadi Faynan, and editor and author of the recent volume *The Early Prehistory of Wadi Faynan, Southern Jordan*.

NATHAN B. GOODALE is currently a visiting instructor in anthropology at Hamilton College. He received his B.A. from Western State College and his M.A. from the University of Montana and is currently a Ph.D. candidate at Washington State University. He has published articles in *American Antiquity*, *Paléorient*, *Archaeology in Montana*, *JONA*, and *Levant*, as well as a chapter in the edited volume *Complex Hunter-Gatherers*. He is interested in using evolutionary theory to explain technological invention and its relationship to population dynamics during the transition to agriculture.

CHERYL HARPER is a graduate student in anthropology at Washington State University and an archaeologist with the U.S. Forest Service. Her research interests focus on understanding landscape use by Archaic populations in the American Southwest.

PETER HISCOCK is a reader at the Australian National University. His work has concentrated on ancient technology, both in Paleolithic Europe and in Australia. His latest book is *The Archaeology of Ancient Australia* (Routledge).

IAN KUIJT specializes in the archaeology of the Near East, Ireland, and Western North America. Completing his doctoral work at Harvard University, he is the editor of *Life in Neolithic Farming Communities:*

Social Organization, Identity, and Differentiation (2000) and the co-editor of *Complex Hunter–Gatherers: Evolution and Organization of Prehistoric Communities on the Plateau of Northwestern North America* (2004). His research has been supported by the National Endowments for the Humanities, the National Science Foundation, the Social Sciences and Humanities Research Council of Canada, and the British Academy.

DOUGLAS H. MACDONALD is an assistant professor at the University of Montana, but worked at GAI Consultants in Pittsburgh during the Skink Rockshelter project. Influenced by his graduate work at Washington State University, he studies lithic technological organization at prehistoric sites in the North American Plains, Rockies, and mid-Atlantic.

SHANE J. MACFARLAN is a Ph.D. student in evolutionary anthropology at Washington State University. He has an M.A. in museum science (Texas Tech University 2003) and a dual B.A. in anthropology and history (University of Pittsburgh 1998). He is currently performing ethnographic research on the island of Dominica and archaeological research in Baja California Sur. His interests are human behavioral ecology, economic reasoning, and cooperation.

MARGARET C. NELSON is a professor in the School of Human Evolution and Social Change and Vice Dean of Barrett, the Honors College at Arizona State University. Her research examines technology and land use in small-scale Puebloan societies in the U.S. Southwest. Her recent book, *Mimbres During the Twelfth Century: Abandonment, Continuity, and Reorganization*, examines the continuities and changes in social and subsistence organization in the Mimbres region of southwest New Mexico from A.D. 1000 to 1250. She has published in numerous journals and books on the organization of prehistoric stone technology.

CURTIS OSTERHOUDT received a dual B.A. in mathematics and physics at Western State College Colorado and a Ph.D. in physics at Washington State University. He is currently a postdoctoral researcher at Los Alamos National Laboratory, focusing on physical acoustics and its uses for measuring various qualities and features of systems. Other academic interests include statistical mechanics and its applications to disparate fields, including archaeology.

MARY E. PRENDERGAST is a doctoral candidate in anthropology at Harvard University, where she has been studying terminal Pleistocene and Holocene hunter–gatherers and transitions to food production. She conducts fieldwork and faunal analyses in China, Kenya, and Tanzania and has published research on Middle Stone Age through Pastoral Neolithic sites in East Africa.

ANNA MARIE PRENTISS earned her B.A. and M.A. degrees in anthropology at the University of South Florida. She completed her Ph.D. in archaeology at Simon Fraser University and is currently associate professor of anthropology at the University of Montana. Her research interests include Pacific Northwest hunter–gatherers, evolutionary theory, and lithic technology. She edited (with Ian Kuijt) *Complex Hunter-Gatherers: Evolution and Organization of Prehistoric Communities on the Plateau of Northwestern North America* (University of Utah Press) and has published articles in a wide range of journals, including *American Antiquity, Current Anthropology, Journal of Anthropological Archaeology*, and *Journal of World Prehistory*.

COLIN PATRICK QUINN is a graduate student at the University of Michigan and has acquired degrees from the University of Notre Dame (B.A.) and Washington State University (M.A.). His research interests include stone tool production and use, experimental techniques, personal adornment production and use, and costly signaling theory in small-scale and middle-range societies in the Near East, western Europe, and northwestern North America.

MICHAEL J. SHOTT teaches archaeology at the University of Akron. His interests include hunter–gatherers, how the archaeological record formed, and lithic analysis. He has written or edited six books and about 80 articles. He works chiefly in the North American Midwest and Great Basin but also has worked in Mexico and Argentina.

JENNIFER WILSON received her M.A. from Washington State University in May 2007. Her thesis focused on technological organization and measuring tool curation from a chert quarry assemblage in the northern Great Basin. Currently, she is employed with Archaeological and Historical Services at Eastern Washington University, and her research focuses on hunter–gatherer organization and lithic technology in the Columbia Plateau and the northern Great Basin.

PART ONE

INTRODUCTION, BACKGROUND, AND REVIEW

I WILLIAM ANDREFSKY, JR.

AN INTRODUCTION TO STONE TOOL LIFE HISTORY AND TECHNOLOGICAL ORGANIZATION

It is relatively easy for most people to understand differences in life histories with organisms such as dragonflies and mollusks, because these organisms undergo dramatic morphological transformations during their life histories. However, if we did not know that *glochidia* living in the gills of fish were the larval phase of mussels, we might classify them as totally different organisms because of their different appearance and different habitat. However, biologists have followed the life histories of these and countless other organisms and have demonstrated the metamorphoses that have taken place. Archaeologists working as taxonomists do not have the benefit of observing the life histories of stone tools. We find and record artifacts in a static state. However, as a result of replication experiments, renewed ethnographic observations, and detailed lithic analytical strategies, it has become apparent to researchers that lithic tools often undergo a series of transformations from the time they are produced or drafted into service until the time they are ultimately discarded. Such transformations relate to all manner of social and economic situations of the tool users. Tools are sharpened when they become dull. They are reconfigured or discarded when they are broken. They are modified to suit a certain task in a certain context. Their uses are often anticipated and they are produced in anticipation of those uses. These and countless other examples of tool transformations can be characterized as part of the life histories of lithic tools.

Lithic tools are dynamic in their morphological configurations because of these life history transformations.

A flake blank originally used as a meat-slicing knife with an acute edge angle may be transformed due to dulling and edge resharpening into a tool that contains a serrated edge used for sawing. This tool can be intentionally chipped and shaped into a projectile point and mounted into a shaft for use as a dart. A single specimen can undergo one or more such transformations during its life history. Such life history transformations not only change the tool form but may also change the tool function, and both formal and functional changes are often associated with forager land-use practices. In this manner, the life histories of stone tools are intimately linked to the organization of stone tool technology.

Lithic technological organization has been defined in a number of different and yet similar ways (Andrefsky 2006; Binford 1973, 1977; Kelly 1988; Koldehoff 1987; Nelson 1991; Shott 1986; Torrence 1983). In all cases, it refers to the manner in which humans organize themselves with regard to lithic technology. Because foraging societies are most often associated with lithic technology, most studies of lithic technological organization deal with forager adaptive strategies. In this context, the manner in which lithic tools and debitage are designed, produced, recycled, and discarded is intimately linked to forager land-use practices, which in turn are often associated with environmental and resource exploitation strategies. I consider lithic technological organization a strategy that deals with the way lithic technology (the acquisition, production, maintenance, reconfiguration, and discard of stone tools) is embedded within the daily lives and adaptive choices and decisions of tool makers and users.

An important component of lithic technological organization concerns the life histories of stone tools. Below I review some of the ways that technological characteristics of lithic artifacts relate to their life histories. I then provide a brief review of the assembled papers in this volume, which address many of the reviewed concepts, such as measuring retouch, recognizing curation, using lithic raw material variability, and understanding tool transformations.

REDUCTION AND REDUCTION SEQUENCES

The life histories of stone tools are often associated with the reduction of stone tools. Because stone tools are produced by reduction or the removal of stone from a nucleus or objective piece, it is easy to equate stone tool life histories to the unidirectional reduction of stone – the farther an objective piece is reduced, the farther the specimen is in its life history. Some of the early thinking in this area can be attributed to William Henry Holmes (1894), who coined the term *lithic reduction sequences*. Stone tool reduction sequences have traditionally been associated with stone tool production phases, stages, or continua. This is particularly true of North American bifacial technology, where the trajectory of reduction begins with raw material acquisition and ends with notching, fluting, or final sharpening of the tip and edges (Callahan 1979; Shott 1993: 94–6; Whittaker 1994: 153–61). Investigators not only examine lithic tools for evidence of reduction sequences but also focus on detached pieces (debitage and debris) in an effort to gain insight into tool production activities (Ahler 1989; Amick and Mauldin 1989; Andrefsky 2001; Bradbury and Carr 1999; Carr and Bradbury 2001; Kalin 1981; Odell 1989; Pecora 2001; Rasic and Andrefsky 2001). Other studies of lithic debitage have examined the source of variation in debitage characteristics in an effort to link those characteristics to broader issues of technological practices. For instance, a series of studies have examined the relationship of debitage striking platform angles to original flake size and production technology (Cochrane 2003; Davis and Shea 1998; Dibble 1997; Dibble and Pelcin 1995; Pelcin 1997; Shott et al. 2000).

The literature on lithic reduction sequences as it relates to technological organization is sometimes complicated by confusing terminology. When talking about the manufacture of "tools" using pressure or percussion flaking methods, I use the term "production." I use the term "reduction" when talking about the removal of detached pieces from cores. In this sense, "reduction" refers to the process of flake removal for the acquisition of detached pieces and "production" refers to the process of flake removal for the purpose of making, shaping, or resharpening a tool. So cores are "reduced" and tools are "produced." I use the term "retouch" as a generic descriptor for removing detached pieces from an objective piece. Essentially, retouch is the process by which flintknappers produce tools and reduce cores.

The recent literature dealing with lithic reduction sequences is not far removed from the concept of chaîne opératoire. Some researchers claim the chaîne opératoire concept "comprises a much wider range of processes than do the English terms reduction sequences or even lithic tool production" (Simek, 1994:119; see also Audouze 1999; Eren and Prendergast, this volume). Inizan and colleagues suggest that chaîne opératoire includes the processes from the procurement of raw materials, through the stages of manufacture and use, and including discard (Inizan et al. 1992; Sellet 1993). Other archaeologists challenge the notion that chaîne opératoire is more encompassing than the concept of "reduction sequences" (Shott 2003). This chapter is not the appropriate venue to explore this discussion. My general opinion is that both concepts are substantially the same thing, and that both are inclusive of the larger issues of procurement, manufacture, use, maintenance, and discard. Furthermore, both concepts are embedded within the larger issues of human land use related to environmental, social, and historical contexts (Andrefsky, this volume; Bleed 1986; Clarkson 2002; Eren et al. 2005; Hiscock and Attenbrow 2003; Hiscock and Clarkson, this volume; Nowell et al. 2003; Pecora 2001; Wilson and Andrefsky, this volume). It is for these reasons that regardless of the terms used, the production of tools and the reduction of cores are central to an understanding of lithic technological organization. Lithic retouch, whether it relates to tool production or maintenance, or the acquisition of blades and flakes, has much to do with the contexts of human land use, and for this reason, understanding reduction sequences and chaîne opératoire allow us to better understand lithic technological organization and the life histories of stone tools.

As lithic analysts begin thinking about the place of stone tools within the framework of life histories, we envision tools in multiple contexts. Stone tools are produced, used, maintained, reconfigured, discarded, reused, discarded, and ultimately discovered by archaeologists and others. These multiple contexts expand our understanding of stone tool reduction from simply the production contexts of tools to a more inclusive understanding of maintenance contexts. Retouch of stone tools not only includes the production stages of tool manufacture, but also includes the chipping of tool edges after use to resharpen or reconfigure the specimen (Brantingham and Kuhn 2001; Flenniken and Raymond 1986; Hiscock and Attenbrow 2003; Morrow 1997; Nowell et al. 2003; Tomka 2001). Recent investigations have shown that some stone tool types such as flake knives have no separate production and use phases. Such tools are retouched as needed, resulting in morphological transformation during the process of use and resharpening (Clarkson 2002; Dibble 1987; Rolland and Dibble 1990). Other stone tool types such as projectile points have very discrete production and maintenance phases; they are not used or maintained until after they have gone through a formal production process (Andrefsky 2006; Hoffman 1985; Shott and Ballenger 2007). Even though stone tools such as projectile points undergo morphological transformation in both the production and use phases as a result of retouch, the production phase is not a good measure of tool use. Such differences in tool types have important implications for measuring reduction as a proxy for curation.

ARTIFACT RETOUCH AND CURATION

In the 1970s Binford (1973, 1979) introduced the curation concept to hunter–gatherer archaeology. Shortly afterward archaeologists began exploring, discussing, and dissecting this concept in some detail (Bamforth 1986; Bleed 1986; Chatters 1987; Close 1996; Gramly 1980; Nash 1996; Odell 1996). One reason the curation concept generated so much discussion was Binford's complicated way of using the term. In my opinion, it was complicated because he did not provide a strict definition and instead used the term in association with a number of interesting ideas. For instance, Binford discussed curation in the context of artifacts being transported from one location to another in anticipation of tasks to be completed at the new location (1973). As a result, some archaeologists associated curation with transported tools (Bettinger 1987; Gramly 1980; Nelson 1991). Binford also linked curation to efficiency of tool use. Bamforth's (1986) paper on technological efficiency and tool curation expanded this concept to include five aspects of tool curation: (1) production in advance of use, (2) implement design for multiple uses, (3) transport of tools to multiple locations, (4) maintenance of tools, and (5) recycling of tools. The notion of tool production effort was added to the definition in the form of complex tools, or tools with haft elements or complex flaking patterns (Andrefsky 1994a; Hayden 1975; Parry and Kelly 1987). Nash's review of the curation concept concludes that the term is ill-defined but already embedded in the literature. He says (Nash 1996:96), "In the absence of such standardization, we should drop the term from the archaeological literature all together." Odell (1996: 75) concludes that for the term "curation" to be useful, "the most parsimonious usage would retain those elements associated with mobility and settlement, and discard the ones associated with tool conservation."

Some of the early lithic analytical practitioners of the curation concept contrasted "curated tools" with "expedient tools" (Andrefsky 1991; Bamforth 1986; Kelly 1988; Parry and Kelly 1987). "Curated" tools were often recognized as having extensive retouch and "expedient" tools were recognized as having very little retouch. This simple way of viewing retouch on tools was sometimes superposed on Binford's model of hunter–gatherer land use, with foragers being residentially mobile and collectors being residentially sedentary or semisedentary. "Curated" tools were often associated with foragers and "expedient" tools were often associated with collectors. This kind of stone tool classification is still popular in the literature. However, most lithic researchers now realize that this one-to-one relationship is not realistic and stone tool configuration is influenced by many other factors, such as raw material availability, shape, and functional considerations (Andrefsky 1994a; 1994b; Bamforth 1991; Bradbury and Franklin 2000; Kuhn 1991; Tomka 2001; Wallace and Shea 2006).

Many early studies of stone tool curation viewed curation as a type of tool. I find the curation concept workable in the context of technological organization if it is recognized as a process associated with tool use rather than a tool type. I refer to it as a process reflecting a tool's actual use relative to its maximum potential use (Andrefsky 2006, this volume; Shott 1996; Shott and Sillitoe 2005). Importantly, then, curation is a process related to tool use. Curation is not a tool type. There are no curated tools, but only tools in various phases of being curated from very low use relative to maximum potential use to very high use relative to maximum potential use. In this way, curation can be measured from low to high, allowing investigators to plug curation into models of human organizational strategies and into the life histories of tools.

For these reasons, it is important to understand that some tools have a production phase discrete from the maintenance phase. Because retouch occurs in both production and maintenance phases, retouch in and of itself may not be a good proxy for curation. Tool curation deals with tool use. Some forms of tools are retouched extensively and never used. As such, they have not undergone curation, even though they are extensively retouched (Andrefsky 2006; Hoffman 1985). This suggests that measures of retouch and reduction must be intimately associated with characteristics such as artifact type and potential artifact function, and even with extramural agencies such as lithic raw material abundance and quality. The collection of papers in this volume demonstrate the importance of these various contextual influences on retouch measures and show how retouch relates to processes such as curation, human land use patterns, and lithic tool functional differences.

HUMAN ORGANIZATION AND LITHIC RAW MATERIAL SELECTION

Another factor that influences lithic technological organization and the life histories of stone tools is lithic raw material availability, abundance, form, and quality. These aspects of lithic raw materials play an important role in the length of time and detail with which a tool is prepared, used, and maintained. Anthropologists studying tool makers and users long understood the importance of lithic raw material availability and abundance to those tool makers and users (Gould 1980, 1985; Gould and Saggers 1985; O'Connell 1977; Weedman 2006). The distribution and availability of lithic raw materials are undeniably important in stipulating how humans manufactured, used, and reconfigured stone tools. Because lithic raw materials can often be provenanced, they provide robust information about the circulation of stone, and by inference, the circulation of people across the landscape. This fact alone makes lithic raw material an important resource for gaining insight into human land use and mobility patterns and relating those to lithic technology. Recent archaeological research has directly linked lithic raw materials to tool production and core reduction technologies (Brantingham and Kuhn 2001; Roth and Dibble 1998) to artifact functional effectiveness (Brantingham et al. 2000; Hofman 1985; Sievert and Wise 2001), to retouch intensity on tools (Andrefsky, this volume; Bradbury et al., this volume; Kuhn 1991, 1992; MacDonald, this volume), and to aspects of risk management (Baales 2001; Braun 2005).

Information gained from lithic raw materials regarding source location, shape, size, durability, and abundance has increased our understanding of stone tool technological organization. Important in this growing knowledge is the fact that lithic raw materials do not play a deterministic role in human organizational decisions, but rather act as one of many factors in how tool makers and users decide to produce, maintain, and discard stone tools.

DISCUSSION

Shott and Nelson (this volume) provide a detailed review of the collection of papers in this volume. I will not repeat their insights here, but instead discuss the multiple linkages among the different papers that bring this volume together into a new synthesis of artifact life histories and lithic technological organization. However, first I must emphasize that the collection of papers covers a broad geographic and temporal span of aboriginal tool maker data. Three papers cover examples from French data sets spanning the Paleolithic. Two papers deal with Near Eastern data during the Neolithic. North American examples are derived from Paleoindian times to historic period aboriginal populations, and from the east coast to the central plains to the west coast, and from Canada to the arid southwest. Other papers touch upon data from Mongolia, Australia, and Italy. The collection of papers as a group illustrate the importance of artifact life history analysis in understanding technology and human organizational strategies.

In the past several decades, lithic artifact production and use experiments have been beneficial in helping researchers understand tool production debris (Amick et al. 1988; Andrefsky 1986; Carr and Bradbury 2004; Kuijt et al. 1995; Titmus 1985), reduction sequences (Ahler 1989; Bradbury and Carr 1999; Magne 1989), and artifact function (Bradley and Sampson 1986; Geneste and Maury 1997; Odell and Cowan 1986; Shea 1993). Several papers in this volume continue the trend of using experiments to generate empirical data for comparison and interpretation of excavated assemblages. Eren and Prendergast (this volume) use a series of retouch experiments to assess various reduction indices. They show that different indices actually measure different aspects of tool retouch. Wilson and Andrefsky (this volume) conduct experiments to show that biface production is analytically separable from biface maintenance after use, and that bifacial retouch related to production is part of the tool's life history but has nothing to do with the curation of the biface. Quinn et al. (this volume) use a suit of experiments to assess retouch on awls and drilling tools. Their results suggest that retouch measures should be designed for specific tool types and assemblage contexts to be most effective for inferring aboriginal behaviors. Bradbury et al. (this volume) use extensive experimental data to isolate raw material influences and hammer type influence in the reduction process. They suggest that lithic raw materials can be partitioned into three broad categories relative to retouch intensity. That is, lithic raw material fracture properties can effectively be segregated into three gross kinds of raw material as opposed to the hundreds and thousands of varieties that exist in chipped stone technology.

Technological organization has been intimately linked to studies of lithic raw material abundance and availability (Ammerman and Andrefsky 1982; Andrefsky 1994b; Daniel 2001; Knell 2004; Larson and Kornfeld 1997) and of suitability for various tool tasks (Amick and Mauldin 1997; Bradbury and Franklin 2000; Ellis 1997; Knecht 1997). Several of the volume contributions focus specifically upon the influence of lithic raw material variability on retouch mechanics or retouch measures. The Bradbury et al. paper (this volume) directly explores the role of raw material type in the flake removal process. MacDonald's paper (this volume) explores raw material abundance and quality as it relates to tool design strategies. His results suggest that aboriginal tool makers and users selected raw material types for their functional qualities. Andrefsky's paper (this volume) uses XRF analysis to locate raw material sources and relates source distances to aspects of tool retouch, resharpening, and discard within the circulation ranges of the tool makers. Harper and Andrefsky (this volume) use lithic raw material analysis to help tease out the life histories of dart points to show how they are recycled in later period occupations in the American southwest. Similarly to Andrefsky's study, Clarkson's paper (this volume) uses raw material diversity to address issues of artifact provisioning and tool stone transport.

Artifact function has always been an important factor in understanding stone tool morphology. Archaeological evidence (Dixon et al. 2005; Elston 1986; Kay 1996; Truncer 1990) and ethnographic analogy

(Greaves 1997; Kelly and Fowler 1986; O'Connell 1977; White 1968) have unquestionably linked tool edge design and tool form to various functions. Several of the papers included in this volume demonstrate the importance of retouch extent and intensity to functional properties of stone tools. Hiscock and Clarkson (this volume) show that reduction of flake tools has much to do with tool form and size and ultimately that reduction state has implications for tool functional interpretations. MacDonald (this volume) shows that stone tool function influenced lithic raw material selection for production of various tool types. Harper and Andrefsky (this volume) use various retouch measures to make a case for the function of recycled dart points as cutting tools and not as projectiles after the introduction of the bow and arrow. Quinn et al. (this volume) demonstrate that artifact function is a critical parameter for selecting or developing a retouch index.

Recently artifact retouch indices have been developed as proxy measures for artifact curation (Davis and Shea 1998; Eren et al. 2005; Hiscock and Clarkson 2005; Shott and Ballenger 2007). Several papers in the volume explicitly test or apply a series of retouch indices or measures to better understand the variability in those indices and the effectiveness of those measures for dealing with curation and forager land use practices. The Eren and Prendergast paper (this volume) initially compared three retouch measures (Clarkson 2002; Eren et al. 2005; Kuhn 1990) in an effort to determine which measured tool mass loss most effectively. They found that retouch was more complicated than they originally anticipated and that each index was effectively measuring different kinds of retouch. Ultimately they dissected various measures to show sources of variability for each retouch index and devised a new display technique to integrate the various indices. Clarkson's paper (this volume) applied the Kuhn index (1990) and the Clarkson index (2002) to an excavated tool assemblage in an effort to link tool morphological transformations to changes in social and environmental conditions. Wilson and Andrefsky (this volume) apply Clarkson's index (2002) to a lithic assemblage from North America and find that the measure is effective for recognizing retouch after use, but it is not effective for measuring retouch on tools that are heavily flaked before use (such as bifaces). As a result, they explore several new techniques for separating production retouch from use retouch. These findings are very similar to Blades's study (this volume), which partitions stone tool production and retouch after use to form a model of "assemblage retouch." In addition to these papers, MacDonald, Hiscock and Clarkson, Andrefsky, Prentiss and Clarke, Harper and Andrefsky, Quinn et al. (all this volume) either apply an existing retouch index or develop new measures to assess retouch.

In the past few years, lithic analysts have been attempting to apply evolutionary approaches to understanding variability in stone tool assemblages (Bamforth 2002; Bamforth and Bleed 1997; Collard et al. 2005; Elston and Brantingham 2002; Ugan et al. 2003). Several papers in the volume add to this effort and attempt to bring various evolutionary approaches into interpretations of artifact life histories. Prentiss and Clarke (this volume) argue that foragers may employ a complex repertoire of inherited technologies in their standard resource gathering activities, and that they also must respond to contingencies, sometimes making alterations to specific tools or creating situational tools to serve in particular circumstances. They suggest that artifact variability is part of a human adaptive response and therefore undergoes selection. Goodale et al. (this volume) also suggest that evolutionary approaches can be used to more effectively understand technological systems. Their study links optimality theory to core reduction strategies by scaling lithic reduction to the concept of diversity. Goodale and company model raw material availability and raw material quality to the ratio of tool producers to tool users, suggesting that diversity of production techniques is a reflection of these three factors. Much like Prentiss and Clarke's study, Clarkson's paper (this volume) documents technological change over a long span of time. Clarkson's study integrates retouch intensity to artifact recycling, raw material selection, and provisioning tactics in an effort to show how aboriginal populations adapted to changing land use patterns. His study goes a step farther by plugging his recognized lithic artifact changes into the social and economic components of risk management and symbolic engagement.

SUMMARY

Archaeologists use stone tools as cultural and temporal markers. Stone tools are also used to infer aboriginal tasks based upon functional information gathered from such tools. These same tools are embedded within aboriginal land use practices and lifeways and, as such, can provide information related to such contexts. As I have stated before (Andrefsky 2005:245),

> It should be obvious to the reader that prehistoric lithic artifacts were made, used, modified, and discarded in cultural contexts unlike any that exists today. Things that were intimately linked to prehistoric activities and tool uses, such as making the tool or searching for the lithic raw material, were probably common chores conducted before an activity was undertaken. Integrating the production of a tool into the process of its use, and then task completion, are all parts of a whole, and differ significantly from modern task accomplishment.

These contexts represent the human framework for the organization of lithic technologies. Understanding stone tool life histories allows researchers to better integrate stone tool assemblages into models of technological organization.

The collection of papers assembled in this volume focus upon the role of stone tool life history within tool makers and users organizational strategies of lithic technologies. In particular, these papers show that tool life histories can be mapped by retouch analysis. However, it is clear that retouch is conducted in complicated ways directly related to the complicated life histories of stone tools. These assembled papers not only demonstrate and explain new techniques for assessing retouch, but also evaluate existing techniques and reveal important associations between retouch characteristics and tool form, function, production, use, and discard and specific situations in which these stone tools are associated.

REFERENCES CITED

Ahler, Stanley A. 1989. Mass Analysis of Flaking Debris: Studying the Forest Rather than the Trees. In *Alternative Approaches to Lithic Analysis*, edited by Donald O. Henry and George H. Odell, pp. 85–118. Archaeological Papers of the American Anthropological Association No. 1. Washington. D.C.

Amick, Daniel S., and Raymond P. Mauldin. 1989. *Experiments in Lithic Technology*. International Series 528, British Archaeological Reports, Oxford.

1997. Effects of Raw Material on Flake Breakage Patterns. *Lithic Technology* 22:18–32.

Amick, Daniel S., Raymond P. Mauldin, and Steven A. Tomka. 1988. An Evaluation of Debitage Produced by Experimental Bifacial Core Reduction of a Georgetown Chert Nodule. *Lithic Technology* 17:26–36.

Ammerman, Albert J., and William Andrefsky, Jr. 1982. Reduction Sequences and the Exchange of Obsidian in Neolithic Calabria. In *Contexts for Prehistoric Exchange*, edited by J. Ericson and T. Earle, pp. 149–72. Academic Press, New York.

Andrefsky, William, Jr. 1986. A Consideration of Blade and Flake Curvature. *Lithic Technology* 15:48–54.

1991. Inferring Trends in Prehistoric Settlement Behavior From Lithic Production Technology in the Southern Plains. *North American Archaeologist* 12:129–144.

1994a. Raw Material Availability and the Organization of Technology. *American Antiquity* 59:21–35.

1994b. The Geological Occurrence of Lithic Material and Stone Tool Production Strategies. *Geoarchaeology: An International Journal* 9:345–62.

2001. *Lithic Debitage: Context, Form, Meaning*. University of Utah Press, Salt Lake City.

2005. *Lithics: Macroscopic Approaches to Analysis*. Second edition. Cambridge University Press, Cambridge.

2006. Experimental and Archaeological Verification of an Index of Retouch for Hafted Bifaces. *American Antiquity* 71:743–57.

Audouze, F. 1999 New Advances in French Prehistory. *Antiquity* 73:167–75.

Baales, Michael. 2001. From Lithics to Spatial and Social Organization: Interpreting the Lithic Distribution and Raw Material Composition at the Final Palaeolithic Site of Kettig (Central Rhineland, Germany). *Journal of Archaeological Science* 28:127–41.

Bamforth, Douglas B. 1986. Technological Efficiency and Tool Curation. *American Antiquity* 51:38–50.

1991. Technological Organization and Hunter–Gatherer Land Use: A California Example. *American Antiquity* 56:216–234.

2000. Core/Biface Ratios, Mobility, Refitting, and Artifact Use-Lives: A Paleoindian Example. *Plains Anthropologist* 45:273–90.

Bamforth, Douglas B., and Peter Bleed. 1997. Technology, Flaked Stone Technology, and Risk. In *Rediscovering Darwin*, edited by C. M. Barton and G. Clark, pp. 109–140. Archaeological Papers of the American Anthropological Association No. 7, Arlington.

Bettinger, Robert. L. 1987 Archaeological Approaches to Hunter-gatherers. *Annual Review of Anthropology* 16:121–42.

Binford, Lewis R. 1973. Interassemblage Variability: The Mousterian and the "Functional" Argument. In *The Explanation of Culture Change: Models in Prehistory*, edited by C. Renfrew, pp. 227–54. Duckworth, London.
—— 1977. Forty-Seven Trips. In *Stone Tools as Cultural Markers*, edited by R. S. V. Wright, pp. 24–36. Australian Institute of Aboriginal Studies, Canberra.
—— 1979. Organization and Formation Processes: Looking at Curated Technologies. *Journal of Anthropological Research* 35:255–73.
Bleed, Peter. 1986. The Optimal Design of Hunting Weapons: Maintainability or Reliability. *American Antiquity* 51:737–47.
Bradbury, Andrew P., and Phillip J. Carr. 1999. Examining Stage and Continuum Models of Flake Debris Analysis: An Experimental Approach. *Journal of Archaeological Science* 26:105–16.
Bradbury, Andrew P., and Jay D. Franklin. 2000. Material Variability, Package Size and Mass Analysis. *Lithic Technology* 25:42–58.
Bradley, Bruce A., and C. Garth Sampson. 1986. Analysis by Replication of Two Acheulian Artefact Assemblages from Caddington, England. In *Stone Age Prehistory: Studies in Memory of Charles McBurney*, edited by G. N. Bailey and P. Callow, pp. 29–45. Cambridge University Press, Cambridge.
Brantingham, P. Jeffrey, and Steven L. Kuhn. 2001. Constraints on Levallois Core Technology: A Mathematical Model. *Journal of Archaeological Science* 28:747–61.
Brantingham, P. Jeffrey, John W. Olsen, Jason A. Rech, and Andrei I. Krivoshapkin. 2000. Raw Material Quality and Prepared Core Technologies in Northeastern Asia. *Journal of Archaeological Science* 27:255–71.
Braun, David R. 2005. Examining Flake Production Strategies: Examples from the Middle Paleolithic of Southwest Asia. *Lithic Technology* 30:107–25.
Callahan, Errett. 1979. The Basics of Biface Knapping in the Eastern Fluted Point Tradition: A Manual for Flintknappers and Lithic Analysts. *Archaeology of Eastern North America* 7(1):1–180.
Carr, Philip J., and Andrew P. Bradbury. 2001. Flake Debris Analysis, Levels of Production, and the Organization of Technology. In *Lithic Debitage: Context, Form, Meaning*, edited by W. Andrefsky, Jr., pp. 126–46. University of Utah Press, Salt Lake City.
—— 2004. Exploring Mass Analysis, Screens, and Attributes. In *Aggregate Analysis in Chipped Stone*, edited by Christopher T. Hall and Mary Lou Larson, pp. 21–44. University of Utah Press, Salt Lake City.
Chatters, James C. 1987. Hunter-Gatherer Adaptations and Assemblage Structure. *Journal of Anthropological Research* 6:336–75.

Clarkson, Chris. 2002. An Index of Invasiveness for the Measurement of Unifacial and Bifacial Retouch: A Theoretical, Experimental and Archaeological Verification. *Journal of Archaeological Science* 29:65–75.

Cochrane, Grant W. G. 2003. On the Measurement and Analysis of Platform Angles. *Lithic Technology* 28:13–25.

Collard, Mark, Michael Kemery, and Samantha Banks. 2005. Causes of Toolkit Variation among Hunter-Gatherers: A Test of Four Competing Hypotheses. *Journal of Canadian Archaeology* 29:1–19.

Daniel, I. Randolph, Jr. 2001. Stone Raw Material Availability and Early Archaic Settlement in the Southeastern United States. *American Antiquity* 66:237–66.

Davis, Zachary J., and John J. Shea. 1998. Quantifying Lithic Curation: An Experimental Test of Dibble and Pelcin's Original Flake-Tool Mass Predictor. *Journal of Archaeological Science* 25:603–10.

Dibble, Harold L. 1987. The Interpretation of Middle Paleolithic Scraper Morphology. *American Antiquity* 52(1):109–17.

——— 1997. Platform Variability and Flake Morphology: A Comparison of Experimental and Archeological Data and Implications for Interpreting Prehistoric Lithic Technological Strategies. *Lithic Technology* 22:150–70.

Dibble, Harold L., and Andrew Pelcin. 1995. The Effect of Hammer Mass and Velocity on Flake Mass. *Journal of Archaeological Science* 22:429–39.

Dixon, E. James, William F. Manley, and Craig M. Lee. 2005. The Emerging Archaeology of Glaciers and Ice Patches: Examples from Alaska's Wrangell–St. Elias National Park and Preserve. *American Antiquity* 70:129–43.

Ellis, Christopher J. 1997. Factors Influencing the Use of Stone Projectile Tips: An Ethnographic Perspective. In *Projectile Technology*, edited by Heidi Knecht, 37–78. Plenum Press, New York.

Elston, Robert G. 1986. Prehistory of the Western Area. In *Handbook of North American Indians, Volume 11: Great Basin*. Edited by Warren L. D'Azevedo. (volume editor), pp. 135–48. Smithsonian Institution Press, Washington, DC.

Elston, Robert G. and P. Jeffrey Brantingham. 2002. Microlithic Technology in Northern Asia: A Risk-Minimizing Strategy of the Late Paleolithic and Early Holocene. In *Thinking Small: Perspectives on Microlithization*, edited by R. G. Elston and S. L. Kuhn, pp. 104–17. Archaeological Papers of the American Anthropological Association, No. 12, Washington, DC.

Eren, Metin I., Manual Dominguez-Rodrigo, Steven L. Kuhn, Daniel S. Adler, Ian Le, and Ofer Bar-Yosef. 2005. Defining and Measuring Reduction in Unifacial Stone Tools. *Journal of Archaeological Science* 32:1190–1206.

Flenniken, J. Jeffrey, and Anan W. Raymond. 1986. Morphological Projectile Point Typology: Replication Experimentation and Technological Analysis. *American Antiquity* 51:603–14.

Geneste, Jean-Michel, and Serge Maury. 1997. Contributions of Multidisciplinary Experiments to the Study of Upper Paleolithic Projectile Points. In *Projectile Technology*, edited by Heidi Knecht, pp. 165–89. Plenum Press, New York.

Gould, Richard A. 1980. Raw Material Source Areas and "Curated" Tool Assemblages. *American Antiquity* 45:823–33.

1985. The Empiricist Strikes Back: A Reply to Binford. *American Antiquity* 50:638–44.

Gould, Richard. A., and S. Saggers. 1985. Lithic Procurement in Central Australia: A Closer Look at Binford's Idea of Embeddedness in Archaeology. *American Antiquity* 50:117–36.

Gramly, R. Michael, 1980. Raw Material Source Areas and "Curated" Tool Assemblages. *American Antiquity* 45:823–33.

Greaves, Russel D. 1997. Hunting and Multifunctional Use of Bows and Arrows: Ethnoarchaeology of Technological Organization among Pume' Hunters of Venezuela. In *Projectile Technology*, edited by Heidi Knecht, pp. 287–320. Plenum Press, New York.

Hayden, Brian. 1975. Curation: Old and New. In *Primitive Art and Technology*, edited by J. S. Raymond, B. Loveseth, C. Arnold, and G. Reardon, 47–59. University of Calgary, Calgary.

Hiscock, Peter, and Val Attenbrow. 2003. Early Australian Implement Variation: A Reduction Model. *Journal of Archaeological Science* 30:239–49.

Hiscock, Peter, and Chris Clarkson. 2005. Experimental Evaluation of Kuhn's Geometric Index of Reduction and the Flat-Flake Problem. *Journal of Archaeological Science* 32:1015–22.

Hoffman, C. Marshall. 1985. Projectile Point Maintenance and Typology: Assessment with Factor Analysis and Canonical Correlation. In *For Concordance in Archaeological Analysis: Bridging Data Structure, Quantitative Technique, and Theory*, edited by C. Carr, pp. 566–612. Westport Press, Kansas City.

Holmes, William H. 1894. Natural History of Flaked Stone Implements. In *Memoirs of the International Congress of Anthropology*, edited by C. S. Wake, 120–39. Schulte, Chicago.

Inizan, M.-L., H. Roche, and J. Tixier. 1992. *Technology of Knapped Stone*. CREP, Meudon.

Kalin, Jeffrey. 1981. Stem Point Manufacture and Debitage Recovery. *Archaeology of Eastern North America* 9:134–75.

Kay, Marvin. 1996. Microwear Analysis of Some Clovis and Experimental Chipped Stone Tools. In *Stone Tools: Theoretical Insights into*

Human Prehistory, edited by George Odell, pp. 315–44. Plenum Press, New York.

Kelly, Isabel T., and Catherine S. Fowler. 1986. Southern Paiute. In *Handbook of North American Indians, Volume 11: Great Basin*. Edited by Warren L. D'Azevedo (volume editor), pp. 368–97. Smithsonian Institution Press, Washington, DC.

Kelly, Robert L. 1988. The Three Sides of a Biface. *American Antiquity* 53:717–34.

Knecht, Heidi. 1997. Projectile Points of Bone, Antler, and Stone: Experimental Explorations of Manufacture and Use. In *Projectile Technology*, edited by Heidi Knecht, pp. 191–212. Plenum Press, New York.

Knell, Edward. 2004. Coarse-Scale Chipped Stone Aggregates and Technological Organization Strategies at Hell Gap Locality V Cody Complex Component, Wyoming. In *Aggregate Analysis in Chipped Stone*, edited by Christopher T. Hall and Mary Lou Larson, pp. 156–83. University of Utah Press, Salt Lake City.

Koldehoff, Brad. 1987. The Cahokia Flake Tool Industry: Socio-Economic Implications for Late Prehistory in the Central Mississippi Valley. In *The Organization of Core Technology*, edited by Jay K. Johnson and Carrol A. Morrow, pp. 151–86. Westview Press, Boulder, CO.

Kuhn, Steven L. 1990. A Geometric Index of Reduction for Unifacial Stone Tools. *Journal of Archaeological Science* 17:585–93.

—— 1991. "Unpacking" Reduction: Lithic Raw Material Economy in the Mousterian of West–Central Italy. *Journal of Anthropological Archaeology* 10:76–106.

—— 1992. Blank Form and Reduction as Determinants of Mousterian Scraper Morphology. *American Antiquity* 57:115–28.

Kuijt, Ian, William C. Prentiss, and David J. Pokotylo. 1995. Bipolar Reduction: An Experimental Study of Debitage Variability. *Lithic Technology* 20:116–27.

Larson, Mary Lou, and Marcel Kornfeld. 1997. Chipped Stone Nodules: Theory, Method, and Examples. *Lithic Technology* 22:4–18.

Magne, Martin P. 1989. Lithic Reduction Stages and Assemblage Formation Processes. In *Experiments in Lithic Technology*, edited by D. S. Amick and R. P. Mauldin, pp. 15–32. International Series 528, British Archaeological Reports, Oxford.

Morrow, Juliet. 1997. End Scraper Morphology and Use-Life: An Approach for Studying Paleoindian Lithic Technology and Mobility. *Lithic Technology* 22:70–85.

Nash, Stephen E. 1996. Is Curation a Useful Heuristic? In *Stone Tools: Theoretical Insights into Human Prehistory*, edited by G. H. Odell, pp. 81–100. Plenum Press, New York.

Nelson, Margaret C. 1991. The Study of Technological Organization. In *Archaeological Method and Theory*, Vol. 3, edited by M. B. Schiffer, pp. 57–100. University of Arizona Press, Tucson.

Nowell, April, Kyoungju Park, Dimitris Mutaxas, and Jinah Park. 2003. Deformation Modeling: A Methodology for the Analysis of Handaxe Morphology and Variability. In *Multiple Approaches to the Study of Bifacial Technologies*, edited by Marie Soressi and Harold L. Dibble, pp. 193–208. University of Pennsylvania Museum of Archaeology and Anthropology, Philadelphia.

O'Connell, James F. 1977. Aspects of Variation in Central Australian Lithic Assemblages. In *Stone Tools as Cultural Markers: Change, Evolution and Complexity*, edited by R. V. S. Wright, pp. 269–81. Australian Institute of Aboriginal Studies, Canberra.

Odell, George H. 1989. Experiments in Lithic Reduction. In *Experiments in Lithic Technology*, edited by D. S. Amick and R. P. Mauldin, pp. 163–98. International Series 528, British Archaeological Reports, Oxford.

—— 1996. Economizing Behavior and the Concept of "Curation." In *Stone Tools: Theoretical Insights into Human Prehistory*, edited by G. H. Odell, pp. 51–80. Plenum, New York.

Odell, George H., and Frank Cowan. 1986. Experiments with Spears and Arrows on Animal Targets. *Journal of Field Archaeology* 13(2):195–212.

Parry, William J., and Robert L. Kelly. 1987. Expedient Core Technology and Sedentism. In *The Organization of Core Technology*, edited by J. K. Johnson and C. A. Morrow, pp. 285–304. Westview Press, Boulder, CO.

Pecora, Albert M. 2001. Chipped Stone Tool Production Strategies and Lithic Debris Patterns. In *Lithic Debitage: Context, Form, Meaning*, edited by William Andrefsky, Jr., pp. 173–91. University of Utah Press, Salt Lake City.

Pelcin, Andrew. 1997. The Formation of Flakes: The Role of Platform Thickness and Exterior Platform Angle in the Production of Flake Initiations and Terminations. *Journal of Archaeolgoical Science* 24:1107–113.

Rasic, Jeffery C., and William Andrefsky, Jr. 2001. Alaskan Blade Cores as Specialized Components of Mobile Toolkits: Assessing Design Parameters and Toolkit Organization through Debitage Analysis. In *Lithic Debitage: Context, Form, Meaning*, edited by Wm. Andrefsky, Jr., pp. 61–79. University of Utah Press, Salt Lake City.

Rolland, Nicolas, and Harold L. Dibble. 1990. A New Synthesis of Middle Paleolithic Variability. *American Antiquity* 55:480–99.

Roth, Barbara, and Harold Dibble. 1998. The Production and Transport of Blanks and Tools at the French Middle Paleolithic Site of Combe-Capelle Bas. *American Antiquity* 63:47–62.

Sellet, Frederic. 1993. Chaîne opératoire: The Concept and Its Applications. *Lithic Technology* 18:106–12.

Shea, John J. 1993. Lithic Use-Wear Evidence for Hunting by Neanderthals and Early Modern Humans form the Levantine Mousterian. In *Hunting and Animal Exploitation in the Later Paleolithic and Mesolithic of Eurasia*, edited by Gail Larson Peterkin, Harvey M. Bricker, and Paul Mellars, pp. 189–98. Archaeological Papers of the American Anthropological Association, Number 4.

Shott, Michael J. 1986. Settlement Mobility and Technological Organization: An Ethnographic Examination. *Journal of Anthropological Research* 42:15–51.

—— 1993. *The Leavitt Site: A Parkhill Phase Paleo-Indian Occupation in Central Michigan.* Memoirs No. 25, University of Michigan Museum of Anthropology, Ann Arbor.

—— 1996. An Exegesis of the Curation Concept. *Journal of Anthropological Research* 52:259–80.

—— 2003. Chaîne opératoire and Reduction Sequence. *Lithic Technology* 28:95–105.

Shott, Michael J., and Jesse A. M. Ballenger. 2007. Biface Reduction and the Measurement of Dalton Curation: A Southeastern Case Study. *American Antiquity* 72:153–75.

Shott, Michael J., Andrew P. Bradbury, Philip J. Carr, and George H. Odell. 2000. Flake Size from Platform Attributes: Predictive and Empirical Approaches. *Journal of Archaeological Science* 27:877–94.

Shott, Michael J., and Paul Sillitoe. 2005. Use Life and Curation in New Guinea Experimental Used Flakes. *Journal of Archaeological Science* 32:653–63.

Sievert, April K., and Karen Wise. 2001. A Generalized Technology for a Specialized Economy: Archaic Period Chipped Stone at Kilometer 4, Peru. In *Lithic Debitage: Context, Form, Meaning*, edited by William Andrefsky, Jr., pp. 188–206. University of Utah Press, Salt Lake City.

Titmus, Gene. 1985. Some Aspects of Stone Tool Notching. In *Stone Tool Analysis: Essays in Honor of Don E. Crabtree.* edited by Marc G. Plew and Max G. Pavesic, 243–64. University of New Mexico Press, Albuquerque.

Tomka, Steve A. 2001. The Effect of Processing Requirements on Reduction Strategies and Tool Form: A New Perspective. In *Lithic Debitage: Context, Form, Meaning*, ed. William Andrefsky, Jr., pp. 207–24. University of Utah Press, Salt Lake City.

Torrence, Robin. 1983. Time Budgeting and Hunter–Gatherer Technology. In *Hunter–Gatherer Economy in Prehistory*, edited by G. Bailey, pp. 11–22. Cambridge University Press, Cambridge.

Truncer, James J. 1990. Perkiomen Points: A Study in Variability. In *Experiments and Observations on the Terminal Archaic of the Middle Atlantic Region*, edited by R. W. Moeller, pp. 1–62. Archaeological Services, Bethlehem, CT.

Ugan, Andrew, Jason Bright, and Alan Rogers. 2003. When Is Technology Worth the Trouble? *Journal of Archaeological Science* 30:1315–29.

Wallace, Ian J., and John J. Shea. 2006. Mobility Patterns and Core Technologies in the Middle Paleolithic of the Levant. *Journal of Archaeological Science* 33:1293–309.

Weedman, Kathryn J. 2006. An Ethnoarchaeological Study of Hafting and Stone Tool Diversity among the Gamo of Ethiopia. *Journal of Archaeological Method and Theory* 13:189–238.

White, J. Peter. 1968. Fabricators, Outils Ecailles, or Scalar Cores? *Mankind* 6:658–66.

Whittaker, John C. 1994. *Flintknapping: Making and Understanding Stone Tools*. University of Texas Press, Austin.

2   MICHAEL J. SHOTT AND MARGARET C. NELSON

LITHIC REDUCTION, ITS MEASUREMENT, AND IMPLICATIONS: COMMENTS ON THE VOLUME

Some years ago one of us wrote, "A glance at a chipped stone tool is enough to see that stone is a subtractive medium" (Shott 1994: 69). The statement bordered on a truism but was worth making in any event. Flakes, the small pieces of stone struck from larger objective pieces, were the subject then; the context was their abundance and diversity as generated in the production, use, and resharpening of tools. Flake analysis makes no sense without understanding the places that flakes occupy in the reduction process.

But the reductive quality of stone also informs the analysis of objective pieces themselves, not least finished tools. Accordingly, tools also are a legitimate subject of reduction studies. A deceptively profound truism worth stating once is worth rephrasing: a glance at a chipped stone tool is enough to see that it was reduced from a larger piece. But the restatement itself requires elaboration. Trivially, tools were reduced from larger objective pieces in the process of production. No one has doubted this since archaeologists demonstrated human agency in the production of stone tools. Yet many tools were further retouched by resharpening, and so continued to experience reduction during use. This is the "reduction thesis" (Shott 2005), which archaeologists did not always appreciate in the past.

This book is a milestone in the development of reduction analysis. Originating in pioneering studies such as Hoffman's (1985), until recently reduction analysis was conducted in isolation by few archaeologists. Clarkson and Lamb's (2005) recent collection demonstrated its value, mostly in Australian flake-tool assemblages. This collection broadens the scope of reduction analysis even farther in geographic and analytical terms. It is about equally divided between North American case studies on the one hand, and Eurasian and Australian ones on the other. North America abounds in bifaces, and their analysis naturally figures more prominently here than elsewhere. Combined with the earlier Australian work, this collection demonstrates that reduction was a truly global process of broad relevance to lithic assemblages, if ever this was doubted.

Here we discuss the importance of reduction analysis in the broadest terms. Then we comment upon chapters separately, and finally we discuss some issues that the book's scope and nature engage.

ESTABLISHING THE THEORETICAL IMPACT: TECHNOLOGICAL ORGANIZATION AND RETOUCH

This treatment of retouch of prehistoric tools, in its broadest sense, is embedded in the study of technological organization (Binford 1979; Nelson 1991). An emphasis on examining the organization of the acquisition of materials, the production, transportation, use, reuse, and discard of tools, and the byproducts of tool manufacture grew from dissatisfaction with debates about the attribution of utilitarian function or style to explain tool form (e.g., Binford 1973; Binford and Binford 1966; Bordes and de Sonneville-Bordes 1970). One of the greatest benefits of an organizational approach to technology is that it allows clearer connections to understanding organization in the economic and social domains of human societies (Andrefsky 1994; Arnold 1987; Bamforth 1991; Bleed 1986; Carr 1994; Johnson and Morrow 1987; Kelly 1988; Parry and Kelly 1987; Shott 1986, 1989b; Torrence 1983, 1989).

The concept of "curation" has been central to studies of technological organization (Bamforth 1986; Binford 1979; Nelson 1991; Parry and Kelly 1987; Shott 1996a). Shott (1996a: 267) has provided a concise and operational definition of curation of tools as "the degree of use or utility extracted, expressed as a relationship between how much utility a tool starts with – its maximum utility – and how much of that utility is realized before discard" (see also Shott 1989a: 24 and 1995). Curation can involve preparation of tools and cores, transportation of those tools and cores, and storage and reuse of tools and cores (Bamforth 1986, Binford 1979; Nelson 1991), and is influenced by the distribution of tool stone in relation to the organization of tool use needs. Stockpiling of materials at regularly used places and transport of prepared core materials, among other strategies, can ameliorate the lack of locally available stone. Thus, the study of degrees of curation aids in understanding the organization of work, the regularity of site occupation, the organization and frequency of movement, and resource scheduling.

Reduction and retouch occur at initial manufacture, during use, and during repair of tools and can therefore yield information about the organization of those activities and the organization of social and economic behaviors. For example, the extent of repair and reshaping of a tool can indicate how long it was curated and possibly transported. High levels of curation and transport indicate frequent mobility (Kelly 1988; Kuhn 1991; Nelson 1991; Torrence 1983). In addition, the distribution of the debris from retouch and the discarded retouched tools indicates the organization of that mobility on the landscape (Andrefsky 2005; Kelly 1988; Nelson 1991). But as Wilson and Andrefsky note in this volume, "retouch amount . . . is a result of several complicated factors that must be considered before it can be applied to measure artifact curation."

In this volume, several aspects of land use are the focus of analysis of stone tool retouch. Andrefsky is interested in circulation ranges and provisioning strategies by pithouse occupants at Birch Creek in southeast Oregon, examining retouch in relation to the distribution of obsidian source materials. Blades examines Old World and New World cases to identify nodes in subsistence–settlement systems based on the characteristics of the retouch on tools at different sites. MacDonald is concerned with the tradeoffs between curation and expediency in a "toolstone-deficient" environment, which he sees as important to understanding risk-minimizing behavior and mobility at the Skink site in West Virginia and the surrounding region. Prentiss and Clark address different aspects of mobility though analysis of retouch. Prentiss and Clark assess mobility and subsistence strategies in their examination of retouch on tools from pithouse villages in Interior British Columbia.

All of the authors agree that understanding how and why retouch varies is essential to higher-order interpretations. "Concepts of reduction, retouching, and resharpening are only important so far as they provide information on the more complex concepts of prehistoric behavior, curation, and tool use-life" (Eren and Prendergast, this volume). The authors of the chapters in this volume move us substantially forward in these interpretations through their focus on methods for recording retouch and interpreting complex relationships among variables influencing retouch. For many years, Shott (1989a, 1995) and others (e.g., Kuhn 1991) have pointed to the importance of retouch studies to the diverse efforts to understand technological organization in varied contexts. This collection answers Shott's call by refining and inventing ways to measure retouch (Eren and Prendergast, Quinn et al., both this volume) and exploring the complexities of the relationships among variables that influence the form, location, and quantity of retouch: raw material availability (Andrefsky, McDonald, both this volume), raw material qualities (Bradbury et al., this volume), aspects of production and repair (Blades; Wilson and Andrefsky, both this volume), function and use (Goodale et al.; Harper and Andrefsky, this volume), and reduction sequences (Clarkson; Hiscock and Clarkson, both this volume). Yet this focus on measurement and methods is driven by concerns for understanding prehistoric behavior.

TYPOLOGY AND THE REDUCTION THESIS

Typologies arrange an abundance of objects or subjects into relatively homogeneous groups. They begin with all specimens as one variable group and ends with types whose members are identical or nearly so, distinguished from one another by size, shape, material, color, or other salient characteristics. Biological taxonomy takes this approach although, of course, it accommodates differences in size and shape between sexes. Yet it knows that animals and plants change by growth from birth to maturity, a difference of proportion by size and age. Difference in proportions as a function of size is allometry, elegantly described in Thompson's (1917) classic study. Because the growth of living things is blindingly obvious, biological taxonomy has no difficulty accommodating the variation it produces within the types – taxa – that it defines.

Paleontology, however, lacks direct observation of growth. It has only fossils, which do not grow or change in any way. Consequently, paleontology risks confusing the variation in size and form that growth creates within a taxon with a difference between taxa. It might, for instance, mistake the ancient equivalent of a tiger cub for an adult cat. Fortunately, it has methods to minimize such risks.

Like biology and paleontology, archaeology uses typology to impose order on the diversity of its subjects. Indeed, typology may be the favorite pastime of lithic analysts. Traditional typologies, unfortunately, assumed that tools were made for use in quite specific sizes and forms and that those qualities of tools never changed during their use. Archaeology, that is, lacks the methods to minimize the risk of confusing the reduced state of tools with their original design.

THE REDUCTION THESIS

In recent years, however, many archaeologists assimilated the reduction thesis (Shott 2005), the understanding that retouched tools vary progressively from first use to discard by decrease in size and change in form depending on extent and pattern of the resharpening that they experience. Not all tools are retouched during use, so the reduction thesis is merely common, not universal. It is amply documented for many tool types from many times and places around the world (e.g., Andrefsky 1997, 2006; Ballenger 2001; Blades 2003; Buchanan 2006; Clarkson and Lamb 2005; Dibble 1995; Ellis 2004; Flenniken and Wilke 1989; Hayden 1977; Hiscock and Attenbrow 2005; Hoffman 1985; Kuhn 1990; Sahnouni et al. 1997; Shott and Ballenger 2007; Shott and Sillitoe 2005; Truncer 1990; Wheat 1974) and in ethnographic sources (e.g., Hayden 1977; Shott and Weedman 2007; Tindale 1965; Weedman 2002).

Arguably, reduction has greater typological implications for flake tools than for bifaces. Whether an artifact is a convergent or transverse scraper, a tula or elouera, can be a matter of degree and pattern of reduction, not necessarily cultural affinity or age (e.g., Dibble 1995; Hiscock and Attenbrow 2005). Bifaces are different because they possess a stem or haft element that rarely changes in use. But points' blades and shoulders, and sometimes even stems (Flenniken and Wilke 1989), can change. With resharpening, length or blade width should decline while thickness changes little if at all (e.g., Andrefsky 2006; Cresson 1990: Fig. 5; Hoffman 1985; Shott and Ballenger 2007; Truncer 1990). Depending upon degree and pattern of reduction, those changes can produce sufficient variation in size and form at discard to make specimens of the same original type seem different (Hoffman 1985). For instance, "Enterline" Paleoindian bifaces in eastern North America may be "more a result of reworking than of deliberate intent" (Cox 1986: 110), i.e., a reduced version of Clovis or Gainey bifaces rather than a distinct type. Fluted points at Debert also were extensively reduced, complicating their technological and typological placement (Ellis 2004). Similarly, Wheat (1974) demonstrated that the "San Jon" type was merely a reduced version of Firstview bifaces. Even bifaces are subject to the typological implications of reduction.

Yet the reduction thesis has implications beyond typology. For instance, patterns of reduction implicate kinds of use by identifying the edges or segments of tools that were retouched. Also, degree of reduction is a measure of curation (Binford 1973; Nelson 1991; Shott 1996a; Shott and Sillitoe 2005), a theoretical quantity of considerable importance in lithic analysis, as several chapters here demonstrate. Although reduction is not identical to longevity, reduction distributions permit archaeologists to calibrate discard rates of different tool types to common scales, and imply different causes of discard. Reduction and curation rate are particularly important to models of hunter–gatherer land use and behavior, for example among North American Paleoindians (e.g., Ellis 1984; Kelly and Todd 1988; Shott 1986; Surovell 2003) but elsewhere too, as several chapters here demonstrate. Reduction distributions that reveal constant discard rates regardless of degree of curation suggest chance as the cause of discard, whereas those that reveal discard rate increasing with curation imply attrition (e.g., Shott and Sillitoe 2005).

On balance, reduction indices do more than just qualify typological inference; integrated with suitable theory, they reveal and quantify degree of curation and ground sophisticated behavioral models in archaeological data. Reduction measurement is a method, but one of great significance for theory. This conclusion warrants the evaluation and use of the range of reduction measures described in this volume, to whose chapters we now turn.

ADDRESSING THE REDUCTION THESIS

Eren and Prendergast's experimental comparison of several reduction measures (this volume) legitimately won the award for best student paper presented at the 2006 Society for American Archaeology meeting. They define reduction as weight loss. Their estimated reduction percentage (ERP), essentially a three-dimensional volumetric extension of Kuhn's (1990) geometric reduction index, is related to an earlier reduction measure partly of Eren's devising (Eren et al. 2005). Comparison omits Dibble and Pelcin's (1995) mass predictor equation and similar measures. As noted, these measures are somewhat ambiguous, but Bradbury et al.'s chapter here suggests that flake thickness remains a useful allometric estimator of original size in most raw materials, and Blades's chapter (this volume) demonstrates that measure in use. ERP emerges as the best general reduction measure, a provisional conclusion that should be further examined in future broader comparisons. The chapter is a fine example of the sort of controlled comparison that more archaeologists should conduct, yet its conclusion is not surprising. If reduction is measured by weight, a good proxy for which is volume, then of course a volume measure such as ERP will perform better than a geometric one such as Kuhn's index. As thorough as it is, the comparison between ERP and IR is not persuasive in all details. For instance, there seems to be more patterned dispersion ("fanning") in Figure 9b than 9d (one outlier there excepted), contrary to Eren and Prendergast's statement (this volume). Moreover, the correlation between the measures in archaeological specimens (their Figure 15c) seems highly dependent upon two outliers. Respectfully, we disagree with Eren and Prendergast that their results moot the "flat-flake" problem merely from the undeniable fact that "different reduction indices measure different attributes" of reduction. Any two-dimensional geometric measure remains vulnerable to flat-flake bias.

Wilson and Andrefsky (this volume) used invasiveness indices to measure degree of reduction in experimental and archaeological specimens. Like Andrefsky's (2006), their study extends to bifaces the measures originally devised for analysis of retouched flakes. Not surprisingly, indices devised for retouched flakes (e.g., Clarkson 2002) did not perform well. As a result, they applied a different method devised by Andrefsky (2006). As a small point, the paper underscores one shortcoming of such indices: the unequal size of the parsed zones of each tool, which are treated for computation as equal in size (e.g., Wilson and Andrefsky's Figure 6). Much more importantly, their chapter is valuable in two respects. First are plots of size measures (area, weight) against resharpening that demonstrate greater reduction effects at earlier resharpening stages (Figure 3; see also Morrow 1997). This alone is a significant observation, which could be enhanced by more extensive experimentation. Once a substantial experimental data set is compiled, archaeologists can better determine which size measure is most sensitive to reduction effects and may be able to define curves mathematically. This is no mere pedantic matter, but a prospect that might enable archaeologists to apply mathematical models to the reduction process to measure it more precisely and to characterize each specimen's location on the continuum. In this respect, Wilson and Andrefsky's experimental approach can integrate the analysis of biface reduction attributes with earlier debris studies (Shott 1994: 91–9; Wilson and Andrefsky 2006). The second original contribution is Wilson and Andrefsky's use of ridges or arrises, a functional equivalent of scar count or density. Experiments showed consistent increases in ridge count up to a possible threshold at five resharpenings. Leave aside possible sampling questions (e.g., whether any variation should be expected in zones near the specimen's center, far from the retouched edge [Figure 7]). Whether subsequent variation is patterned or merely random (at a glance, Figure 8 suggests stochastic variation beyond the threshold), it is no surprise that this variant of resharpening has different effects at different points in the reduction continuum. Reduction is a constant, but its effects can be variable for allometric and other reasons.

The reduction thesis applied to Australian assemblages casts doubt on traditional typology owing largely to the work of Hiscock and his students (e.g., Clarkson and Lamb, 2005; Hiscock and Attenbrow 2005). Australians are particularly strong advocates of Kuhn's reduction index, which generally works well to measure reduction in the retouched flakes common there. As in an earlier study (Hiscock 1996), Hiscock and Clarkson now apply both the thesis and Kuhn's measure to French Paleolithic assemblages, the setting for Dibble's (e.g., 1995) statement of the reduction thesis. They properly qualify their view, acknowledging technology, individual preference, and other factors besides reduction that contribute to the size, form, and retouch patterns of flake tools. In this connection, Hiscock and Clarkson's treatment of other reduction perspectives becomes something of a straw man, because reduction alone never was claimed as the sole determinant of tool size and form. One interesting result is the high correlation that Hiscock and Clarkson calculate between Kuhn's index and mass loss, contra Eren and Prendergast. Such conflicting results call for further experimentation, a point to which we return below. Hiscock and Clarkson conclude, reasonably, that reduction is one factor among several that determine tool size and form. In a Paleolithic context, Dibble (1995) argued that Bordean types had descriptive, not analytical, value (i.e., that they legitimately described modes in pattern and degree of reduction). Results here suggest that they have none at all.

Too often, semantic differences obscure substantive affinities between North American and European approaches. Americans ordinarily call the resharpening that tools experience "reduction." From his European Paleolithic perspective, Blades (this volume) distinguishes the reduction of individual tools ("retouch intensity") from the gross reduction patterns of entire assemblages or industries ("reduction"). Despite the semantic distance, the result of different contexts of customary use, Blades's chapter illustrates some strengths of the European approach. Where most contributors emphasize the reduction of individual tools, he legitimately treats assemblages as his analytical subjects. Like other Paleolithic archaeologists, Blades measures assemblage reduction using flake–core ratios, the size of cores or blanks, and the amount of cortical cover. He compares measures between North American and European assemblages, drawing inferences about the mode or organization of stone acquisition from degrees of assemblage-level reduction.

The most suitable reduction measure depends upon tool type, industry, context, and research question. Rightly, Quinn et al. (this volume) eschew the search for universal measures. Instead, they introduce a new geometric reduction index, a good example of the need to devise methods and measures that are as diverse as the types and research problems to which they are applied. Quinn et al.'s "curation index for el-Khiam points" (EKCI) includes a proper name that renders it unsuitable for general use. Perhaps Quinn et al. might consider renaming their measure the "curation index for tip resharpening" (CITR), the simple "sharpness" (or "tip sharpness") label that they use alternatively if briefly, or a suitably generic alternative? The index could prove valuable in the study of other types such as Folsom bifaces, whose tip form and sharpness are implicated in reduction analysis (e.g., Ahler and Geib 2000; Buchanan 2006) and generally in evolutionary studies of biface design and function (e.g., Hughes 1998). As have others, Quinn et al. want to estimate the original size of specimens found at reduced size. Their simple regression of blade axial length upon thickness (Figure 5) is statistically significant but, as elsewhere (e.g., Davis and Shea 1998; Shott et al. 2000) somewhat diffuse. As a result, it has only modest predictive value. Using it, Quinn et al. (this volume) fashion an allometric reduction measure that subtracts estimated original length from reduced length as observed upon recovery. They rescale the resulting measure when appropriate, but might consider further rescaling to account for haft size or thickness and the constraint it imposes on usable length of stone tools. For instance, North American Paleoindian flakeshavers were depleted when resharpened not to the haft, but to a point a centimeter or more above it, because, as a consequence of haft thickness and kinetics of use, the remaining exposed length of the tools was too short for the retouched bit to reach the worked material (Grimes and Grimes 1985). On balance, Quinn et al.'s (this volume) geometric and allometric measures, validated by experiments, deserve a place in the toolkit of reduction analysis.

Harper and Andrefsky (this volume) explore the metric correlates and functional contexts of dart and arrow use. Like others (e.g., Shott 1996b; VanPool 2006), they conclude that darts continued in use – or at least were recycled for later use – after the introduction of arrow technology. Unlike others, Harper and Andrefsky (this volume) conclude that Pajarito Plateau dart points were recycled principally as knives, not as projectile tips. As a minor point, it would take Harper and Andrefsky little time to validate their identification of specimens as darts or arrows against classification functions (Shott 1997b; Thomas 1978). They cite convincing evidence for the reuse of dart points, and reason that reuse involved "sawing and cutting" rather than the tipping of projectiles, because dart points are more extensively reduced than are arrow points. This logic is reasonable but arguable, for two reasons. First, they measure degree of resharpening indirectly, but might instead use the direct measures (e.g., invasiveness indices) documented in other studies, notably in this volume. Second, they assume that dart points would not undergo resharpening during their use as projectile tips. Therefore, all resharpening noted on them is attributed to reuse as knives. Yet dart points typically are of a size and form that accommodates some amount of resharpening while used on projectiles (see, for instance, Hunzicker's [2005] extensive resharpening of experimental Folsom points used exclusively as projectile tips). Also, as Harper and Andrefsky state, dart points might have been used simultaneously as both tips and knives, especially if hafted to foreshafts that themselves could be detached easily from mainshafts. Use of dart points as knives neither precludes nor necessarily follows use as projectile tips. Therefore, Pajarito Plateau dart points may be much reduced not because they originated as Late Archaic projectile tips and only later were reused as knives, but because during Archaic or later times they were reduced in original use. Whatever the case on the Pajarito Plateau, Harper and Andrefsky's study (this volume) illustrates the interpretive value of patterns of reduction in bifaces.

Andrefsky's chapter (this volume) exemplifies the approach taken by several contributors to this volume. He uses obsidian distance-to-source data from the northern Great Basin and reasonable, ethnographically informed threshold values to distinguish local from nonlocal scales of acquisition. Andrefsky's detailed analysis clearly documents patterning between distance-from-source and degree of reduction, measured using his own (Andrefsky 2006) hafted biface retouch index. Reasonably, he attributes the patterning to supply effects. It could, however, be influenced by manufacturing cost or other factors. Although supply patterns well with degree of reduction in some cases, the relationship is by no means universal, as Ballenger (2001) shows. As valid as Andrefsky's (this volume) analysis is, like most such approaches it nevertheless reduces continuous variables (source distance, degree of reduction) to dichotomous attributes, a treatment that does not fully exploit the potential of reduction measures. As small points, Tables 2 and 3 are somewhat underspecified (i.e., several expected values are less than five) and Figure 8's bottom-heavy scatter shows considerable variation in retouch values at a narrow and low range of distance values.

MacDonald (this volume) examines the influence of raw-material quality, abundance, and distribution by comparing the use of Upper Mercer and Kanawha cherts at Skink Rockshelter. His thesis is that tool design and curation rate are determined largely by stone abundance and how this varies with distance and time. In this respect,

MacDonald's approach resembles that of European Paleolithic archaeologists who interpret patterns and degrees of reduction in terms of material supply. The Paleolithic approach recruits traditional tool typology to the measurement of reduction. MacDonald uses Andrefsky's (2006) invasiveness index along with assemblage measures (even if assemblage sizes are quite low). The result both links the two levels or scales of reduction that Blades (this volume) discussed and illustrates the value of reduction measures in testing higher-level theory. MacDonald argues that tools of more-distant Upper Mercer should be more heavily reduced than those made of nearer Kanawha. Essentially, he extends the logic of fall-off curves to reduction as a function of distance. The result confirms his prediction, but the very small sample size qualifies it. Elsewhere MacDonald, like Prentiss and Clarke (this volume) and like Clarkson (this volume), equates number of tools with occupational intensity (e.g., Table 1, showing Late Woodland tools outnumbering Archaic ones, Woodland "intensity" thus being higher), which elides both deposition spans (all else equal, admittedly, Late Woodland deposition span at Skink might be less than Archaic spans, but this point must be demonstrated) and rates of use and discard.

Bradbury et al. (this volume) examine the effect of raw material upon reduction, independent of amount of use and resharpening. This subject is important because many reduction measures, especially in flake tools, require knowing or inferring original flake size for comparison to the discarded (and presumably reduced) form. Thus, Bradbury et al.'s chapter concerns estimation of tools' original size, not measurement of their reduction. Although chert sources vary greatly in mechanical attributes and applications, Bradbury et al. find that a tripartite division accommodates most variation. If the conclusion is borne out in further experiments, then lithic analysts need not measure the precise mechanical properties of each source, which, in any event, are apt to vary within the source formation (even within the cobble) depending upon context, degree of weathering, and intrinsic factors. Instead, analysts may apply Bradbury et al.'s tripartite scale and thereby reserve precious analytical time and talent for more advanced tasks. The conclusion has a pleasant implication for the several chapters here that compare assemblages or contexts without controlling for differences in raw material. One of Bradbury et al.'s most significant findings is that hammer type has only a slight effect on flake allometry and, by extension, reduction. Despite variation by material, however, platform thickness correlates significantly with flake weight. (Nevertheless, Bradbury et al.'s Figures 2 and 3 suggest some threshold above which platform thickness only weakly constrains weight. Further results might determine the boundary conditions for the effects of platform thickness.) This conclusion strengthens the validity of Pelcin's (1996) pioneering research and of reduction measures based on comparing observed size at discard to inferred original size.

Prentiss and Clarke (this volume) use ratios of tool types within assemblages to measure reduction. They combine these with reduction measures of individual tools. Partly, their chapter is an effort to encompass technological organization (Nelson 1991) within evolutionary archaeology. This intriguing prospect might help invigorate lithic analysis as it struggles to increase its relevance to broader theoretical currents. Although the argument further illustrates the service that reduction analysis can provide to theory development, it is largely beyond the scope of present discussion. Prentiss and Clarke's (this volume) account of changing patterns of land use, technological practices, and reduction in the Mid-Fraser valley is one example of the great breadth of reduction analysis. Their two case studies are widely separated in time and cultural context, a point that can be ignored in a heuristic study such as Prentiss and Clarke's. Their interpretation is reasonable, but their functional classification groups distinct types and includes at least one default category ("All other flake tools and light retouched scrapers"). Also, Prentiss and Clarke equate archaeological frequency with frequency of ancient use, without considering the intervening role of use life. It also is unclear if the "sudden shift" to light-duty tools and heavy reduction at Hidden Falls is an absolute or proportional change. The complex interactions of activity patterns, occupation span, and use life (e.g., Shott 1997a) urge caution in attributing change in assemblage composition to one factor only.

Clarkson (this volume) links three reduction measures – core platform rotation (little known in North America but common in Australia), Kuhn's (1990) geometric index, and his own invasiveness index – to changing scales and patterns of hunter–gatherer land use in the Wardaman country of northern Australia. As above, a qualified demurral on the validity of the geometric and invasiveness indices: they will not work on all varieties of flake blanks and tools. Neither, for instance, works well on the hafted endscrapers that are common in North American assemblages: the geometric index because retouch is concentrated on the distal edge and these tools have nearly rectangular longitudinal sections that vary as little during resharpening in t as in T (to use Clarkson's nomenclature), and invasiveness because that concentrated retouch consistently affects only one or few zones so produces little if any change in index values as reduction advances. Allometric measures are suitable for endscrapers and similar tools (e.g., Shott and Weedman 2007). Thus, Clarkson's chosen indices are not always the best reduction measures, but they are sensible choices for the retouched-flake industries that he studied.

A possible complication to Clarkson's analysis (this volume) involves artifact (including stone-tool) density as a measure of occupational intensity, which is a composite of population, rate of tool use, and duration. But the quantity also depends on two factors not controlled in Clarkson's treatment. One is sedimentation rate, a geological matter. The other, however, is rate of discard, which depends not just on rate of tool use but also on curation. No matter their use rate, highly curated tools are discarded and so contribute to artifact density at lower rates than equally used poorly curated tools. Clarkson measures reduction and links it to curation, but does not link either to discard rate or, by extension, occupational intensity. He might consider calibrating artifact density to curation rate, which may bear upon analysis and interpretation. This chapter's value lies in demonstrating the relevance of reduction to a range of cultural properties and practices connected with land use. The reduction thesis may or may not have "stale" implications for tool typology, but Clarkson clearly shows that reduction and its measurement are relevant to other issues as well.

Goodale and colleagues (this volume) equate variation in reduction processes with diversity, and inversely correlate both with efficiency. They model reduction as a function of material supply, quality, and, less convincingly, the producer:consumer ratio. This original approach has uncertain relevance to reduction measurement, and more to the inference of reduction sequences and their variation. This is less a criticism than an observation that places the model in a different analytical perspective. The model begs the question of efficiency, as though there were a single, unambiguous reduction sequence equally suitable to a wide range of material supply, core size and shape, and contexts, and risks conflating inherent technological variation with social causes. Also, Bradbury et al.'s chapter suggests that material quality may be an ordinal more than the interval variable that Goodale et al. assume. Their several case studies are heuristic examples rather than detailed analyses.

DISCUSSION

These chapters encompass a wide range of geographic space and archaeological time, as well as of technological and assemblage contexts. Yet they cover only a narrow range of subsistence and land use because most chapters study hunter–gatherer assemblages. Although hunter–gatherers made, used, and curated stone tools, nearly all prehistoric people used stone tools. Thus, concern with reduction and curation analysis should be broadly relevant.

Archaeologists used to, and perhaps still do, compile fall-off curves to map the use or discard of tool stone across landscapes. Generally, they expected such curves to fall off either rapidly or gradually, but in any case to fall off as a regular function of distance. This assumption not only ignores the possibility of transport over long distance but also emphasizes entropy among the factors that determine rates of use and discard. In effect, it assumes that ancient people did not organize their stone consumption, but, in fact, allowed circumstance to disorganize it. A strolling child who eats cookies from a bag scatters crumbs in his path. As he walks, first he eagerly gobbles handfuls of cookies. Many crumbs fall behind him. As his appetite wanes, he eats fewer cookies and so trails fewer crumbs. There is little organization to the child's cookie consumption, and the result is a cookie-crumb fall-off curve similar to many tool stone curves. It is unreasonable to suppose that ancient people could do no better than a hungry child with a sweet tooth, that their technologies were governed chiefly by the entropy of declining supply with distance from source. This view ignores the reality that ancient people did organize their technologies and manage tool supply with their land-use practices to prevent supply of materials and appropriate tools from inexorably decreasing with distance from sources of stone. Distance-to-source arguments also engage a scale problem when similar degrees of reduction occur with different distances from sources.

A universal reduction measure equally suitable to all assemblages and tool types is a chimera. Land-use practices and the organization of tool use in relation to these practices are highly variable and somewhat context-dependent, particularly with regard to stone availability. Tools, even their stone parts alone, are complex objects. Reduction alters different parts of tools to different degrees, introducing further complexity. Add to this the different purposes of various analyses and the suitability of various reduction measures to different kinds or parts of tools, and the complexities increase again. No single measure will serve all analytical purposes. Accordingly, in recent years we have witnessed a dramatic increase in the number of reduction measures proposed for various types of stone tools, and in the development and testing of those measures, as this volume shows. This collection is a major contribution both to the array of reduction measures and to reduction measurement generally, extending its analytical breadth and the range of assemblages to which reduction measurements are applied. All of this work is a testament to the importance of the reduction thesis and its accommodation to the great diversity of stone-tool types.

Many reduction measures are inevitable, and perhaps even desirable to some extent. No doubt still more will be devised, and then tested with experimental and archaeological data. Yet too many measures hamper comparison and broad application. Beyond new measures themselves, we see several urgent tasks that confront reduction analysis.

The first challenge concerns our concepts of appropriate or relevant dimensions or characteristics. Perhaps only weight is both a measure of size and a unitary character that is measured in only one way. So simple a dimension as length can be parsed by component or orientation (e.g., total, axial, blade, stem). There is no universal length measure suitable for all purposes, nor should there be. Instead, different measures record different aspects of reduction. Most reduction measures are calculated from orthogonal dimensions (e.g., maximum length, maximum width) or ratios among them. Orthogonal dimensions are perfectly legitimate and have the added virtue of wide use. Yet they reduce complex wholes to (usually) a few linear dimensions. They are no more a full description of tool size and design than are stick-figure caricatures adequate depictions of the human form. Archaeologists should consider measuring two-dimensional or, ideally, three-dimensional form and size using attribute schemes such as Buchanan's (2006) that are more detailed and therefore better approximations of actual size and form.

Second, as work advances in the development of new reduction measures, the need for their comparative evaluation and integration grows acute. We need well-controlled experiments that apply different measures to the same specimens. Several specimens of the same type can be fashioned, and then repeatedly dulled and resharpened. At each resharpening stage, variables such as dimensions, mass, and edge angles sufficient to calculate several reduction measures can be recorded and compared for their validity and accuracy in estimating degree of reduction. But other reduction measures may be devised in the future. Because specimens continue to experience reduction in size and change in shape at each resharpening, it is not always possible to make observations necessary for new reduction measures from the same original specimens. Size and form at, say, second resharpening are lost once specimens are resharpened a third time, and so on. To control comparisons among the number of measures, which is apt to grow, accurate casts of each specimen should be made at each resharpening episode to serve as archival controls for the later testing of even further reduction measures.

Third, archaeologists must determine each measure's fidelity to underlying causes and patterns of reduction. Lithic analysts might emulate paleodemographers, who confront similar problems in using several estimates of age at death in skeletal populations (Shott 2005:120). In a comparative study, Meindl et al. (1982: 75–6) calculated an average of several independent estimators' values weighted by each one's score on the first component of a principal-components analysis of all estimators. They interpreted this quantity as an aggregate estimator, and found that estimators varied in their correlation with it. All estimators were not equal. Although archaeologists have begun to compare reduction measures (e.g., Clarkson, this volume; Eren and Prendergast, this volume; Hiscock and Clarkson 2005), much more must be done.

Finally, this collection shows that archaeologists understand the connection between reduction as a physical process and curation as a behavioral one. Yet some archaeologists continue to treat curation as a qualitative state or condition that sometimes stands opposed to the equally qualitative condition of "expediency." Reduction and curation both are continuous processes; curation is no more a qualitative state than is "shortness." People do not ask, What is your shortness? They ask how tall you are. Archaeologists should not ask, Was this tool curated? They should ask, *How much* was this tool curated? Both curation and height are continuous variables. Simple height measurement makes that point with respect to height; reduction measures and distributions make the same point with respect to curation.

CONCLUSIONS

Reduction is integral to determining the form of used stone tools, just as are stylistic and functional aspects of design. No one doubts that size and form can be measured, nor that their measurement and analysis bear on many archaeological questions. No less is true of reduction.

Reduction is also a key aspect of the organization of stone tool use. Access to suitable materials, movement, resource scheduling, and work group composition, among other aspects of land use, can be understood through analysis of stone artifacts, including the reduction process.

This volume is the latest and among the best of recent reduction research. It marks the growing maturity of these approaches and their expanding scope. The volume also demonstrates the relevance of reduction analysis to more than typology; it includes curation distributions, land-use patterns, and most broadly technological organization. Its chief conclusion is unambiguous: stone-tool analysis makes no sense without understanding the places that tools occupy in the reduction process.

REFERENCES CITED

Ahler, Stanley A., and Phil R. Geib. 2000. Why Flute? Folsom Point Design and Adaptation. *Journal of Archaeological Science* 27:799–820.
Andrefsky, William. 1994. The Geological Occurrence of Lithic Material and Stone Tool Production Strategies. *Geoarchaeology: An International Journal* 9:345–62.
———. 1997. Thoughts on Stone Tool Shape and Inferred Function. *Journal of Middle Atlantic Archaeology.* 13:125–44.
———. 2005. *Lithics: Macroscopic Approaches to Analysis.* Second edition. Cambridge University Press, Cambridge.

2006. Experimental and Archaeological Verification of an Index of Retouch for Hafted Bifaces. *American Antiquity* 71:743–58.

Arnold, Jeanne E. 1987. Technology and Economy: Macroblade Core Production from the Channel Islands. In *The Organization of Core Technology*, edited by J. K. Johnson and C. A. Morrow, pp. 207–37. Westview Press, Boulder, CO.

Ballenger, Jesse A. M. 2001. *Dalton Settlement in the Arkoma Basin of Eastern Oklahoma*. Sam Noble Oklahoma Museum of Natural History Monographs in Anthropology No. 2, Norman, OK.

Bamforth, Douglas B. 1986. Technological Efficiency and Tool Curation. *American Antiquity* 51(1):38–50.

—— 1991. Technological Organization and Hunter-Gatherer Land Use: A California Example. *American Antiquity* 56(2):216–34.

Binford, Lewis R. 1973. Interassemblage Variability – The Mousterian and the "Functional" Argument. In *The Explanation of Culture Change*, edited by Colin Renfrew, pp. 227–54. Duckworth Press, London.

—— 1979. Organization and Formation Processes: Looking at Curated Technologies. *Journal of Anthropological Research* 35(3):255–73.

Binford, Lewis R., and Sally R. Binford. 1966. A Preliminary Analysis of Functional Variability in the Mousterian of Levallois Facies. *American Anthropologist* 68(2):238–95.

Blades, Brooke S. 2003. End Scraper Reduction and Hunter–Gatherer Mobility. *American Antiquity* 68:141–56.

Bleed, Peter. 1986. The Optimal Design of Hunting Weapons: Maintainability and Reliability. *American Antiquity* 51(4):737–47.

Bordes, François, and D. de Sonneville-Bordes. 1970. The Significance of Variability in Palaeolithic Assemblages. *World Archaeology* 2:61–73.

Buchanan, Briggs. 2006. An Analysis of Folsom Projectile Point Resharpening Using Quantitative Comparisons of Form and Allometry. *Journal of Archaeological Science* 33:185–99.

Carr, Philip J. (editor). 1994. *The Organization of Prehistoric North American Chipped Stone Tool Technologies*. International Monographs in Prehistory, Ann Arbor, MI.

Clarkson, Chris. 2002. An Index of Invasiveness for the Measurement of Unifacial and Bifacial Retouch: A Theoretical, Experimental, and Archaeological Verification. *Journal of Archaeological Science* 25:603–10.

Clarkson, Christopher, and Lara Lamb. 2005. *Lithics "Down Under": Australian Perspectives on Lithic Reduction, Use and Classification*. BAR International Series 1408. Oxbow, Oxford.

Cox, Steven L. 1986. A Re-analysis of the Shoop Site. *Archaeology of Eastern North America* 14:101–70.

Cresson, Jack. 1990. Broadspear Lithic Technology: Some Aspects of Biface Manufacture, Form, and Use History with Insights towards Understanding Assemblage Diversity. In *Experiments and Observations on the Terminal Archaic of the Middle Atlantic Region*, edited by R. W. Moeller, pp. 105–30. Archaeological Services, Bethlehem, CT.

Davis, Zachary J., and John J. Shea. 1998. Quantifying Lithic Curation: An Experimental Test of Dibble and Pelcin's Original Flake-Tool Mass Predictor. *Journal of Archaeological Science* 25:603–10.

Dibble, Harold L. 1995. Middle Paleolithic Scraper Reduction: Background, Clarification, and Review of Evidence to Date. *Journal of Archaeological Method and Theory* 2:299–368.

Dibble, Harold L., and Andrew W. Pelcin. 1995. The Effect of Hammer Mass and Velocity on Flake Mass. *Journal of Archaeological Science* 22:429–39.

Ellis, Chris. 1984. "Paleo-Indian Lithic Technological Structure and Organization in the Lower Great Lakes Area: A First Approximation." Ph.D. diss., Department of Archaeology, Simon Fraser University.

——— 2004. Understanding "Clovis" Fluted Point Variability in the Northeast: A Perspective from the Debert Site. *Canadian Journal of Archaeology* 28:205–53.

Eren, Metin, Manuel Dominguez-Rodrigo, Steven L. Kuhn, Daniel S. Adler, Ian Le, and Ofer Bar Yosef. 2005. Defining and Measuring Reduction in Unifacial Stone Tools. *Journal of Archaeological Science* 32:1190–1201.

Flenniken, J. Jeffrey, and Philip J. Wilke. 1989. Typology, Technology, and Chronology of Great Basin Dart Points. *American Anthropologist* 91:149–58.

Grimes, J. R., and B. L. Grimes. 1985. Flakeshavers: Morphometric, Functional and Life Cycle Analyses of a Paleoindian Unifacial Tool Class. *Archaeology of Eastern North America* 13:35–57.

Hayden, Brian. 1977. Stone Tool Functions in the Western Desert. In *Stone Tools as Cultural Markers: Change, Evolution, and Complexity*, edited by R. Wright, pp. 178–88. Australian Institute of Aboriginal Studies, Canberra.

Hiscock, Peter. 1996. Transformations of Upper Paleolithic Implements in the Dabba Industry from Haua Fteah (Libya). *Antiquity* 70:657–64.

——— 1999. Revitalising Artefact Analysis. In *Archaeology of Aboriginal Australia: A Reader*, edited by T. Murray, pp. 257–65. Allen and Unwin, St. Leonards, New South Wales.

Hiscock, Peter, and Val Attenbrow. 2005. *Australia's Eastern Regional Sequence Revisited: Technology and Change at Capertee 3*. BAR International Series No. 1397. Oxbow, Oxford.

Hiscock, Peter, and Chris Clarkson. 2005. Measuring Artefact Reduction: An Examination of Kuhn's Geometric Index of Reduction. In *Lithics "Down Under": Australian Perspectives on Lithic Reduction, Use and*

Classification, edited by C. Clarkson and L. Lamb., pp. 7–20. BAR International Series 1408. Oxbow, Oxford.

Hoffman, C. Marshall. 1985. Projectile Point Maintenance and Typology: Assessment with Factor Analysis and Canonical Correlation. In *For Concordance in Archaeological Analysis: Bridging Data Structure, Quantitative Technique, and Theory*, edited by C. Carr, pp. 566–612. Westport Press, Kansas City, MO.

Hughes, Susan S. 1998. Getting to the Point: Evolutionary Change in Prehistoric Weaponry. *Journal of Archaeological Method and Theory* 5:345–408.

Hunzicker, David A. 2005. "Folsom Hafting Technology: An Experimental Archaeological Investigation into the Design, Effectiveness, Efficiency and Interpretation of Prehistoric Weaponry." M.A. thesis, Department of Museum and Field Studies, University of Colorado, Boulder.

Johnson, Jay K., and Carol A. Morrow (editors). 1987. *The Organization of Core Technology*. Westview Press, Boulder, CO.

Kelly, Robert L. 1988. The Three Sides of a Biface. *American Antiquity* 53(4):717–34.

Kelly, Robert L., and Lawrence C. Todd. 1988. Coming into the Country: Early Paleoindian Hunting and Mobility. *American Antiquity* 53:231–44.

Kuhn, Steven L. 1990. A Geometric Index of Reduction for Unifacial Stone Tools. *Journal of Archaeological Science* 17:583–93.

———. 1991. "Unpacking" Reduction: Lithic Raw Material Economy in the Mousterian of West-Central Italy. *Journal of Anthropological Archaeology* 10:76–106.

Meindl, Richard S., C. Owen Lovejoy, and Robert P. Mensforth. 1982. Skeletal Age at Death: Accuracy of Determination and Implications for Human Demography. *Human Biology* 55:73–87.

Morrow, Juliet E. 1997. Scraper Morphology and Use-Life: An Approach for Studying Paleoindian Lithic Technology and Mobility. *Lithic Technology* 22:70–85.

Nelson, Margaret C. 1991. The Study of Technological Organization. *Archaeological Method and Theory* 3:57–100.

Parry, William J., and Robert L. Kelly. 1987. Expedient Core Technology and Sedentism. In *The Organization of Core Technology*, edited by Jay K. Johnson and Carol A. Morrow, pp. 285–304. Westview Press, Boulder, CO.

Pelcin, Andrew W. 1996. "Controlled Experiments in the Production of Flake Attributes." Ph.D. diss., Dept. of Anthropology, University of Pennsylvania, Philadelphia.

Sahnouni, Mohamed, Kathy Schick, and Nicholas Toth. 1997. An Experimental Investigation into the Nature of Faceted Limestone "Spheroids" in the Early Palaeolithic. *Journal of Archaeological Science* 24:701–13.

Shott, Michael J. 1986. Technological Organization and Settlement Mobility: An Ethnographic Examination. *Journal of Anthropological Research* 42:15–51.

———. 1989a. Diversity, Organization, and Behavior in the Material Record: Ethnographic and Archaeological Examples. *Current Anthropology* 30:283–301.

———. 1989b. On Tool-Class Use Lives and the Formation of Archaeological Assemblages. *American Antiquity* 54(1):9–30.

———. 1994. Size and Form in the Analysis of Flake Debris: Review and Recent Approaches. *Journal of Archaeological Method and Theory* 1:69–110.

———. 1995. How Much Is a Scraper? Uniface Reduction, Assemblage Formation, and the Concept of "Curation." *Lithic Technology* 20:53–72.

———. 1996a. An Exegesis of the Curation Concept. *Journal of Anthropological Research* 52:259–80.

———. 1996b. Innovation and Selection in Prehistory: A Case Study from the American Bottom. In *Stone Tools: Theoretical Insights into Human Prehistory*, edited by G. Odell, pp. 279–313. Plenum, New York.

———. 1997a. Stones and Shafts Redux: The Metric Discrimination of Chipped-Stone Dart and Arrow Points. *American Antiquity* 62:86–101.

———. 1997b. Activity and Formation as Sources of Variation in Great Lakes Paleoindian Assemblages. *Midcontinental Journal of Archaeology* 22:197–236.

———. 2005. The Reduction Thesis and Its Discontents: Overview of the Volume. In *Lithics "Down Under": Australian Perspectives on Lithic Reduction, Use and Classification*, edited by C. Clarkson and L. Lamb, pp. 109–25. BAR International Series 1408.

Shott, Michael J., and Jesse A. M. Ballenger. 2007. Biface Reduction and the Measurement of Dalton Curation: A Southeastern Case Study. *American Antiquity*, 72: 153–75.

Shott, Michael J., Andrew P. Bradbury, Philip J. Carr, and George H. Odell. 2000. Flake Size from Platform Attributes: Predictive and Empirical Approaches. *Journal of Archaeological Science* 27:877–94.

Shott, Michael J., and Paul Sillitoe. 2005. Use Life and Curation in New Guinea Experimental Used Flakes. *Journal of Archaeological Science* 32:653–63.

Shott, Michael J., and Kathryn Weedman. 2007. Measuring Reduction in Stone Tools: An Ethnoarchaeological Study of Gamo Hidescraper Blades from Ethiopia. *Journal of Archaeological Science* 34: 1016–35.

Surovell, Todd A. 2003. "Behavioral Ecology of Folsom Lithic Technology." Ph.D. diss., Department of Anthropology, University of Arizona, Tucson.

Thomas, David H. 1978. Arrowheads and Atlatl Darts: How the Stones Got the Shaft. *American Antiquity* 43:461–72.

Thompson, D'Arcy. 1917. *On Growth and Form*. Cambridge University Press, Cambridge.

Tindale, Norman. 1965. Stone Implement Making among the Nakako, Ngadadjara and Pitjandjara of the Great Western Desert. *Records of the South Australian Museum* 15:131–64.

Torrence, Robin. 1983. Time Budgeting and Hunter–Gatherer Technology. In *Hunter-Gatherer Economy in Prehistory: A European Perspective*, edited by G. Bailey, pp. 11–22. Cambridge University Press, Cambridge.

　1989. Retooling: Towards a Behavioral Theory of Stone Tools. In *Time, Energy, and Stone Tools*, edited by R. Torrence, pp. 57–66. Cambridge University Press, Cambridge.

Truncer, James J. 1990. Perkiomen Points: A Study in Variability. In *Experiments and Observations On the Terminal Archaic of the Middle Atlantic Region*, edited by R. Moeller, pp. 1–62. Archaeological Services, Bethlehem, CT, USA.

VanPool, Todd L. 2006. The Survival of Archaic Technology in an Agricultural World: How the Atlatl and Dart Endured in the North American Southwest. *Kiva* 71:429–52.

Weedman, Kathryn J. 2002. On the Spur of the Moment: Effects of Age and Experience on Hafted Stone Scraper Morphology. *American Antiquity* 67:731–44.

Wheat, Joe Ben. 1974. Artifact Life Histories: Cultural Templates, Typology, Evidence and Inference. In *Primitive Art and Technology*, edited by J. Raymond, B. Loveseth, and G. Reardon, pp. 7–15. University of Calgary Department of Archaeology, Calgary.

Wilson, Jennifer Keeling, and William Andrefsky, Jr. 2006. The Debitage of Bifacial Technology: An Application of Experimental Data to the Archaeological Record. Paper presented at the 59th Annual Northwest Anthropological Conference, Seattle, WA.

PART TWO

PRODUCTION, REDUCTION, AND RETOUCH

3 METIN I. EREN AND MARY E. PRENDERGAST

COMPARING AND SYNTHESIZING UNIFACIAL STONE TOOL REDUCTION INDICES

Abstract

Intensity of stone tool reduction has important implications for understanding hominid behavior, tool use and modification, mobility, and cognitive ability. There are a variety of reduction indices available to the lithic analyst. While each has strengths and weaknesses, different index values obtained on the same stone tools do not necessarily correlate with each other. Significantly different interpretations of an assemblage may be made depending on the analyst's choice of reduction index. In this paper we demonstrate this point by presenting different reduction indices calculated for both an experimental assemblage and a sample from the La Colombière Perigordian assemblage. Additionally, this paper presents models for combining different indices in order to better understand retouch and resharpening on unifacial stone tools.

INTRODUCTION

Archaeological quantification allows comparison between groups or attributes of artifacts that may otherwise be difficult to understand.

We would like to thank William Andrefsky, Jr., for inviting us to present this paper in the symposium "Artifact Life-Cycle and the Organization of Lithic Technologies" at the seventy-first Society for American Archaeology (SAA) conference in San Juan, Puerto Rico, and for including it in this volume. Thanks to Ofer Bar-Yosef, Manuel Dominguez-Rodrigo, David Meltzer, and C. Garth Sampson for their support and suggestions for the presentation of this paper. Thanks to Michael Shott, David Meltzer, William Andrefsky, and the 2006 SAA Student Paper Award committee for comments and suggestions that greatly improved this paper. Thanks also to Peter Hiscock, whose informative and kind suggestions in San Juan motivated the reanalysis of Kuhn's index

Additionally, it can organize and simplify data, as well as reveal hidden patterns in the archaeological record. Although quantification of unifacial stone tool retouching and resharpening has received considerable attention over the past twenty years (e.g., Andrefsky 2006; Clarkson 2002; Davis and Shea 1998; Dibble 1995, 1998; Dibble and Pelcin 1995; Eren et al. 2005; Hiscock and Clarkson 2005; Kuhn 1990, 1992; Pelcin 1998; Shott 2005; Shott et al. 2000; see also Clarkson, Hiscock and Clarkson, Quinn et al., all this volume), there remain several unresolved issues. Attempts to measure the abstract concept of "reduction" from a flake or blade blank have become ever more complex. Indices sometimes correlate reduction with different stone tool attributes that change through retouching or resharpening: retouch invasiveness, edge thickness, or volume loss. Other times, reduction is defined as mass or weight lost from a blank (Pelcin 1998) – in these cases different variables act as proxies for mass loss.

At the beginning of this study, our goal was to determine which reduction index most accurately measured mass loss. We applied three reduction indices – Kuhn's (1990) index of reduction (IR), Clarkson's (2002) index of invasiveness (II), and Eren et al.'s (2005) estimated reduction percentage (ERP) – to the same replicated assemblage of scrapers that served as the control assemblage. We also applied the indices to an archaeological assemblage from the Upper Paleolithic site of La Colombière in France. However, as we proceeded with our analysis, we began to develop an altogether new understanding of tool reduction. In particular, we began to gain a new appreciation of the ways in which the use of reduction indices was influencing our perception of the reduction sequence.

Though we discovered that some reduction indices did, indeed, gauge mass loss better than others, we also learned that "reduction" is

in this paper. We would like to thank Diana Loren, Ofer Bar-Yosef, and the Peabody Museum of Archaeology and Ethnology for providing the La Colombière Assemblage. Finally, one of us (Eren) would like to thank Mustafa Eren, Kathleen Eren, and Nimet Eren for support and financial assistance.

Research space was provided by the following institutions: the Stone Age Laboratory at Harvard University's Peabody Museum of Archaeology and Ethnology; the Department of Anthropology at Harvard University; the Department of Anthropology at Southern Methodist University; and the Department of Archaeology at the Cleveland Museum of Natural History. Thanks to Think Computer Corporation for technical/computer assistance.

Any mistakes or shortcomings in this paper are our own.

too complex an idea to be simply defined by one variable, whether that variable is edge thickness, scar invasiveness, or volume loss – or going further, whether any of these variables can be used as a proxy for mass loss. Through the presentation of an experimental analysis and the analysis of an archaeological assemblage, we will demonstrate that the use of different reduction indices on the same assemblage can actually produce differing results (graphically and quantitatively) as to how "reduced" that particular assemblage is. Although this result on its own has major theoretical implications for how an archaeologist interprets a lithic assemblage, what we also intend to suggest with this analysis is that reduction cannot simply be described by a single variable, and perhaps a combination of variables would more accurately depict "reduction." We feel that a multiple-index approach to reduction encompasses the complexity of the term much better than any single index on its own.

Some scholars may question why our analysis does not include the application of the mass predictor equation (Dibble and Pelcin 1995). First, time and space constraints permit only the presentation of three reduction indices in this paper. Second, many scholars (Davis and Shea 1998; Dibble 1998; Pelcin 1998; Shott et al. 2000) have shown that the mass predictor's applicability is dependent on raw material type. Third, even when raw material type is accounted for, the mass predictor has been shown to be inaccurate in its calculation of a retouched tool's original mass. Nevertheless, despite its drawbacks, the mass predictor model is extremely valuable and may potentially contribute significantly to the understanding of blank retouching and resharpening.

THE ERP, IR, AND II

Three reduction indices used here, and a brief explanation of each is in order. The first reduction index used in this experiment is the estimated reduction percentage (ERP, Eren et al. 2005) (Figure 3.1). Unlike most retouching indices, this approach treats artifact size and shape in three, rather than two dimensions. The ERP method quantifies volume loss due to retouch/resharpening relative to the original unmodified blank by reconstructing the original volume of a modified blank, thereby allowing a realistic percentage of volume loss to be calculated. In an experimental test, the ERP quantified overall mass

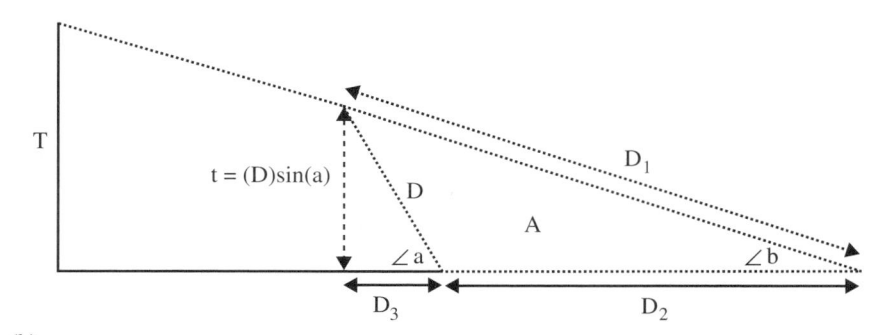

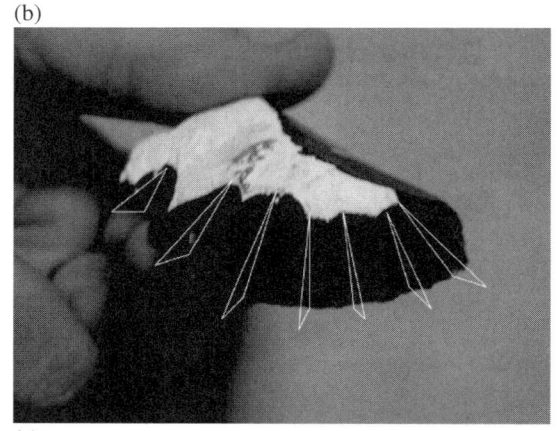

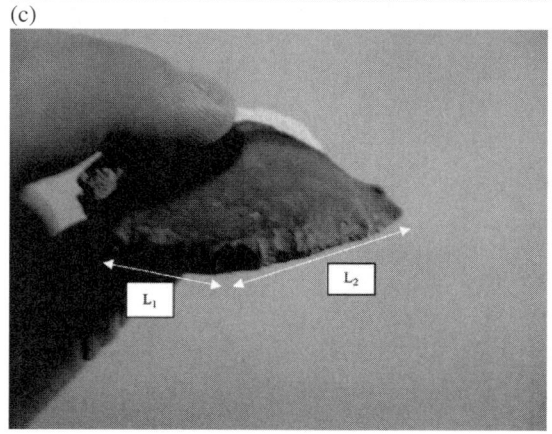

FIGURE 3.1. Estimated reduction percentage. (a) Imaginary triangles with sides D_1, D_2, D and area A are constructed onto the cross section of a unifacial stone tool, with the variables described in the text; (b) the area of each triangle is calculated and averaged; (c) The averaged area is then multiplied by the length of retouched edge, L, to get the volume missing from the tool (Eren et al. 2005: 1193, 1195).

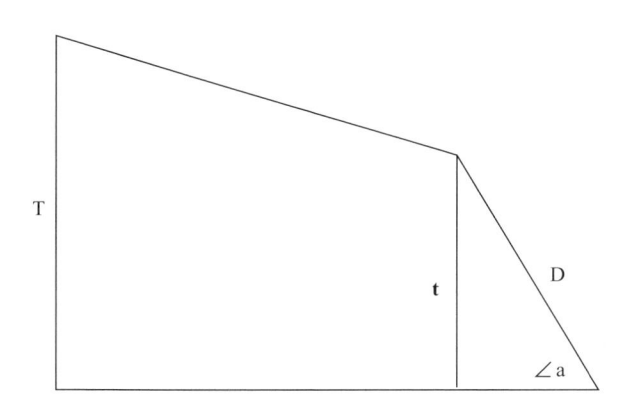

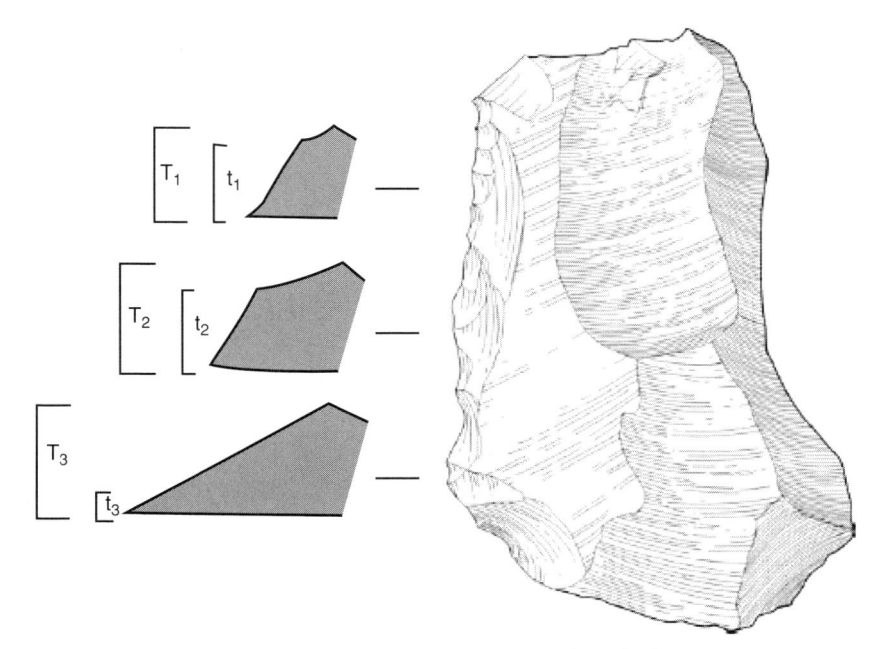

FIGURE 3.2. Index of reduction. (a) Measurements required for calculating Kuhn's (1990) index of reduction (Eren et al. 2005: 1192); (b) demonstration of how the index of reduction changes as retouching progresses (figure from Hiscock and Clarkson 2005: 1016).

loss better than other reduction indices (Eren et al. 2005). To measure volume lost due to retouch, the volume of debitage removed from a unifacial tool is found with the reduction equation (RE),

$$V = L\frac{D^2}{2}(\sin^2(a)\cot(b) - \sin(a)\cos(a)),$$

where b is the dorsal plane angle, a is the retouched edge angle, D is the retouch length, and L is the length edge retouched. The value V estimates the volume of debitage removed from a unifacial tool (*VolumeEstimatedDebitage*). To calculate the percentage of volume loss, called the estimated reduction percentage (ERP), in relation to the original unmodified blank, one has to measure the volume of the retouched tool by putting it in water and measuring the volume displacement (*VolumeRetouchedPiece*). This enables one to solve the equation

$$\frac{VolumeEstimatedDebitage}{VolumeEstimatedDebitage + VolumeRetouchedPiece}.$$

The second reduction index used in this experiment is the index of reduction (IR, Kuhn 1990) (Figure 3.2). Kuhn's (1990) equation for the IR,

$$\text{IR} = \frac{(D)\sin(a)}{T},$$

quantifies the ratio of the maximum medial thickness of the unifacial tool (T) to the vertical thickness of the flake at the retouch terminations (t). Trigonometry equates t with the depth of retouch scars D multiplied by the angle of retouch a. Kuhn (1990) shows that as edges are progressively retouched, the IR increases in value. Though Kuhn (1990) in his experiments never used the IR to quantify tool mass/weight loss, Hiscock and Clarkson (2005: 1021, 1022) suggest that Kuhn's IR "is strongly positively related to log(%weight loss)" and that it is "a robust indicator of the extent of reduction when retouching patterns are suited to the calculation of the index". They also note that the IR is not linearly scaled and should be calibrated if it is to indicate weight lost from a specimen accurately (Hiscock and Clarkson 2005: 1019). However, Eren et al. (2005) demonstrate that without calibration the IR does not gauge mass or volume loss accurately. Further, data presented below cast doubt on the accuracy of the IR for gauging mass loss, even with calibration. For this experiment average values of a and D were calculated and used in the equation $(D)\sin(a)$, which was then divided by the single medial thickness value T.

The third reduction index used in this experiment is the index of invasiveness (II, Clarkson 2002) (Figure 3.3). Dividing a stone tool into 16 segments (8 segments on both the dorsal and ventral sides) and two zones (an outer zone and an inner zone), Clarkson (2002) assigns a score to each segment according to the invasiveness of retouch. A score of 0 is assigned to a segment exhibiting no retouch. A score of 0.5 is assigned to a segment exhibiting retouch invading only the outer zone. A score of 1 is assigned to a segment exhibiting retouch invading the inner zone. The scores are then summed to give a total figure for the invasiveness of the stone tool. Dividing this sum by the number of segments gives a result ranging between 0 and 1. Clarkson's formula for calculating the Index of Invasiveness is

Index of Invasiveness = $\Sigma S_s / 16$,

where ΣS_s is the summed total of segment scores (Clarkson 2002: 68). Clarkson correlates his index with weight loss and retouch blows on the basis of experimental work. He notes that the II is not linearly scaled and should be calibrated if it is to indicate weight lost from a specimen accurately.

As shown above, each index is correlated with mass/weight lost. Researchers using each index argue for the importance of mass/weight loss for understanding the reduction concept (Clarkson 2002: 74; Eren et al. 2005: 1191; Hiscock and Clarkson 2005: 1020). Clarkson (2002: 74) goes so far as to call percentage weight lost from a specimen an "absolute measure of reduction."

EXPERIMENTAL ASSEMBLAGE

The experimental assemblage consists of 49 blanks knapped by Eren using hard hammer direct percussion. (Due to a typographical error that was discovered after we completed our analysis, we eliminated one artifact from our original sample of 50.) The amount and location of unifacial retouch on each specimen differed substantially, resulting in a diverse assemblage. Summary data of the unmodified blanks are presented in Table 3.1, whereas reduction data are presented in Table 3.2. Box plots of the ERP, IR, and II values are shown in Figure 3.4.

Table 3.1. Summary statistics of the experimental assemblage before retouching

	m	V	L	W	T	MD	L:W	SA	PL	PD	PA	EPA
n	49	49	49	49	49	49	49	49	48	48	48	48
mean	42.49	19.58	68.80	42.51	11.55	73.18	1.75	2,994.60	21.80	8.20	211.46	70.00
sd	32.25	14.90	13.69	13.80	4.77	14.41	0.53	1,340.9	9.82	3.44	178.4	10.50
min	5.50	2.53	43.75	19.14	5.02	47.27	0.76	837.38	4.89	1.77	8.66	33.00
q1	18.60	8.57	59.10	32.66	7.74	61.82	1.37	2,020.64	16.09	5.56	89.46	64.00
med	29.40	13.55	67.73	39.09	10.68	73.09	1.66	2,534.35	20.18	7.85	158.41	71.00
q3	57.30	26.41	77.23	50.71	14.04	82.40	2.13	3,896.68	29.22	10.61	310.01	77.00
max	140.20	64.61	103.66	77.46	24.77	109.98	3.45	6,502.55	42.02	17.02	715.18	93.00
range	134.70	62.07	59.91	58.32	19.75	62.71	2.69	5,665.17	37.13	15.25	706.53	60.00
iqr	38.70	17.83	18.13	18.05	6.30	20.58	0.76	1,876.03	13.13	5.05	220.56	13.00

Notes: m = mass; V = volume; L = length; W = width; T = medial thickness; MD = maximum dimension; L:W = length-to-width ratio; SA = surface area; PL = platform length; PD = platform depth; PA = platform area; EPA = exterior platform angle.
(m) is in grams. (L), (W), (T), (MD), (PL), and (PD) are in millimeters. (SA) and (PA) are in millimeters. (EPA) is in degrees.

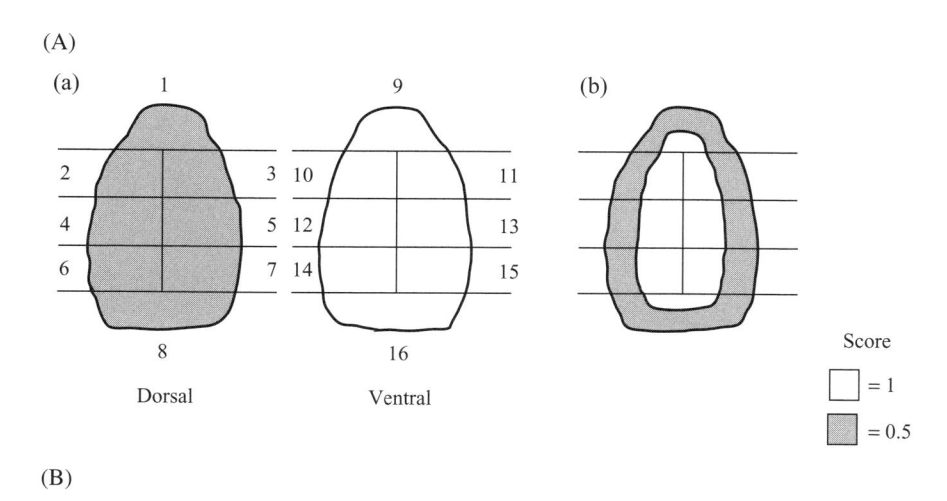

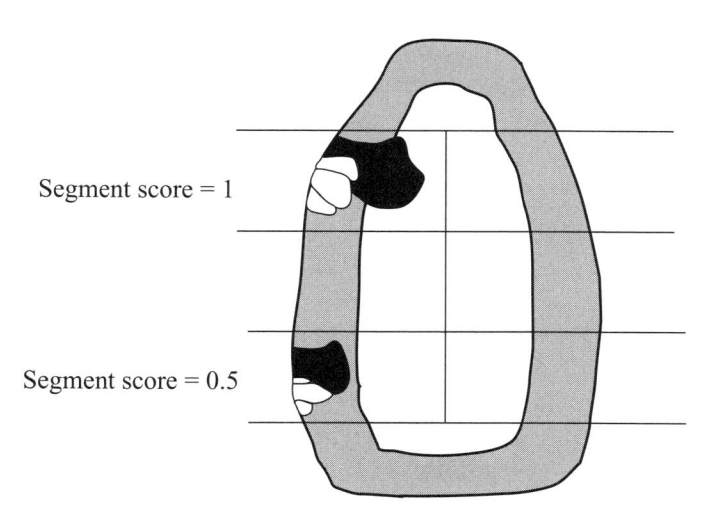

$$\text{Index} = \frac{\text{Total segment score } (1 + 0.5)}{\text{Total segment } (16)} = 0.093$$

FIGURE 3.3. Index of invasiveness. (A) Method for constructing the sixteen segments (a) and the inner and outer zones (b) required for calculating Clarkson's (2002) index of invasiveness (figure from Clarkson 2002: 67); (B) demonstration of how the index of invasiveness is applied to a stone tool (figure from Clarkson 2002: 68).

Table 3.2. Summary statistics of experimental assemblage reduction data

	ERP	II	IR
n	49	49	49
mean	0.08	0.13	0.49
sd	0.04	0.05	0.10
min	0.02	0.06	0.27
q1	0.05	0.09	0.42
med	0.08	0.09	0.49
q3	0.11	0.19	0.55
max	0.21	0.25	0.73
range	0.19	0.19	0.46
iqr	0.06	0.09	0.14

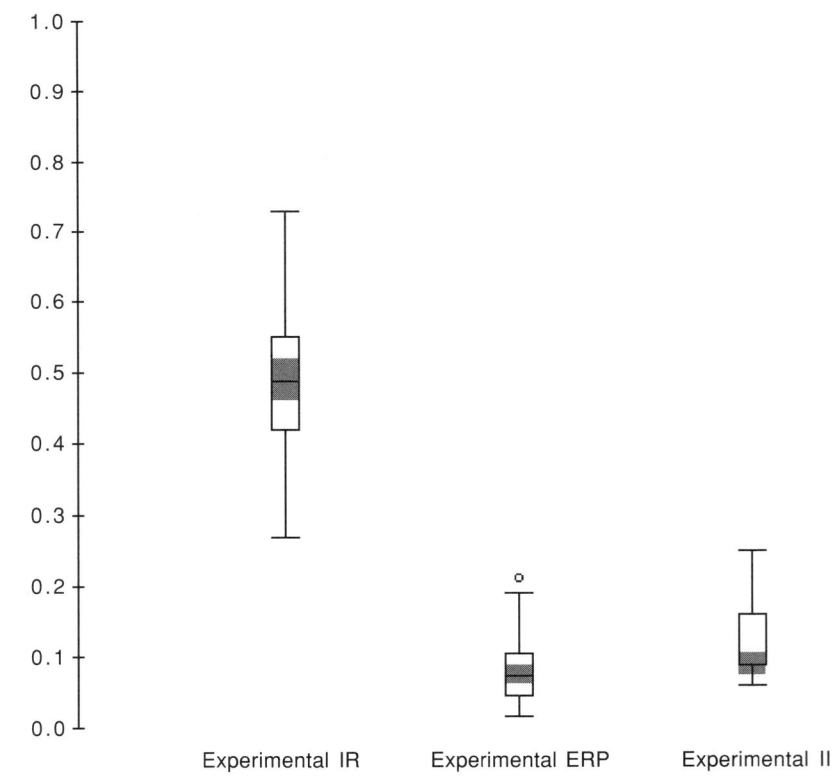

FIGURE 3.4. Experimental assemblage box plots of the IR, ERP, and II ($n = 49$).

Table 3.3. Experimental assemblage correlations of each reduction index with the others and with the actual percentage mass lost ($n = 49$)

	ERP	IR	II	Percentage actual mass lost
ERP	1.0000	—	—	—
IR	.4997	1.0000	—	—
II	.4202	−.0883	1.0000	—
Percentage actual mass lost	.7308	.2200	.3549	1.0000

The means, medians, and ranges differ drastically between each reduction indices, though the ERP and II are more similar than either is to the IR. Based on Figure 3.4 alone, it is apparent that, if using a single reduction index in isolation, a researcher may draw different conclusions on how "reduced" or "exhausted" the lithic assemblage is. Despite the differences among them, if the IR, II, and ERP are each attempting to calculate the amount of material removed from the tool, one might expect some degree of overlap among the reduction indices, whether calculations are based on two dimensions (in the case of IR and II) or three (in the case of ERP). However, the side-by-side box plots show large differences between the indices in terms of their medians and ranges.

A better means of comparison among the indices is correlation. For example, a heavily reduced scraper should have a high degree of reduction whether calculated by IR, II, or ERP, so we expect these three indices to move in tandem. However, correlations among these indices (Table 3.3, Figures 3.5–3.7) are generally weak, with the exception of a moderate correlation between the IR and ERP ($r = 0.49$).

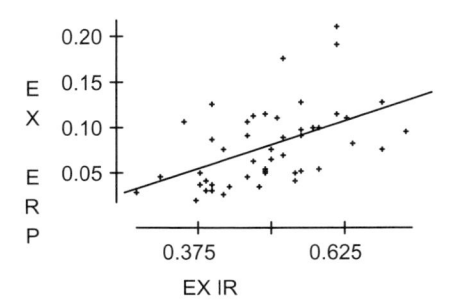

FIGURE 3.5. Experimental assemblage correlation between the ERP and IR ($r = .4997$).

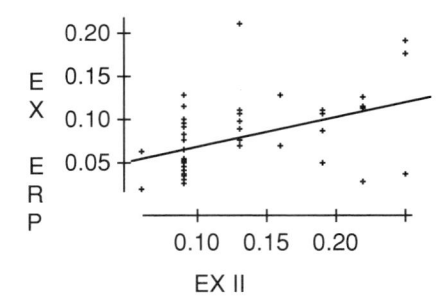

FIGURE 3.6. Experimental assemblage correlation between ERP and II ($r = .4202$).

Interestingly, given that the box plots showed the ERP and the II to have closer medians and ranges than either has to the IR, the correlation between the ERP and II is quite low. This indicates that although the ERP and II ranges appear similar at first, in fact, when each specimen is analyzed individually the ERP and II calculate very different values.

The ERP, IR, and II were then compared to the actual percentage of mass lost. Mass loss was calculated by measuring a specimen's mass on a scale before and after retouching and then subtracting the values. The correlations between the ERP, IR, and II and percentage mass lost are shown in Table 3.3. It is evident that the ERP gauges mass lost better than either the IR or the II, which supports the conclusions of Eren et al. (2005). Yet this does not mean that the IR or II is obsolete. On the contrary, as will be discussed below, each index is responding to different aspects of reduction, resulting in variability among indices.

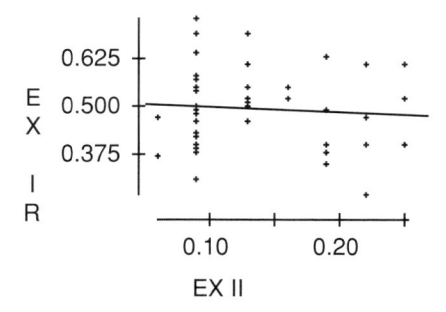

FIGURE 3.7. Experimental assemblage correlation between IR and II ($r = -.0883$).

Table 3.4. La Colombière tool types present in the sample ($n = 113$)

Type	n	Percentage
Retouched blade	44	38.93%
Backed blade	33	29.20%
Retouched flake	17	15.04%
End scraper	11	9.73%
Borer	5	4.42%
Retouched core trimming element	2	1.77%
Retouched core tablet	1	0.88%

The best way to examine this variability is by comparing indices calculated on individual tools. Because the archaeological assemblage has a higher diversity of tool types than the experimental assemblage, we turn to our analysis of it.

ARCHAEOLOGICAL ASSEMBLAGE

The archaeological assemblage consists of the Perigordian unifacial stone tools from La Colombière (LC), France. Excavated on multiple occasions since the 1870s, La Colombière is a rock shelter with multiple components. In 1948 Harvard University professors Hallam L. Movius, Jr. and Kirk Bryan (Movius and Judson 1956) identified these components as Neolithic, Magdalenian, and Perigordian. The Perigordian assemblage consists of a variety of unifacial stone tools, such as scrapers, notched pieces, and backed blades. Summary data for the La Colombière assemblage are presented in Tables 3.4 and 3.5.

Table 3.5. La Colombière fragmentation types present in the sample ($n = 113$)

Fragmentation	n	Percentage
Complete	54	47.79%
Distal	34	30.09%
Mid section	15	13.27%
Proximal	10	8.85%

Table 3.6. Summary statistics of La Colombière assemblage reduction data.

	ERP	II	IR
n	110	113	113
mean	0.1272	0.1103	0.5979
sd	0.1260	0.0668	0.2575
min	0.0008	0.0313	0.2296
q1	0.0288	0.0625	0.4065
med	0.0885	0.0938	0.5997
q3	0.1943	0.1563	0.7188
max	0.6207	0.3750	2.2482
range	0.6199	0.3438	2.0186
iqr	0.1655	0.0938	0.3123

The ERP, IR, and II were calculated for the LC assemblage (Table 3.6). As shown in Figure 3.8, the range of variation in each of these indices is much higher than that in the experimental assemblage. As in the experimental assemblage, each index's median and range are significantly different from the others. Differences are also reflected by the low degree of correlation the reduction indices have with each other (Table 3.7). As noted above, we expect the indices to move in tandem for a specimen that is highly reduced. Yet this is not the case, as two of the three correlations are quite low. An important exception exists: the apparently high correlation between the IR and ERP. This is negated by a distinct J-shaped pattern in the plotted data, with a fanning pattern in the residuals of regression analysis (Figure 3.9). These patterns show that there is a nonlinear relationship between these indices: as the two-dimensional IR increases on individual tools,

Table 3.7. La Colombière assemblage correlations of each reduction index with the other

	ERP	IR	II
ERP	1.0000	—	—
IR	0.6688	1.0000	—
II	0.2266	−0.04210	1.0000

Note. ERP-IR, $n = 110$; ERP-II, $n = 110$; II-IR, $n = 113$.

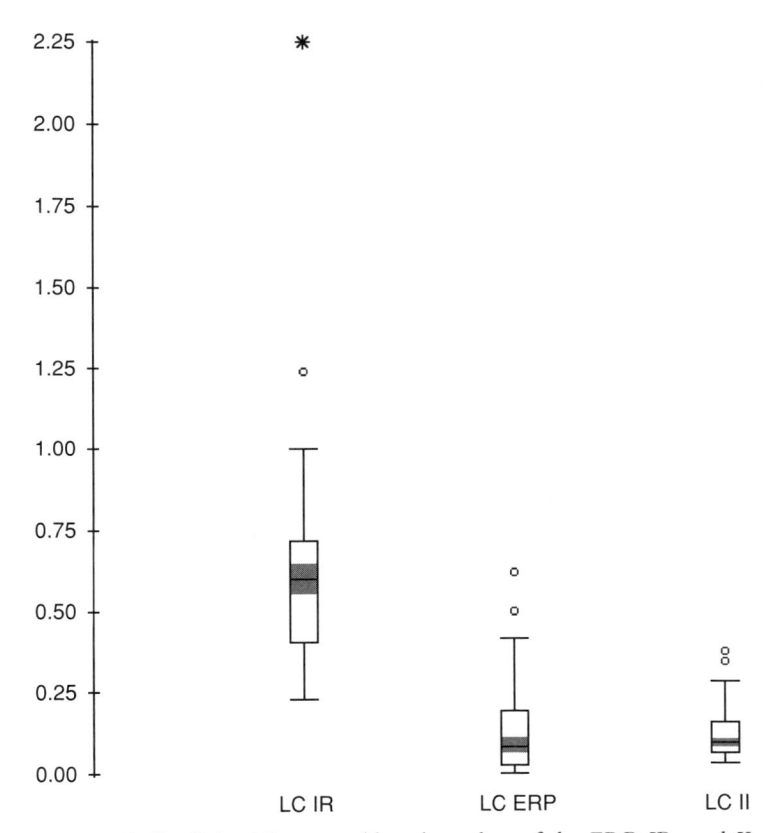

FIGURE 3.8. La Colombière assemblage box plots of the ERP, IR, and II ($n = 113$). Specimens measuring greater than 1 with the IR are specimens where the retouched areas are thicker than the medial thickness.

the three-dimensional ERP increases by much more, creating wide discrepancies between the two indices on heavily reduced tools.

Box-plots of individual tool types' reduction data show the same patterns as the experimental and archaeological assemblages (Table 3.8, Figures 3.10–3.14). Medians and ranges for each reduction index on each tool type differ drastically. Yet the comparison of different tool types provides interesting results.

The backed blades (Figure 3.10) and borers (Figure 3.11) are one such example. The IR and ERP for the backed blades have higher values than the IR and ERP for the borers. However, the borers have

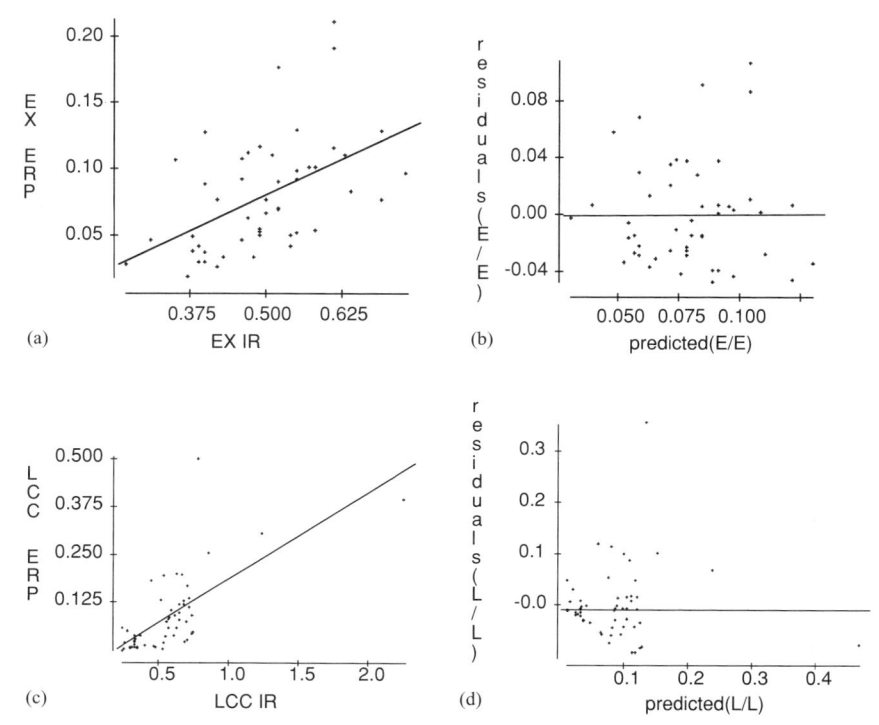

FIGURE 3.9. Scatterplots of IR versus ERP, with regression lines (a, c) and residuals of regression analysis (b, d), for the experimental assemblage ("EX") (a, b) and the La Colombière archaeological assemblage of 55 *complete tools only* ("LCC") (c, d). Fanning residuals (d) in the archaeological assemblage, in contrast to the experimental assemblage (b), suggest that the relationship between ERP and IR in the archaeological assemblage is nonlinear: as IR increases slightly, ERP increases greatly, suggesting that IR may not be as effective as ERP on heavily reduced tools. Specific traits of the outlying tools may be influencing this pattern.

a higher II value than the backed blades' II value. If each reduction index measures the same variable (e.g., mass/weight loss, or, alternatively, simply "reduction"), there would be a major discrepancy. However, when each index is understood to measure a different variable, the result makes sense: backed blades would be expected to have an exhausted edge (which would give a high IR value) and a lot of mass would be removed (which would give a high ERP value). However, from a dorsal viewpoint, backing is not invasive (which explains the low II value). On the other hand, a borer only requires a small

Table 3.8. La Colombière reduction data medians and ranges of each tool type

n	Backed blades 33	End scrapers 11	Borers 5	Core trimming element 2	Core tablet 1	Retouched blade 44	Retouched flake 17
IR med	0.72	0.74	0.33	0.56	0.59	0.48	0.59
IR range	0.66	0.80	0.45	0	0	0.53	2.01
ERP med	0.24	0.15	0.04	0.06	0.08	0.04	0.05
ERP range	0.59	0.27	0.07	0	0	0.23	0.39
II med	0.09	0.06	0.16	0.06	0.03	0.09	0.09
II range	0.16	0.31	0.13	0	0	0.34	0.22

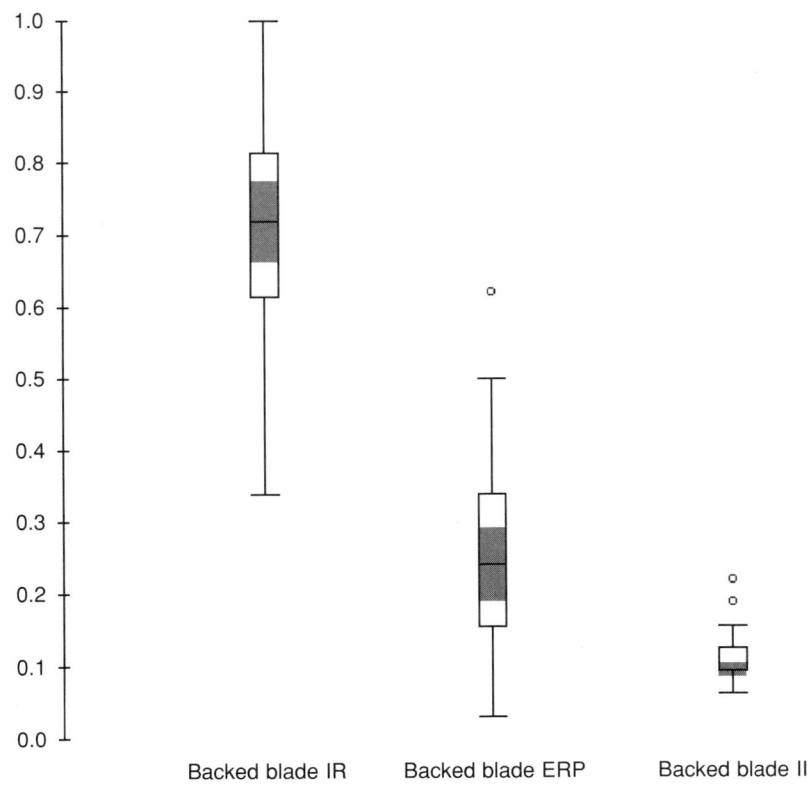

FIGURE 3.10. La Colombière backed blades' box plots of the ERP, IR, and II.

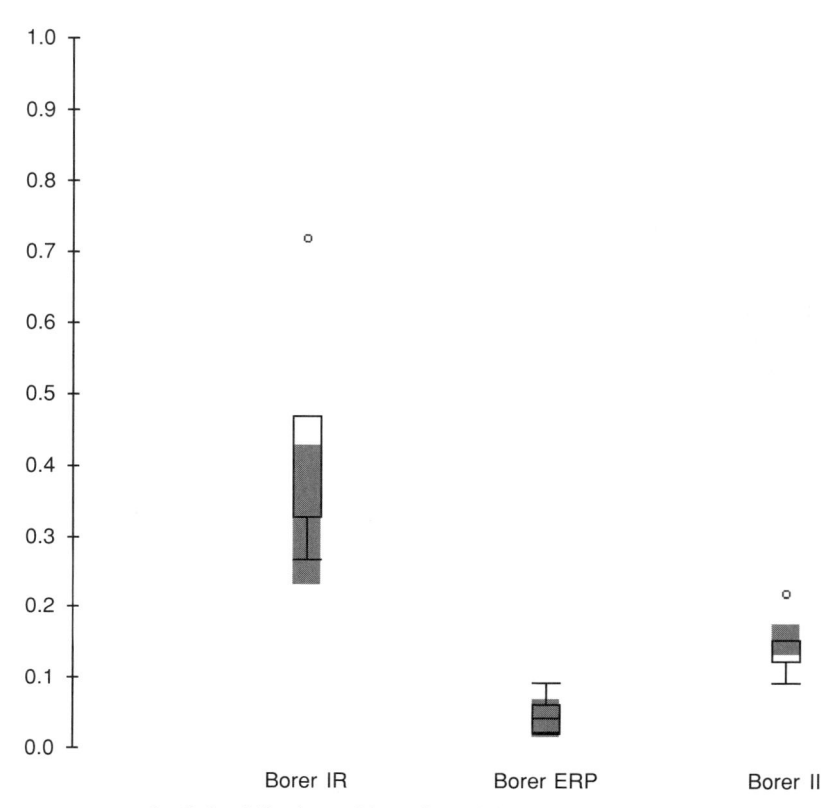

FIGURE 3.11. La Colombière borers' box plots of the ERP, IR, and II.

pointed section on a specimen. This would result in little overall mass lost (shown by the low ERP value), an edge that was not exhausted (shown by a low IR value), and retouch that was somewhat invasive on the functioning part of the tool (shown by the II). Comparisons such as the one above can be made between any of the tool types presented in Table 3.8.

Some researchers looking at Table 3.5 may question our use of reduction indices on broken tools. For this reason, we have provided Figure 3.15, showing differences in reduction values between complete and broken LC assemblages. Although there are some differences, the point of our paper is further confirmed: whether the indices are used on a complete specimen or a broken one, they still show divergent medians, ranges, and values.

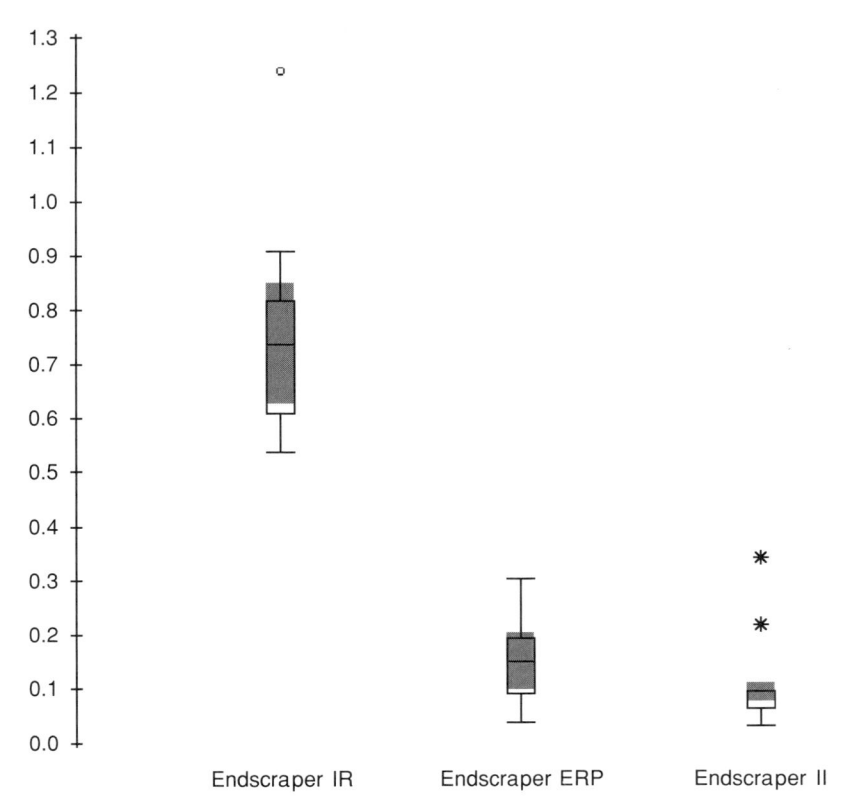

FIGURE 3.12. La Colombière end scrapers' box plots of the ERP, IR, and II.

ANOTHER LOOK AT THE IR

After the symposium "Artifact Life-Cycle and the Organization of Lithic Technologies" at the 71st Society for American Archaeology Conference in San Juan, Puerto Rico, informative and helpful conversations with Peter Hiscock regarding how lithic analysts employ Kuhn's (1990) IR inspired us to briefly reanalyze the experimental assemblage.

Kuhn (1990) uses the equation $(D)\sin(a)$ to estimate t. He estimates t at three points along a tool's edge. Then he divides t by T, which is calculated as a single value recorded at the blank's longitudinal midpoint (Kuhn 1990: 587). In our analysis, we followed Kuhn's (1990) methodology, which in the experimental assemblage resulted in a low correlation ($r = .2200$) between IR values and percentage mass lost.

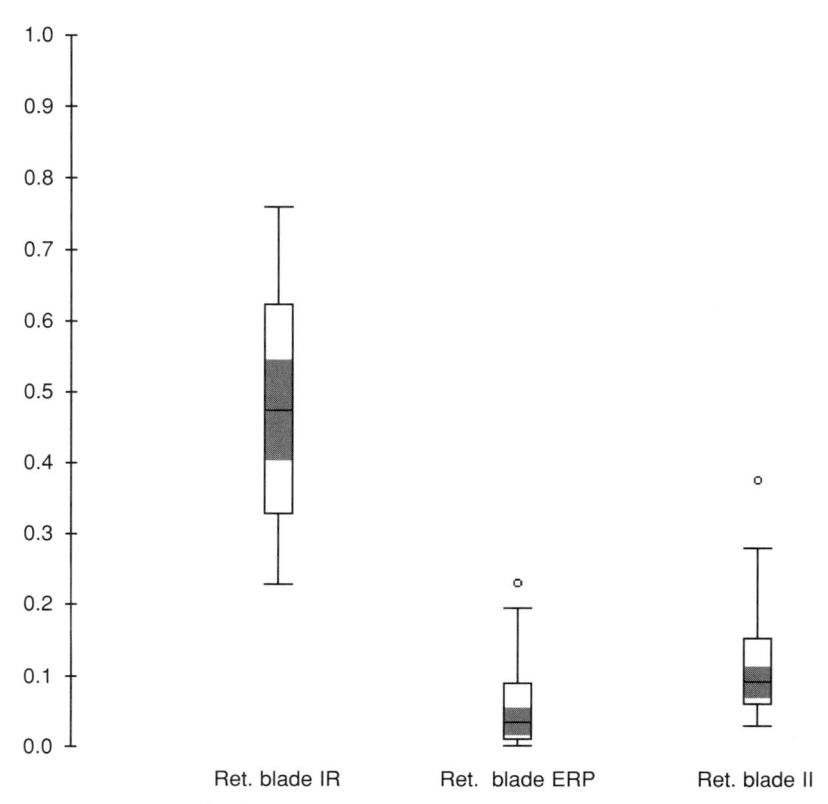

FIGURE 3.13. La Colombière retouched blades' box plots of the ERP, IR, and II.

Peter Hiscock suggested that we apply his own "average Kuhn reduction index" (Hiscock and Attenbrow 2005: 60) to our experimental assemblage. Hiscock and Attenbrow's (2005) methodology differs from Kuhn's (1990) methodology in that multiple T values are recorded at each point where D and a are calculated. So instead of

$$\text{IR} = ((t_1 + t_2 + t_3)/3)/T,$$

Hiscock and Attenbrow calculate:

$$\text{IR} = (t_1/T_1 + t_2/T_2 + t_3/T_3)/3.$$

This second formula was used to calculate the IR on the experimental assemblage. The values, median, and range are shown in Table 3.9.

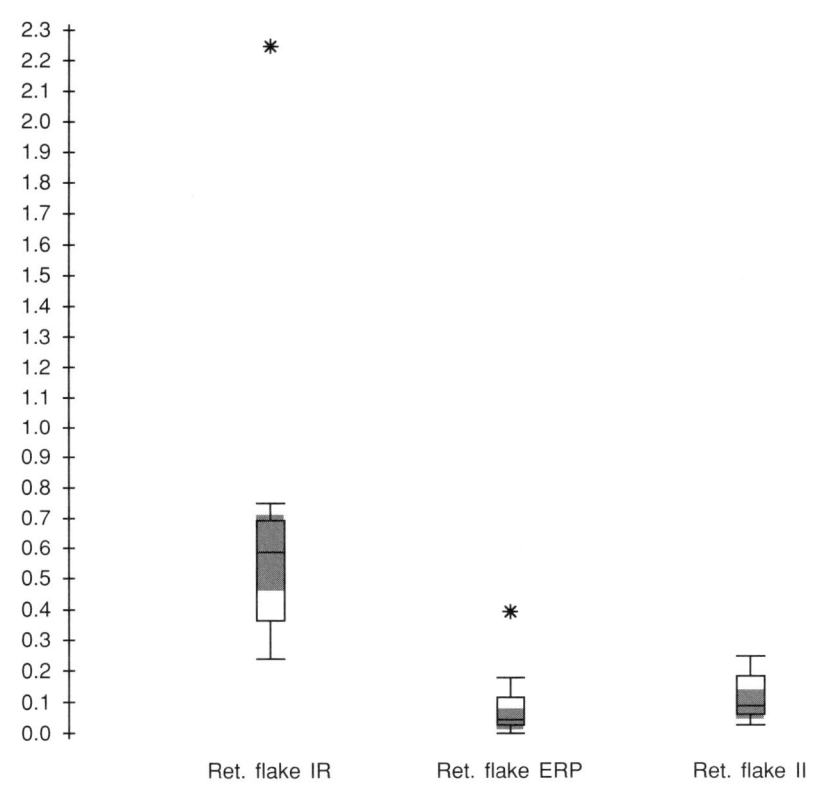

FIGURE 3.14 La Colombière retouched flakes' box plots of the ERP, IR, and II.

When compared to Kuhn's original methodology (Table 3.9, columns 2 and 3), it appears that the equations' results do not differ too drastically (in terms of precision, rather than accuracy). Next, the average Kuhn reduction index values were correlated with percentage mass lost. This correlation ($r = .5160$) is stronger than the original calculated correlation ($r = .2200$), but still not nearly as strong as the correlation between the ERP and percentage mass lost ($r = .7308$). As noted above, it must be stressed that the experimental assemblage was quite diverse regarding how much retouch was applied to each specimen as well as where that retouch was applied on each specimen. The lower correlation values (when compared to Eren et al. 2005) seen in this paper involving actual mass lost (for both the IR and ERP) might be

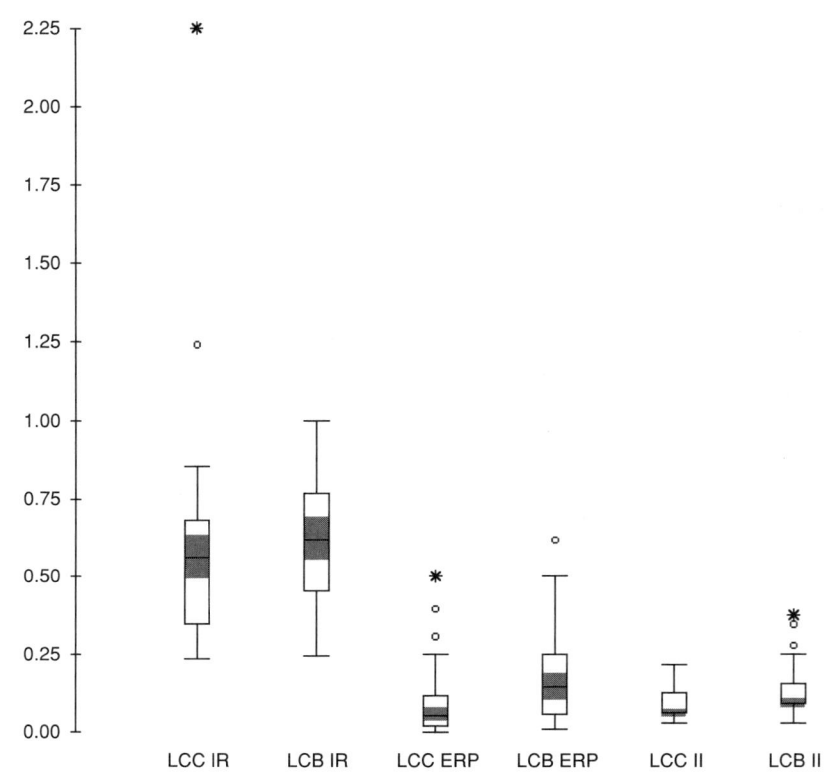

FIGURE 3.15. Box plots show the means and ranges for the index of reduction (IR), estimated reduction percentage (ERP), and index of invasiveness (II), calculated on the La Colombière (LC) Upper Perigordian assemblage of 113 retouched pieces. Here, the results are separated into groups of broken ($n = 58$, LCB) and complete ($n = 55$, LCC) tools.

due to the challenge posed by the experimental assemblage's diversity. If anything, this may show that all indices are lacking somewhat when applied to diverse assemblages.

Interestingly, when the average Kuhn reduction index values recorded from the experimental assemblage are calibrated to a logarithmic scale as Hiscock and Clarkson (2005) suggest, the correlation to percentage mass lost is even lower ($r = .5033$). The IR values calculated from Kuhn's original methodology were also calibrated to a logarithmic scale, with no improvement in correlation with percentage mass lost ($r = .1828$).

Table 3.9. Data comparison between Kuhn's (1990) original index of reduction original methodology and Hiscock and Attenbrow's (2005) average Kuhn reduction index

n	IR (Hiscock and Attenbrow 2005) 49	IR (Kuhn 1990) 49	Difference na
mean	0.53	0.49	0.04
sd	0.12	0.10	0.02
min	0.30	0.27	0.03
q1	0.43	0.42	0.01
med	0.53	0.49	0.04
q3	0.60	0.55	0.05
max	0.85	0.73	0.12
range	0.55	0.46	0.09
iqr	0.23	0.14	0.09

These results are peculiar. One must wonder why Hiscock and Clarkson (2005: 1021) found such a high correlation between the IR and mass/weight lost ($r = .933$, with Log transformation), whereas in two separate tests (Eren et al. 2005; this paper), we failed to repeat their results. Examining the different experimental methodologies may provide a clue. Sample sizes for both methodologies are statistically robust: Hiscock and Clarkson begin with a sample of 30 flakes, whereas we begin with a sample of 50 blanks. Additionally, both Hiscock and Clarkson's unretouched sample and our own demonstrate variability in original blank size, mass, and linear variables (length, width, thickness). However, the experimental methods diverge when it comes to blank modification. In their experiment, Hiscock and Clarkson retouched each flake multiple times on one lateral edge until they were exhausted, recording the average Kuhn reduction index at each "retouching event." We, on the other hand, retouched each flake once: retouch varied on each flake in intensity and location (one lateral edge, two lateral edges, a distal and lateral edge, etc.). We believe that herein lies the discrepancy. It is already known that the IR can increase with mass loss (Kuhn 1990). Thus, if a flake is retouched multiple times on a single edge, then a particular amount of mass loss will occur at each retouching event. In this specific scenario, the dependent variable, the IR value, has nowhere to go but up. In other words, by analyzing only modification on one lateral edge, all Hiscock

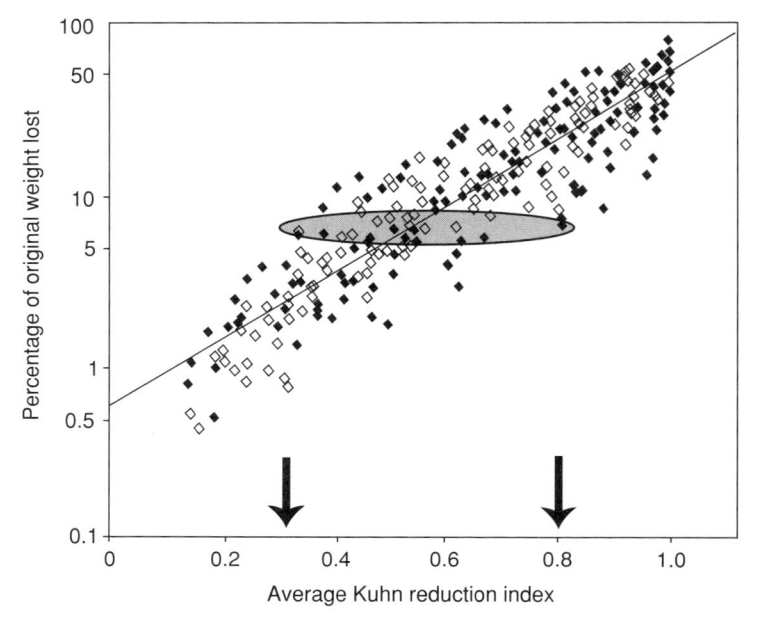

FIGURE 3.16. Graph from Hiscock and Clarkson (2005:1020) depicting the relationship between the average Kuhn reduction index and percentage weight lost. The oval encircles data points representing blanks with 6–7% weight lost. Notice, despite the similar values in weight lost, that the range of the oval (represented by the arrows) is quite large.

and Clarkson show is that IR *can* increase with weight loss, not that it *must*. Measuring the IR on the same flake at successive retouching events might be driving their correlation.

Nevertheless, although methodological differences may produce differences between Hiscock and Clarkson's results and our own, even when dealing with a single lateral edge, the strong correlation between the IR and mass lost suggested by Hiscock and Clarkson is undermined by their Figure 5 (Hiscock and Clarkson 2005: 1020, reproduced here in Figures 3.16 and 3.17). The oval in Figure 3.16 encircles their data points, representing blanks that have lost about six to seven percent of their original weight. As indicated by the arrows, these blanks give IR values ranging from 0.3 to 0.8, despite the fact that these blanks have lost similar amounts of weight. Alternatively, the oval in Figure 3.17 encircles their data points representing blanks that possess

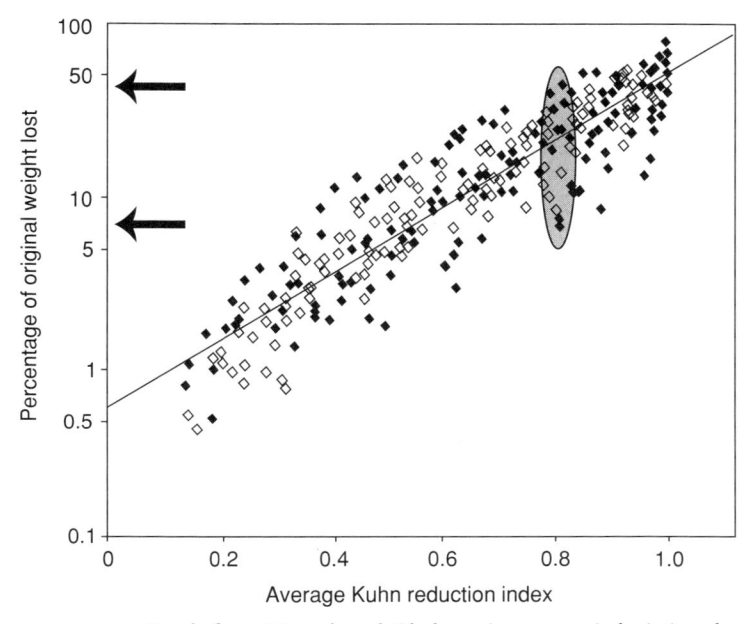

FIGURE 3.17. Graph from Hiscock and Clarkson (2005: 1020) depicting the relationship between the average Kuhn reduction index and percentage weight lost. The oval encircles data points representing blanks with an average Kuhn reduction index of about 0.8. Despite the similar reduction values, the data points range from about 6–7% weight lost to almost 50% weight lost.

an IR value approximating 0.8. Despite similar IR values, these blanks have lost anywhere from six percent weight to almost fifty percent weight! We wonder if this sort of variability would be even greater if more than one edge were retouched and each blank were only retouched once. Thus, we are forced to disagree with Hiscock and Clarkson (2005: 1020) when they state, "it is reasonable to assert that, at least in single margin reduction of the type experimentally tested, the percentage of weight lost could be reliably predicted from the value of the Kuhn Reduction Index that can be measured on specimens." If the IR were a "robust" indicator of "progressive loss of weight from a retouched flake worked on a single lateral margin" (Hiscock and Clarkson 2005: 1020), then it would be good to know what specific weight loss amount equaled what specific IR value, and visa versa.

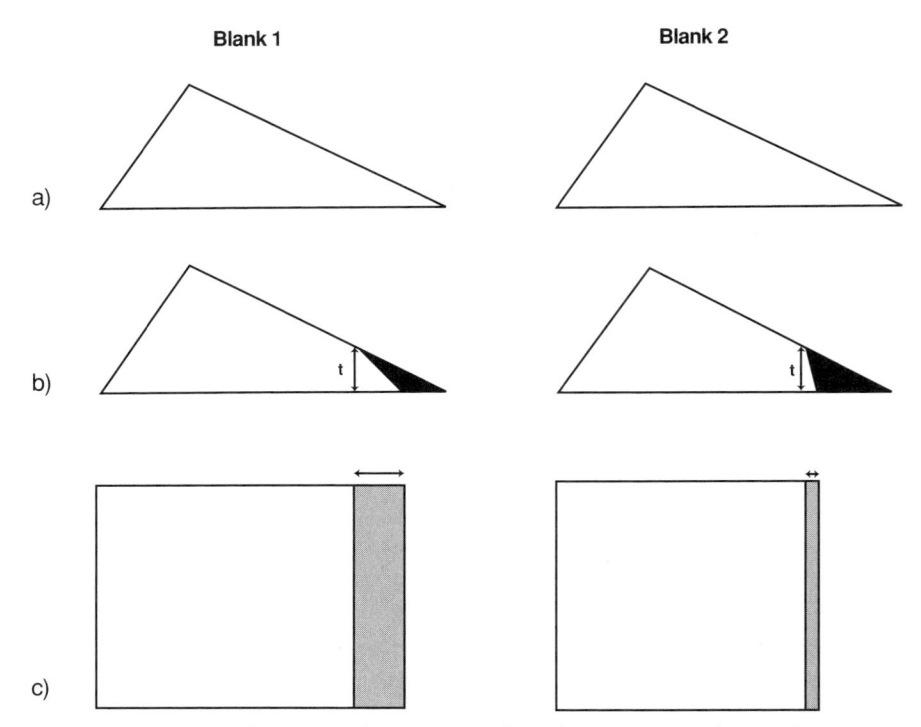

FIGURE 3.18. Illustration of scenario one from the text. Despite having the same IR value, blank two has lost more mass (as represented by the black space). From a dorsal viewpoint (c), the retouch on blank one appears much more invasive.

Why is the IR a poor indicator of mass loss? Shott (2005) and Eren et al. (2005) already provide some examples. However, two more hypothetical cases dealing with retouch on a single lateral edge are presented here.

(1) Imagine two blanks, similar in mass and morphology (Figure 3.18a). The first blank is retouched on a single lateral edge with feather retouch, while the second blank is backed on a single lateral edge (Figure 3.18b). Yet, if the thickness at the retouch terminations t is the same on each blank, the IR value for each blank will be equal, despite the fact that the backed blank has lost more mass. (If the II is employed, the blank with the feathered retouch will have a higher II value than the blank with the backed retouch (Figure 3.18c). This quantitatively differentiates the morphology of two similarly exhausted

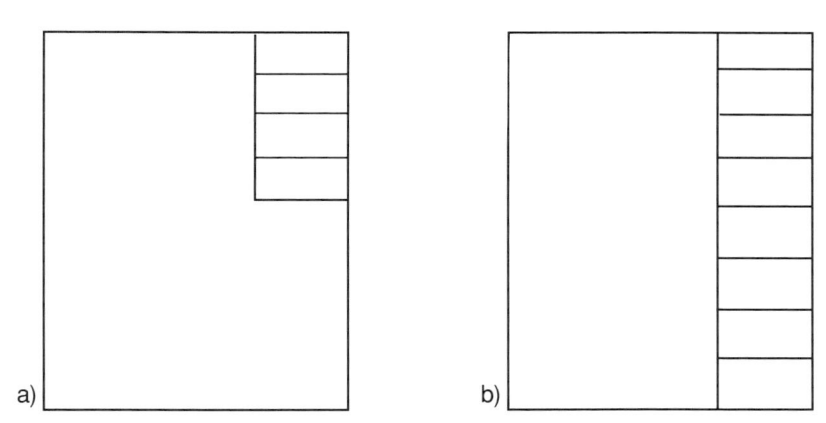

COMPARING AND SYNTHESIZING RETOUCH INDICES 75

FIGURE 3.19. Illustration of scenario two from the text. The large square in "a" has retouch scars (represented by the rectangles) on only one part of its lateral edge. If the rest of the lateral edge is retouched, as shown in "b," the IR value will not change despite more mass being lost.

edges (i.e., two edges with the same IR value, or in other words, the same edge thickness t).

(2) Imagine one blank with retouch on only 50% of a single lateral edge (Figure 3.19a). This retouch has an IR value of 0.6. If the rest of the lateral edge is retouched and exhausted to an IR value of 0.6 (Figure 3.19b), then the IR for the blank will remain at 0.6 despite the fact that even more mass was lost. Of course, even more difficulties arise regarding the IR and mass lost if the two hypothetical scenarios are combined in some fashion (as is likely to be the case in the archaeological record).

The reanalysis presented in this section supports the original analysis of the experimental and archaeological assemblages presented above and suggests that the IR should not be used as a proxy for mass (or weight) lost due to retouch. Yet, as demonstrated in the archaeological assemblage above, and as will be explained below, the IR still plays an important part for understanding reduction.

DISCUSSION

The concepts of reduction, retouching, and resharpening are only important insofar as they provide information on the more complex concepts of prehistoric behavior, curation, and tool use-life (e.g., Bamforth 1986; Binford 1973; Nash 1996; Odell 1996; Shott 1989, 1995). Curation, as explained by Bamforth (1986) and Odell (1996), involves production, design, transportation, recycling, and maintenance of tools. Even if researchers agree with Odell's (1996: 75) omission of tool conservation from the definition of curation (focusing instead on mobility and settlement), retouching and resharpening would still play an important role in understanding how prehistoric groups move across landscapes and settle in certain environments.

Determinants of use-life (Shott 1989: 17–19) might be seen to covary with different stages of tool reduction. Additionally, analysis of how reduced certain tools are indicates the time, effort, and manufacturing cost of producing those tools, which in turn is directly proportional to the tools' use-life (Shott 1989: 20). As Eren et al. (2005: 1200) note, reduction indices do not portray the life histories of stone tools, only the end results of those life histories. However, looking at the end of a life history might indicate how long that life actually was. This is an important contribution, because use-life probably is best expressed as a function of time (Shott 1989: 10).

Shott (1995: 67; 2000) states that the degree of tool-using behavior of interest to archaeologists registered in archaeological specimens, such as curation and use-life, can be determined only when stone tools are properly quantified. The data above are presented to illustrate a single point: though supposedly quantifying the same concept (i.e., reduction, specifically retouching/resharpening by mass loss), different reduction indices provided drastically different values, means, and ranges, even for specimens of the same type category. We hope that this analysis helps to clarify the proper role each index plays in the quantification of lithic reduction. Though each index has its own weaknesses, the ERP index is best suited for estimating how much overall mass has been removed from a specimen, because the variable quantified by the ERP, volume, is directly proportional to mass. In two separate experiments (Eren et al. 2005 and this chapter), the ERP has quantified mass loss better than the IR and the II. The IR should be used for understanding issues dealing with edge exhaustion. The ratio the IR uses, that of the retouched edge to the spine of the tool, is ideal for quantifying this attribute. Finally, the II should be used for answering questions dealing with scar invasiveness, another varying

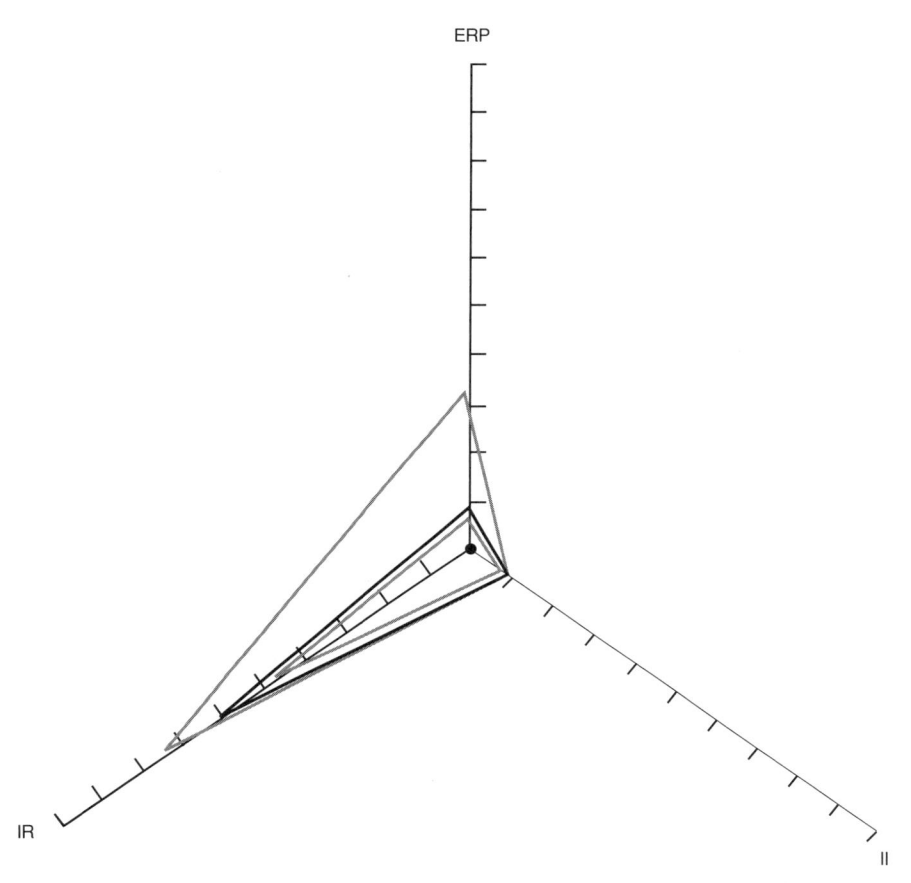

FIGURE 3.20. A possible tri-index approach: retouched blades (innermost triangle); backed blades (outermost triangle); retouched flakes (central triangle).

result of tool reduction. In this sense, studies that argue that one index is better than another (Eren et al. 2005; Hiscock and Clarkson 2005) are really arguing that one index is better than another at quantifying a single variable of reduction, not reduction itself.

Because each reduction index measures a different aspect of reduction, different questions should be posed and analyzed with each index. Further, combining reduction indices can portray differences in overall assemblages or individual tool types. For example, Figures 3.20 and 3.21 demonstrate one way that a tri-index approach might be portrayed (though there may be better ways to do so). These figures show that differences in tool categories can be depicted clearly.

Yet illustrating differences among tool types is not the only value a tri-index approach might have. Comparisons of lithic assemblages manufactured from local and foreign raw materials may show interesting patterns dealing with how different aspects of retouching and resharpening interact depending on distance. Different raw material types might be another aspect that would be interesting to analyze regarding the interactions between how much mass was removed, where, and how.

Though the tri-index approach is valuable, it is also somewhat limited in its static nature. Although it can clearly and precisely depict stone tool types and assemblages as they are, it does not explain or portray how they came to be. Dynamic approaches for understanding lithic technology, such as the chaîne opératoire (e.g., Sellet 1993), can provide different information than looking at typological end products or metric measurements alone. Six major analytical categories exist within the lithic chaîne opératoire:

(1) Procurement (e.g., direct, indirect)
(2) Core reduction (e.g., core preparation, blank removal, core repreparation)
(3) Tool reduction (e.g., blank or tool modification through retouch and resharpening)
(4) Transport (e.g., quarry to site, base camp to kill site)
(5) Use (e.g., cutting, scraping, shooting)
(6) Discard (e.g., exhaustion, breakage, cache, interment)

During the life history of a stone tool, these categories do not occur in any rigid order, but instead may interact in a fluid manner, perhaps depicted best by Conard and Adler (1997). Categories (2) (core reduction) and (3) (tool reduction) make up the *reduction sequence*. On many occasions, lithic analysts have successfully reconstructed category (2). Well-known examples include Levallois blank production (e.g., Bar-Yosef and Dibble 1995, Van Peer 1992), prismatic blade production (e.g., Collins 1999), and even small flake production (Dibble and McPherron 2006). Category (3) has not yet been successfully reconstructed in such a dynamic fashion. Yet, in the same way that general reduction sequences such as Levallois blank production or prismatic blade production are modeled, a dynamic understanding of unifacial

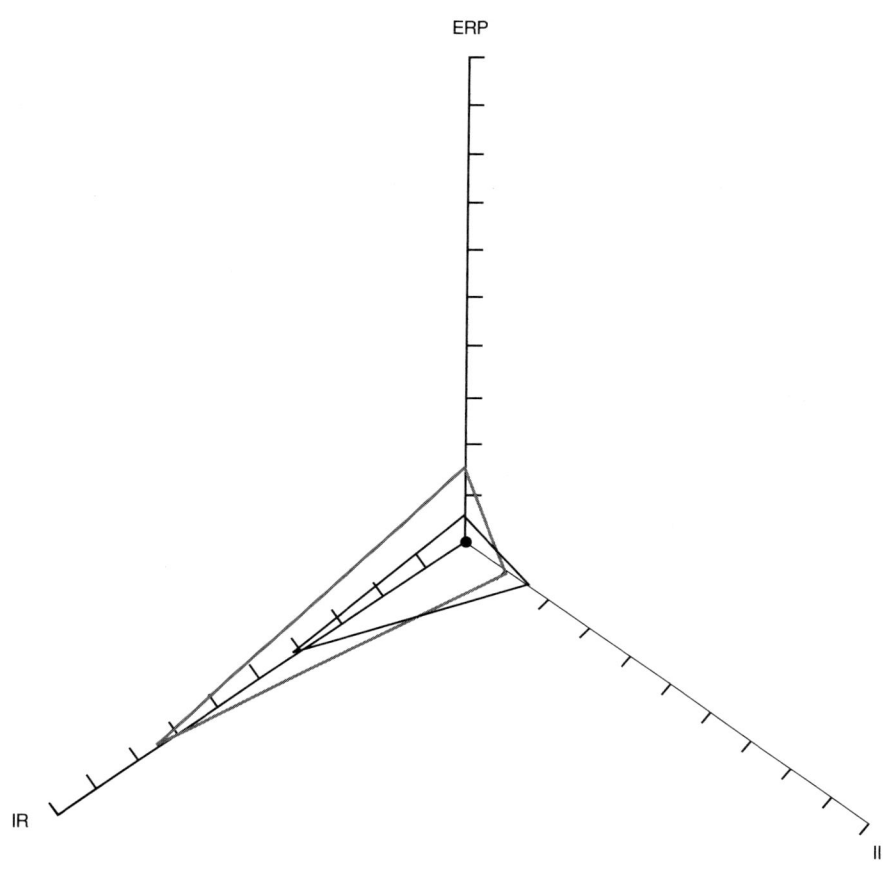

FIGURE 3.21. A possible tri-index approach: end scrapers (gray); borers (black).

stone tool retouching and resharpening processes might be possible through the construction of "retouch-tion" sequences. Below, this new methodology is introduced that combines three reduction indices into a single model for assessing how unifacial stone tools change through retouching and resharpening.

The basic principle of this methodology is as follows: because different reduction indices measure different aspects of retouching and resharpening, they can be manipulated and arranged so that a retouching sequence for an "average" or "common" unifacial stone tool in an assemblage can be illustrated. A methodology for constructing lithic

Table 3.10. Hypothetical ERP, IR, and II values discussed in the text

ERP	0% mass lost	0–5% mass lost	5–10% mass lost	10–15% mass lost	15–20% mass lost	20–25% mass lost
Sample size	10	14	26	29	28	8
Averaged IR values (right lateral edge)	0	0.11	0.20	0.43	0.74	0.80
Averaged II values (right lateral edge)	0	0.13	0.05	0.35	0.04	0.19
Average IR values (left lateral edge)	0	0	0	0	0	0.49
Averaged II values (left lateral edge)	0	0	0	0	0	0.47

Note: Scrapers are grouped into increments of mass lost, provided by the ERP calculation. The IR and II values are recorded on multiple edges (presented here are the right and left lateral edges) and then averaged in each grouping, so a mean IR or II value is calculated for scrapers with 0–5% mass lost, 5–10% mass lost, etc. By knowing these averaged values, it may be possible to visualize the evolution of a scraper edge as mass is lost, as depicted in Figure 3.22.

retouching sequences for unifacial stone tools of the same type category might progress as follows:

(1) Using the ERP, quantify mass loss on each uniface in an assemblage.
(2) Arrange the unifaces from least mass loss to most mass loss. It may be helpful to group the unifaces by increments of 5% or 10% mass loss.
(3) Quantify edge exhaustion and retouch invasiveness by applying the IR and II, respectively, to each uniface at different edge points (i.e., distal end, lateral edge).
(4) Analyze how edge exhaustion and retouch invasiveness change as mass loss increases. By analyzing the unaveraged sixteen II segments, it is possible to see at what point during the retouching sequence (i.e., at what mass loss increment) particular sections of a stone tool are retouched or resharpened. Additionally, by applying the IR to different sections of a stone tool, it is possible to understand how quickly or slowly edge exhaustion takes place at that particular point as mass continues to be lost.

In sum, looking at a single tool type at different mass lost increments might provide a general sequence for how that type was retouched over time.

Table 3.10 and Figure 3.22 provide a hypothetical interpretation of the retouching sequence. Suppose we are presented with 115 unifacial tools of the same type category. By applying the ERP to each uniface,

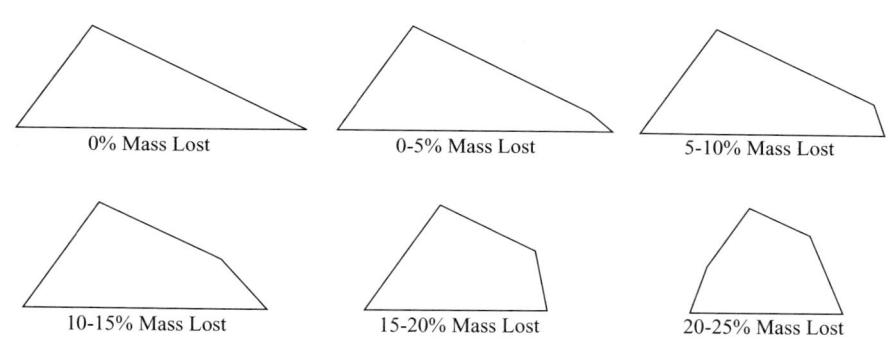

FIGURE 3.22. A hypothetical "retouching sequence" portrayed by cross sections of unifacial tools as described in the text and Table 3.10.

we can determine the mass lost. Once this is known, the unifaces can be grouped into categories of mass lost (Table 3.10). Next, IR and II values are recorded on one edge and averaged within each group (Table 3.10). For example, the 14 unifaces in the 0–5% Mass Lost category have a mean IR value of 0.11 and a mean II value of 0.13 on their right lateral edges. Once these averages are calculated for each category, visual depictions of the numbers (here shown in cross section) can be constructed (Figure 3.22):

- The 0% Mass Lost category shows no edge modification.
- The 0–5% Mass Lost category has only slight modification, as evidenced by the small IR and II values. This may be indicative of rejuvenating a cutting edge.
- The 5–10% Mass Lost category has greater edge modification. The IR increases with retouch, but the II decreases as the edge angle becomes steeper. At this stage the tool's function may involve rough scraping, woodworking, or heavy-duty cutting.
- Stronger retouch blows increase both the IR and II values in the 10–15% Mass Lost category. A cutting edge may again be desired at this stage.
- The IR increases dramatically in the 15–20% Mass Lost category, but the II decreases. This stage in the retouching sequence may indicate that scrapers were backed so that the opposite edge could be used for cutting.
- Finally, the 20–25% Mass Lost category shows only small increases in the IR and II values. This may illustrate a "last gasp" of the modified edge before the tool is discarded or another edge is used. Notice that it is only at this final stage that the left edge exhibits modification.

By applying the retouching sequence methodology to a unifacial stone tool assemblage, other dynamic questions can be addressed regarding how unifacial stone tools change with mass loss:

- At what point (at what mass loss category) in a tool's life history does basal thinning occur?
- Are there particular retouching strategies for extending the life of a stone tool?
- Which sections of a stone tool are modified first? Are edges exhausted simultaneously or one at a time in sequence?
- How is a "retouch-tion" sequence affected by lithic procurement distance or raw material type?
- When does heat damage occur in the life history of a unifacial stone tool?
- Do specific tool types depend on particular amounts of mass loss, edge exhaustion, retouch invasiveness, or some combination of all three?
- How are metric measurements affected as mass is lost, as edges are exhausted, or as retouch invasiveness varies?
- During the life of a uniface, when is the edge angle acute (perhaps for cutting or slicing) and when is it steep (perhaps for backing or scraping)?

Although it is understood that each uniface is probably not reduced in exactly the same way, a retouching sequence would attempt to understand how a "common" or "average" uniface changes with mass loss. Retouching sequences from different sites or layers could then be compared against distance, raw material, or other factors. Additionally, reasons for tool discard might be possible to assess: was a tool discarded because of size, mass lost, edge exhaustion, scar length, or some other factor? A retouching sequence is currently being constructed for the unifacial stone tool assemblage from the Paleo Crossing Site, Ohio (Eren 2005).

CONCLUSIONS

The results of this paper have three important implications. First, when researchers decide to quantify retouching and resharpening processes, they should decide which questions they wish to ask and choose an appropriate reduction index, because each index measures different aspects of reduction. Second, for the reasons just stated, comparison of data using one reduction index to data using another index is not possible because each index measures different variables. Third, published data dealing with stone tool retouching and resharpening should be carefully reexamined, and, if used for reference, carefully cited because conclusions might differ with the application of a different index.

We hope that this paper has shown that reduction should not be understood or measured as simply mass lost, but instead as a complex concept that also includes how and where on a specimen that mass has been removed. We encourage others to continue experimenting with reduction indices. Simple models combining and calibrating the results from different indices hold promise for quantifying and describing lithic assemblages in great detail and in dynamic fashion, which will in turn allow for succinct, useful, and, perhaps, standardized presentations of lithic data around the world.

REFERENCES CITED

Andrefsky, William, Jr. 2006. Experimental and Archaeological Verification of an Index of Retouch for Hafted Bifaces. *American Antiquity* 71:743–57.

Bamforth, Douglas B. 1986. Technological Efficiency and Tool Curation. *American Antiquity* 51(1):38–50.

Bar-Yosef, O., and H. L. Dibble. 1995. *The Definition and Interpretation of Levallois Technology*. Prehistory Press, Madison.

Binford, Lewis R. 1973. Interassemblage Variability – The Mousterian and the "Functional" Argument. In *The Explanation of Culture Change: Models in Prehistory*, edited by C. Renfrew, pp. 227–54. Duckworth, London.

Clarkson, Chris. 2002. An Index of Invasiveness for the Measurement of Unifacial and Bifacial Retouch: A Theoretical, Experimental, and Archaeological Verification. *Journal of Archaeological Science* 29:65–75.

Collins, M. B. 1999. *Clovis Blade Technology*. University of Texas Press, Austin.

Conard, N. J., and D. S. Adler. 1997. Lithic Reduction and Hominid Behavior in the Middle Paleolithic of the Rhineland. *Journal of Anthropological Research* 53(2):147–75.

Davis, Z. J., and J. J. Shea. 1998. Quantifying Lithic Curation: An Experimental Test of Dibble and Pelcin's Original Flake-Tool Predictor. *Journal of Archaeological Science* 25:603–10.

Dibble, H. L. 1995. Middle Paleolithic Scraper Reduction: Background, Clarification, and Review of the Evidence to Date. *Journal of Archaeological Method and Theory* 2(4):299–368.

———. 1998. Comment on "Quantifying Lithic Curation: An Experimental Test of Dibble and Pelcin's Original Flake-Tool Mass Predictor," by Zachary J. Davis and John J. Shea. *Journal of Archaeological Science* 25:611–13.

Dibble, H. L., and S. P. McPherron. 2006. The Missing Mousterian. *Current Anthropology* 47(5):777–803.

Dibble, H. L., and A. W. Pelcin. 1995. The Effect of Hammer Mass and Velocity on Flake Mass. *Journal of Archaeological Science* 22:429–39.

Eren, M. I. 2005. "Northeastern North American Paleoindian Uniface Types: Techno-Typological Analyses of Unifacial Stone Tools from Paleo Crossing (33ME274), Ohio." A.B. thesis, Harvard College, Cambridge, MA.

Eren, M. I., M. Dominguez-Rodrigo, S. L. Kuhn, D. S. Adler, I. Le, and O. Bar-Yosef. 2005. Defining and Measuring Reduction in Unifacial Stone Tools. *Journal of Archaeological Science* 32:1190–1201.

Hiscock, P., and V. Attenbrow. 2005. *Australia's Eastern Regional Sequence Revisited: Technology and Change at Capertee 3*. British Archaeological Reports, International Monograph Series 1397. Oxford: Archaeopress.

Hiscock, P., and C. Clarkson. 2005. Experimental Evaluation of Kuhn's Geometric Index of Reduction and the Flat-Flake Problem. *Journal of Archaeological Science* 32:1015–22.

Kuhn, S. L. 1990. A Geometric Index of Reduction for Unifacial Stone Tools. *Journal of Archaeological Science* 17:583–93.

———. 1992. Blank Form and Reduction as Determinants of Mousterian Scraper Morphology. *American Antiquity* 57(1):115–28.

Movius, H. L., and S. Judson. 1956. *The Rockshelter of La Colombière: Archaeological and Geological Investigations of an Upper Perigordian Site Near Poncin (Ain)*. Peabody Museum, Cambridge, MA.

Nash, S. E. 1996. Is Curation a Useful Heuristic? In *Stone Tools: Theoretical Insights into Human Prehistory*, edited by G. H. Odell, pp. 81–99. Plenum Press, New York.

Odell, G. H. 1996. Economizing Behavior and the Concept of "Curation." In *Stone Tools: Theoretical Insights into Human Prehistory*, edited by G. H. Odell, pp. 51–80. Plenum Press, New York.

Pelcin, A. W. 1998. The Threshold Effect of Platform Width: A Reply to Davis and Shea. *Journal of Archaeological Science* 25:615–20.

Sellet, F. 1993. Chaîne Opératoire: The Concept and Its Applications. *Lithic Technology* 18:106–12.

Shott, M. J. 1989. On Tool-Class Use Lives and the Formation of Archaeological Assemblages. *American Antiquity* 54(1):9–30.

1995. How Much Is a Scraper? Curation, Use Rates, and the Formation of Scraper Assemblages. *Lithic Technology* 20(1):53–72.

2005. The Reduction Thesis and Its Discontents: Overview of the Volume. In *Lithics "Down Under": Australian Perspectives on Lithic Reduction, Use, and Classification*, edited by C. Clarkson and L. Lamb, pp. 109–25, BAR International Series 1408, Archaeopress, Oxford.

Shott, M. J., A. P. Bradbury, P. J. Carr, and G. H. Odell. 2000. Flake Size from Platform Attributes: Predictive and Empirical Approaches. *Journal of Archaeological Science* 27: 877–894.

Van Peer, Phillip. 1992. The Levallois Reduction Strategy. Prehistory Press, Madison.

4 JENNIFER WILSON AND WILLIAM ANDREFSKY, JR.

EXPLORING RETOUCH ON BIFACES: UNPACKING PRODUCTION, RESHARPENING, AND HAMMER TYPE

Abstract
Measuring retouch amounts on stone tools has been helpful for understanding human organizational strategies. Multiple retouch indices geared toward assessing retouch amounts on flake tools and unifaces have been developed, but few have been developed to evaluate retouch exclusively for bifaces. For this study, a retouch index was developed and evaluated on an experimental assemblage of bifaces. It is shown that reduction activities on bifaces may create extensive amounts of retouch that are contingent upon a number of factors from both the production and resharpening events that must be taken into consideration before understanding a biface's life history.

INTRODUCTION

Tool curation has been defined as the relationship between a tool's potential utility and its actual usage (Andrefsky 2005; Bamforth 1986; Shott 1996), or its "life history" (Eren et al. 2005). This curation concept has been linked to studies of hunter–gatherer organizational strategies in understanding issues of land use, economy, and, mobility. For stone tools, retouch amount has been used as an effective measure to assess the degree to which a tool has been curated (for discussion of curation see Andrefsky 2006; Barton 1988; Binford 1973, 1979; Blades 2003; Clarkson 2002; Davis and Shea 1998; Dibble 1997; Nelson 1991; Shott 1989, 1996).

However, assessing retouch amount may not be as universal as we might initially believe. We define retouch as the deliberate modification of a stone tool edge created by either percussion or pressure-flaking techniques (Andrefsky 2005). As such, retouch takes place in the beginning production stages of a tool as well as in the subsequent episodes of resharpening and reshaping of a tool's cutting edge. Therefore, we would expect the amount of retouch to progressively increase throughout the production and the use-life of a tool. To evaluate retouch in terms of degree of curation, analysts have created indices that quantify retouch for comparisons of stone tools.

Previous retouch measures have been effective for different kinds of stone tool forms. Barton (1988) and Clarkson (2002) measure retouch on flake tools based upon progressive use of the original flake blank. Kuhn (1990) measured retouch on scraper edges. Andrefsky (2006) and Hoffman (1985) measured retouch on hafted bifaces (see also Eren and Prendergast; Hiscock and Clarkson; Quinn et al., this volume). We feel that North American bifaces represent a different tool type than some bifaces from other parts of the world. Bifaces are stone tools that have two surfaces (or faces) that meet to form an edge around the entire perimeter and usually have flake scars that extend from the edge to the midline of the surface (Andrefsky 2005). North American bifaces tend to undergo a production phase and a subsequent use-life phase (Callahan 1979; Whitaker 1994). In some areas of the world, bifaces are produced from flake blanks as a result of their being used and resharpened extensively (cf. Clarkson 2002). However, we feel that some bifaces do not become bifaces as a result of this use and resharpening process. We feel that some bifaces are shaped by extensive retouch before they are even used. Some bifaces, particularly those in parts of North America, are extensively retouched during their production phase, and thus, the retouch amount on the biface has little or no meaning with regard to curation. We suggest that retouch indices should be specifically tailored to different kinds of tools and that we need to consider the differences between bifacial production and bifacial resharpening after use.

To gather information on biface retouch, we conducted a series of production and use experiments attempting to replicate bifaces similar to those recovered from Chalk Basin, a chert quarry workshop area on the Owyhee River in southeastern Oregon. Our experiment systematically gathered attribute information from both the bifaces being produced and used and the debitage resulting from the experiment.

THE EXPERIMENT

Our experimental study involved the production of three "quarry bifaces" made from high-chipping-quality chert. Information on each biface was recorded after six arbitrary production and use-life events. The first two events were arbitrary production events and the last four were associated with resharpening episodes after tool use events. When the biface was reduced by approximately half of its starting weight during the production process, we arbitrarily stopped and collected debitage shatter for that production event. Resharpening episodes occurred when the edges of the biface were retouched enough so that it could be used as a tool with a cutting edge around the entire perimeter. The biface edges were then dulled and resharpened again to create a series of resharpening episodes of each biface.

One of the authors performed all of the flintknapping over a drop cloth using either a hard hammer or a soft hammer percussor, while the other author recovered and numbered each flake as it was removed. All production and resharpening were done with percussion flaking (no pressure flaking). The greatest number of flakes collected was from the first production event, which yielded an average of twenty-nine flakes per biface. This makes intuitive sense, because the biface was reduced by the greatest amount during this episode. The smallest number of flakes collected was from the first resharpening event, with an average of 12 flakes collected for each biface. The resharpening events had the greatest amount of variability amongst all three bifaces. The average number of flakes collected for each biface, after the first resharpening event, was nineteen flakes per event. The cores chosen for the experiment were all roughly the same shape and size (approximately weighing 1,000 g each). However, one biface had about twenty more flakes removed from it during the experiment than the other two bifaces. This was due to the presence of material flaws that had to be removed in order to maintain an effective cutting tool.

After each event, all of the shatter was collected and the biface was photographed and measured. For consistency, all of the measurements were done by just one of the authors throughout the experiment. Initially, all three bifaces were reduced using a quartzite hard hammer to remove most of the cortex from the objective piece. After the first half-life, a siltstone hard hammer and a soft hammer (i.e., antler billet) were used to shape and thin the bifaces. Gradually, throughout the experiment, the percentage of hard hammer flakes decreased whereas the percentage of flakes made by a soft hammer increased.

DEBITAGE PATTERNS WITH BIFACE PRODUCTION AND RESHARPENING

In a previous study (Wilson and Andrefsky 2006), we explored the variability found in debitage characteristics between biface production and biface resharpening events from the experiment. From 256 proximal flakes analyzed, we found that debitage characteristics were significantly associated with differences in production and resharpening events. Metric variables sensitive to these different retouch activities include maximum length, width, thickness, weight, and platform area (maximum platform width multiplied by maximum platform thickness) (Table 4.1). Nominal attributes that were sensitive to retouch activities were platform type and presence of cortex.

Using platform types previously defined (Andrefsky 2005), we found that flakes made from biface production exhibit more flat or cortical platforms than flakes made during resharpening events (Figure 4.1). Most of the flakes also had dorsal cortex and have a smaller width-to-thickness ratio, heavier weight, and larger platform area (Table 4.1).

In contrast, flakes that are the by-products of resharpening events tend to have more complex and abraded platforms. The flakes produced from resharpening events also weigh relatively less, with a smaller platform area, and have a higher width-to-thickness ratio. Even though the two comparative groups in the debitage study did have some overlap, the average sizes of the two groups were significantly different.

Given the results of differences noted in the debitage attribute analysis, we could expect to see a positive correlation between flake weight and platform area, and also between flake weight and width-to-thickness ratio. Intuitively, if the weight increases, there should

Table 4.1. Comparison of attributes recorded from proximal flakes

	Attribute	Max	Min	Mean	Std. deviation
Production	Weight (g)	99.5	0.7	12.784	15.935
	Width (mm)	93.8	1.5	41.245	16.635
	Length (mm)	113.1	4.6	42.132	17.916
	Thickness (mm)	26.9	2.8	7.870	4.222
	Platform width (mm)	58.8	7.4	19.499	10.622
	Platform thickness (mm)	16.1	1.0	5.852	3.085
	Platform area (mm)	946.7	7.4	133.822	138.801
	Width to thickness (mm)	15.37	0.48	5.8859	2.5206
Resharpening	Weight (g)	5.3	5.3	1.193	1.054
	Width (mm)	44.4	7.2	18.736	7.054
	Length (mm)	65.2	8.5	24.315	11.237
	Thickness (mm)	7.5	0.7	2.308	.873
	Platform width (mm)	20.5	2.1	8.363	3.729
	Platform thickness (mm)	5.3	0.4	1.892	.852
	Platform area (mm)	91.2	1.7	17.603	15.556
	Width to thickness (mm)	22.50	3.85	8.6019	3.0650

FIGURE 4.1. Platform types identified on proximal flakes.

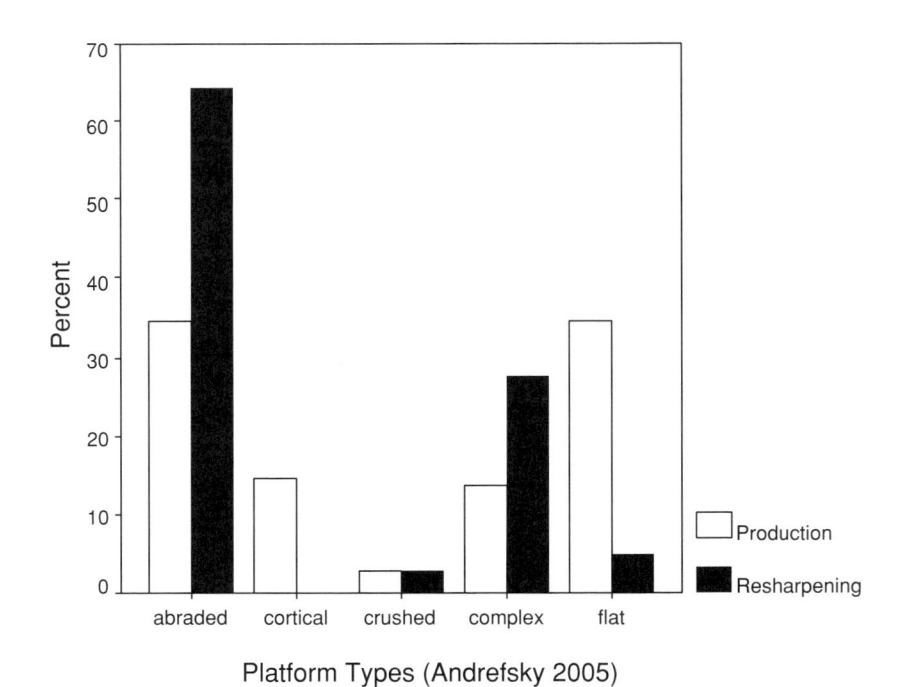

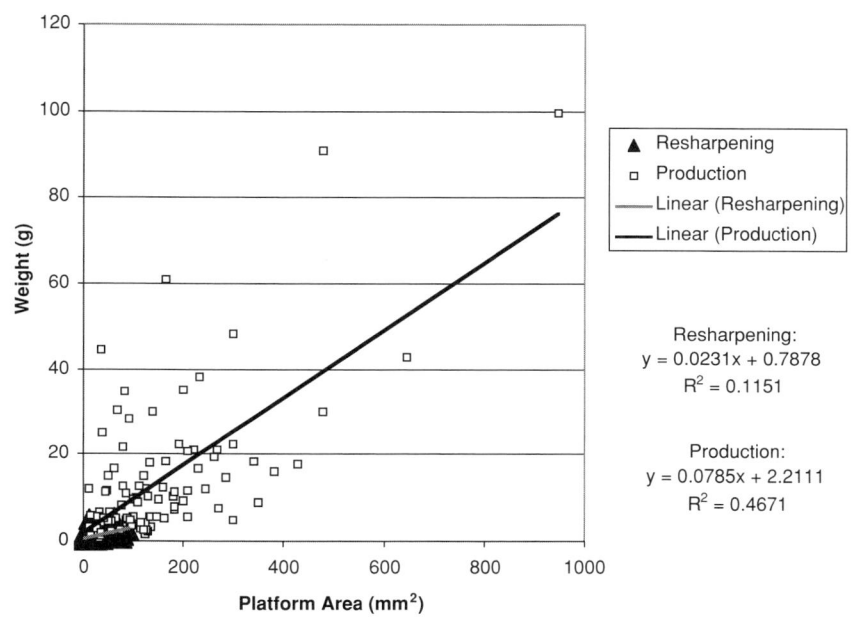

FIGURE 4.2. Platform area of proximal flakes plotted against their weight.

also be an increase in platform area, given that bigger flakes usually have larger platforms, and there should also be a decrease in the width-to-thickness ratio, assuming that the more a flake weighs the larger it should be in size, which is expressed as a ratio. Figure 4.2 displays a scattergram that shows a strong and significant ($R^2 = .4671$, $F = 94.979$, $p < .001$) relationship between increasing flake weight and platform area with production flakes. From the resharpening episodes, flakes clustered together around the lower weights and smaller platform areas. This relationship was not as strong (Pearson's $r = 0.336$) as with production flakes but was still statistically significant ($R^2 = .1151$, $F = 18.027$, $p < .0005$).

When flake weights were plotted against the width-to-thickness ratio, it appeared that the weight of the flake increased as the ratio began to decrease (Figure 4.3). On closer examination, however, this was a significant ($R^2 = .0403$, $F = 6.781$, $p = .010$) but weak correlation for resharpening flakes. For the production events, there was an insignificant relationship between the variables, where only 0.7% of the variance could be explained ($R^2 = .0074$, $F = 0.880$, $p = .350$).

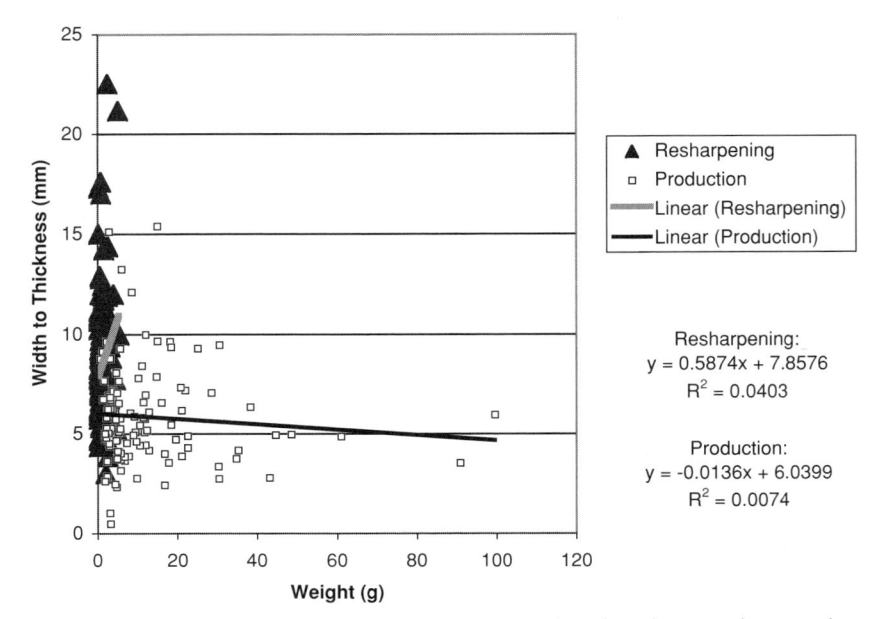

FIGURE 4.3. Width-to-thickness ratio of proximal flakes plotted against their weight.

These correlations have shed light on how the weights of production and resharpening flakes relate to platform area and width-to-thickness ratio. There is more variation between the groups in regards to size (width to thickness) and weight, in comparison to the stronger correlation with weight and platform area (i.e., as the weight of the flake increases, so does the platform area for production and resharpening events).

GENERAL BIFACE PATTERNS OF PRODUCTION AND RESHARPENING

Based upon results gathered from our debitage pattern study, we expected that biface size, shape, and flake removal patterns would also reveal differences between retouch associated with production and retouch associated with resharpening. When graphed, it is apparent that all three bifaces show a continual decrease in both surface area and weight throughout the use-life events (hereafter called use-life events) (Figures 4.4 and 4.5). This is what would be expected given the fact that the use-life events follow a reductive process, resulting in

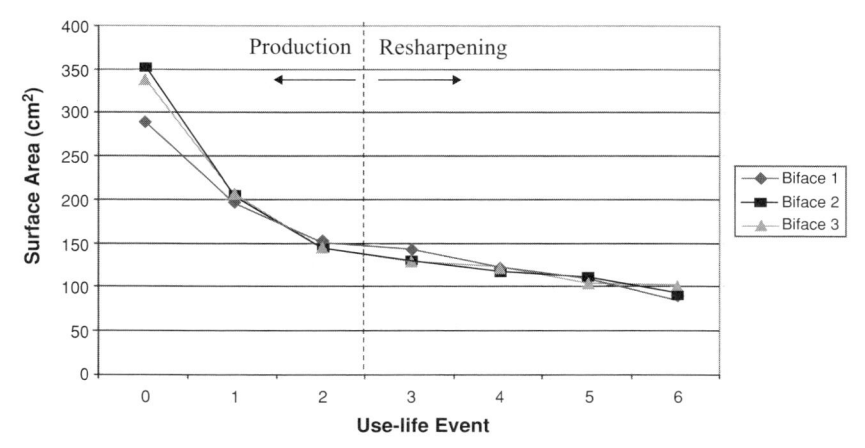

FIGURE 4.4. Total surface area of the bifaces throughout the experiment.

progressively smaller bifaces. However, these data also suggest that the biface use-life events 1 and 2 are responsible for the greatest amount of size reduction and that biface size reduction is significantly less during the resharpening events (3–6).

This pattern is clear when we graph the amount of surface area lost during each use-life event. Even though the amount of total surface area of bifaces progressively decreased during the use-life events, the amount of surface area lost stabilized after the production events 1 and 2 (Figure 4.6). Essentially, the resharpening events (3–6) show very

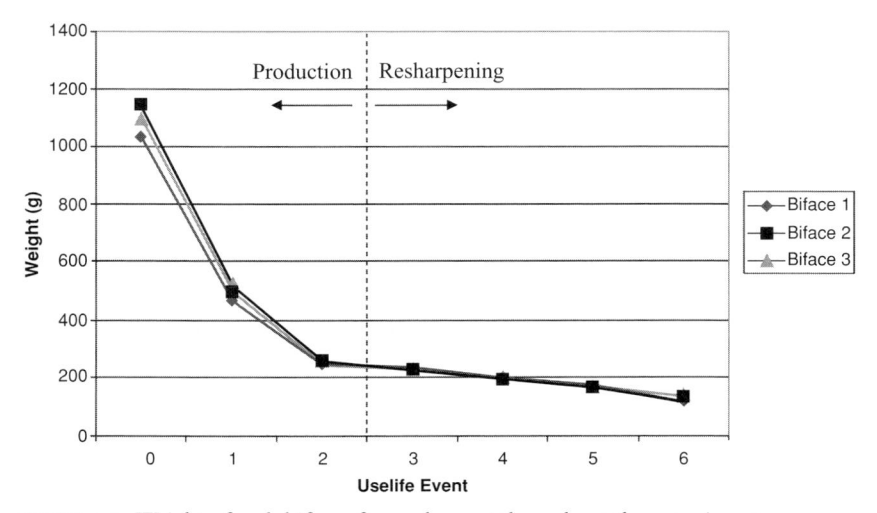

FIGURE 4.5. Weight of each biface after each event throughout the experiment.

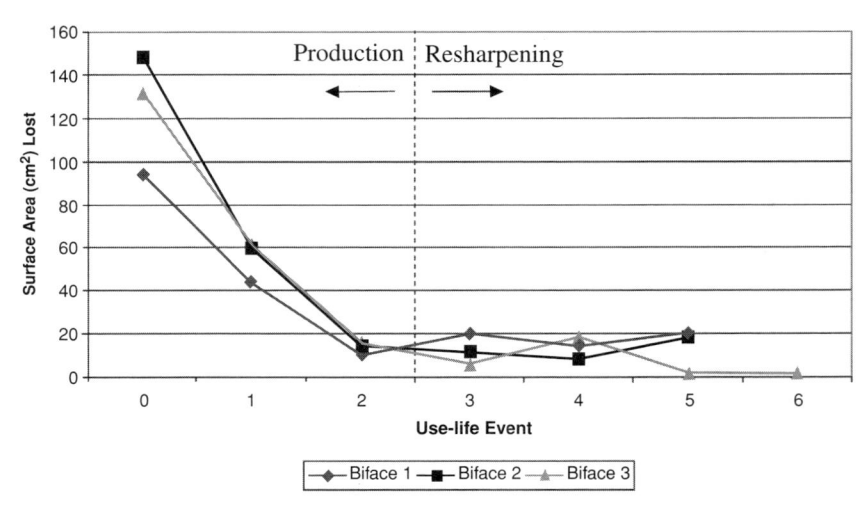

FIGURE 4.6. Surface area lost for each biface throughout the reduction sequence.

little lost surface area; the average surface area lost for each biface is roughly 50 cm^2, compared to about 200 cm^2 lost during production. This pattern also suggests that there may be some observable differences in biface characteristics between production and resharpening events. However, lost surface area is only effective for discriminating such events in a controlled experimental setting. It is not possible to effectively use such a measure on excavated assemblages, because surface area lost can only be calculated based upon knowing the original size of the biface. However, like the change in debitage attributes, it does suggest that other biface characteristics might help assess differences between production and resharpening events.

RETOUCH INTENSITY

Other studies have shown that retouch intensity has been an effective measure of curation on stone tools (Clarkson 2002; Eren and Prendergast, this volume, Quinn et al., this volume; Hiscock and Clarkson 2005; Kuhn 1990). We suggest that bifaces have a unique production life and use life and thus, retouch amount has to account for these two phases of a biface life cycle. To assess our assumption, we applied Clarkson's (2002) index of invasiveness to our experimentally produced bifaces.

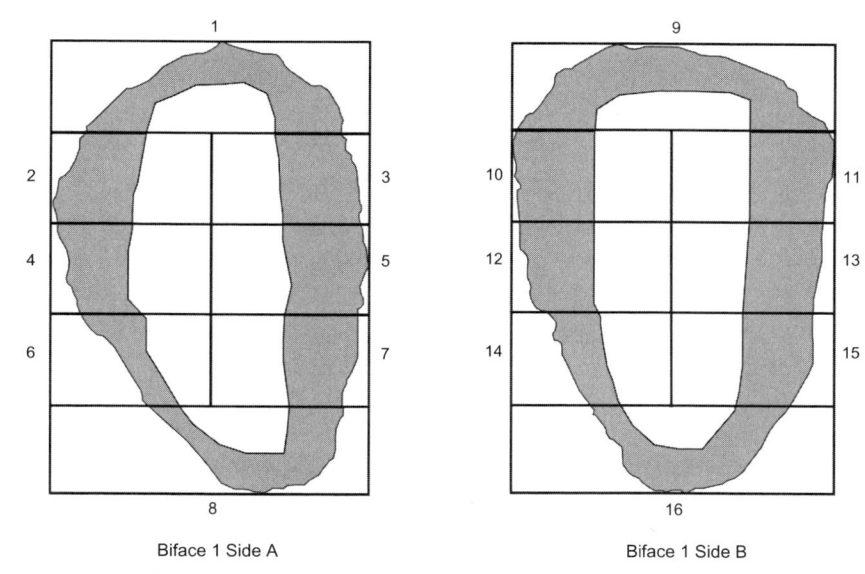

FIGURE 4.7. Illustration of Clarkson's (2002) grid used to calculate his index of invasiveness with numbered squares on each side of the biface. The gray areas on the biface figure indicate the midpoint between the midline of the biface and the edge. Flake scars originating from the edge that do not reach the midpoint would receive a 0.5 and flakes that extend past the midpoint would score a 1 for that segment.

Each side of the biface was partitioned into eight equal segments (cf. Clarkson 2002), each one accounting for 12.5% of the total area (Figure 4.7). After a reduction event, each segment was given a score of either 0, 0.5, or 1. Segments exhibiting no retouch would receive a score of 0. If the flake patterning was evident but did not reach the midpoint area of the artifact, defined as the arbitrary line from the midline of the biface to the edge, that square would have a value of 0.5. A score of 1 was given to squares where retouch extended from the edge of the biface and past the midpoint area. The scores from each square were added up and then divided by 16 for the average retouch amount, which was the index of invasiveness score. If the invasiveness score was close to 0, the biface would be considered to exhibit little to no retouch. When the invasiveness score approached 1, the tool is classified as being completely retouched.

We found that retouch amount using this technique is not sensitive to resharpening after the production phase. Using this index, the bifaces were scored as heavily retouched after the second use-life event

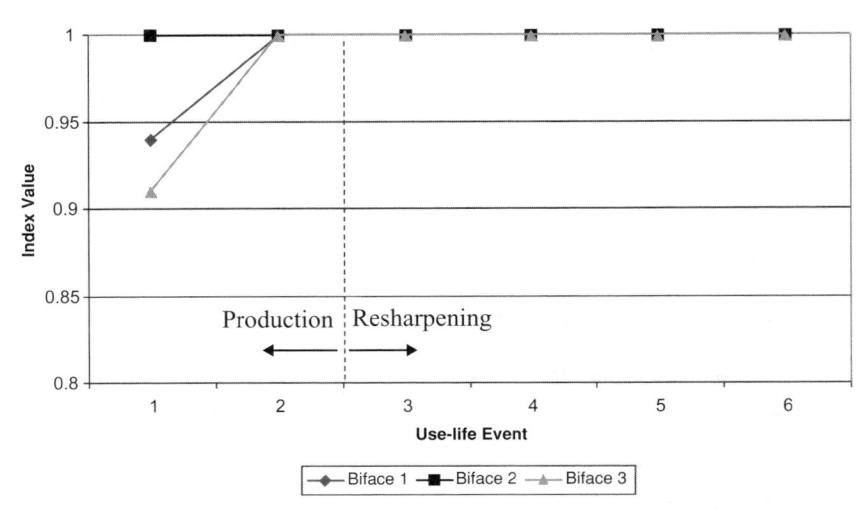

FIGURE 4.8. Results of applying Clarkson's index of invasiveness (2002) to our experimental bifaces.

(Figure 4.8). One of the bifaces even reached a value of one (maximum retouch amount) after the first use-life event. This is interesting because we know the bifaces were never used. However, the index reveals a maximum level of retouch and subsequently a maximum level of curation. This suggests that the index of invasiveness may not be a good indicator of bifacial retouch as it relates to curation, and also that bifaces are produced, used, and resharpened differently than artifacts such as flake tools (cf. Andrefsky 2006). The outcome of this method is not surprising, as Clarkson (2002: 72) does warn about the potential shortcomings of the index of invasiveness when applied to artifacts that have been "fully retouched." Clarkson's index was intended for application to bifacially retouched flakes.

RIDGE COUNT RETOUCH INDEX

Clarkson's index of invasiveness does not adequately segregate biface production from biface resharpening after use. These two use-life events are important in measuring retouch on bifaces. One of the things that intuitively appear to be occurring on the surface of our experimental bifaces is a progressive increase in the number of

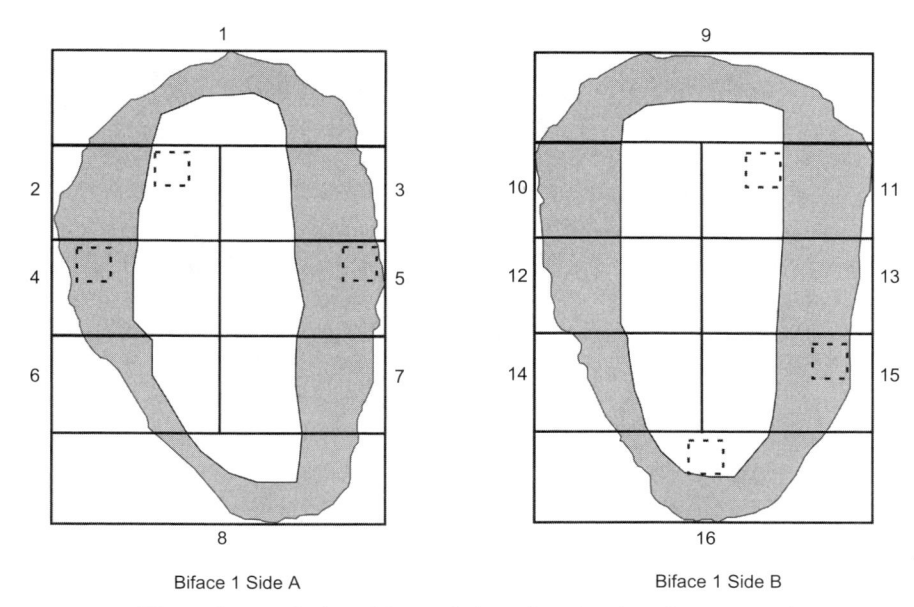

FIGURE 4.9. The analysis grid adapted from Clarkson (2002) and used to count ridges systematically for our study. The squares on the biface are 1 × 1 cm in size and indicate the locations where ridge counts were analyzed throughout the experiment.

flake removal scars from early production events to final resharpening events. To explore flake removal scar counts, we developed a retouch index based upon a sample of the flake removal patterns found on the surface of each biface. The average number of ridge counts was used as a proxy for flake removals to derive this index.

To test this retouch index, we collected biface data after each biface use-life event (weight, maximum length, width, thickness, and flake ridge count). The flake ridges were recorded in a systematic way that involved scanning the biface at a high resolution (600 dpi) and then sampling the bifacial surface image using Deneba's Canvas 8 drafting program. The analysis of each biface image was partitioned using Chris Clarkson's grid for evaluating retouch invasiveness, which partitioned each side of the biface into eight segments. Once the grid was digitally superimposed on the biface, six 1 × 1 cm squares were drawn on the biface and positioned in the same location after each use-life event (Figure 4.9). Three 1 × 1 cm squares were sampled on each face of

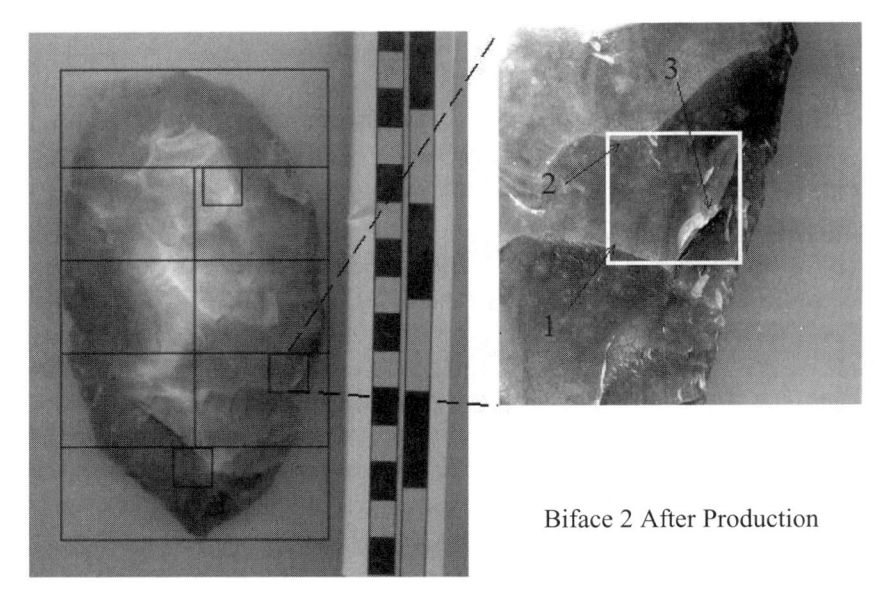

FIGURE 4.10. Image of one of the analysis squares from one of the experimental bifaces, showing how flake ridges were counted.

the specimen. By using a standardized size (1 × 1 cm) box, the same amount of surface area was evaluated from the beginning production stages through the usage and resharpening episodes, regardless of biface shape.

Dorsal flake ridges, or arrises, were counted in each of the sampled boxes. Dorsal ridges were defined as the raised areas that form between the intersections of flakes that were removed from the biface (Figure 4.10). Flake ridges that form as a result of platform preparation, which were present around the biface edge, were not included in this analysis. Flake ridges were identified with the aid of a magnification lens (16×) and by examining the scanned image of the biface. By using the scanned image of the biface to supplement the analysis, it was easier to determine the number of ridges present in the analysis square by focusing in on a particular grid and by adjusting the brightness and contrast of the image. Because the biface surface is not smooth, changing the brightness and contrast levels of the image allowed particular ridges to become more pronounced with different combinations of

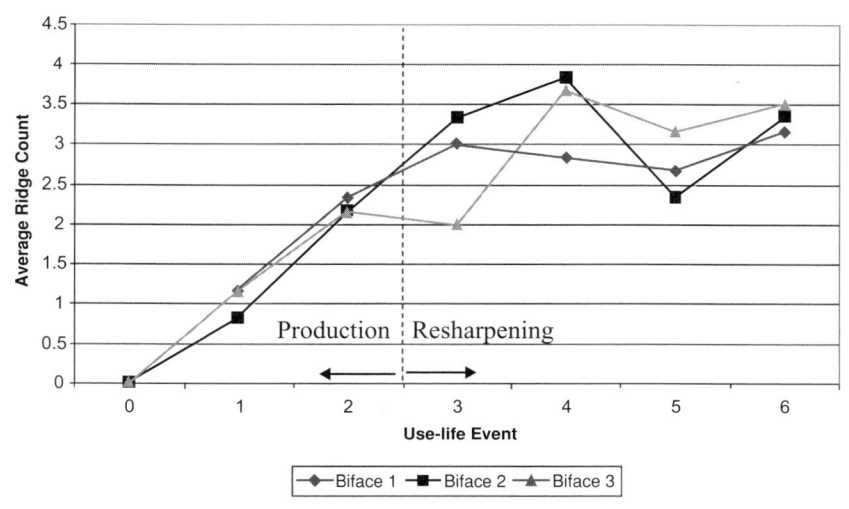

FIGURE 4.11. Average ridge count for each biface throughout the experiment.

light and contrast. The ridges identified on the scanned image were checked on the actual biface to ensure that the lines observed were not biface fissures or ripple marks but actual flake ridges. Once the number of ridges for each square was confirmed, all six ridge counts were added up and divided by six. This resulted in an average ridge count for each biface after each use-life event.

This retouch index was applied to our assemblage of replicated bifaces, with expectations that there would be significant differences between production and retouch use-life events, as seen in the debitage data, and in the amount of surface area lost on the experimental bifaces. The average ridge count associated with each experimentally produced biface use-life event illustrates that the ridge counts increase throughout all of the use-life events before dropping at use-life event 5 during resharpening (Figure 4.11). This pattern reveals some interesting aspects of biface production and resharpening after use. First, the ridge count measure seems to work as an effective tool to assess use-life events from the beginning of the production cycle through the fourth use-life event, and retouch seems to increase as each use-life event increases. However, this progressive pattern ends at use-life event 5, where there is a drop in the retouch index. We

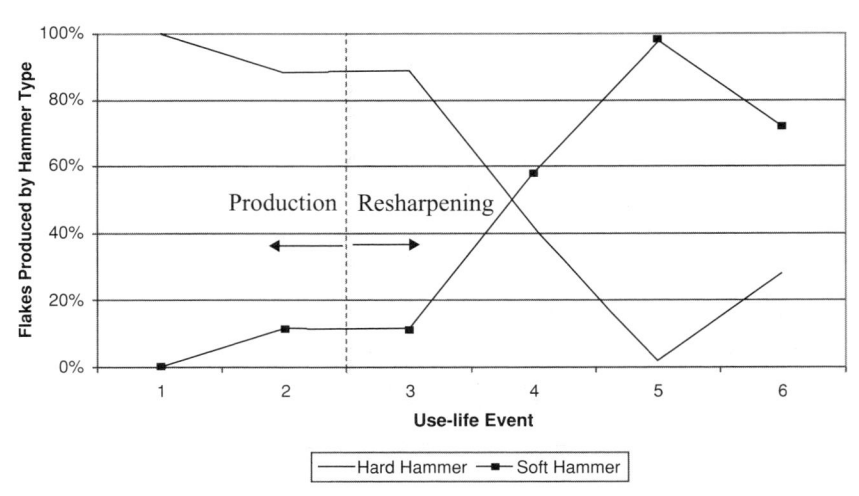

FIGURE 4.12. Graph of the percentages of flakes produced by soft and hard hammer percussion.

also see that the retouch progression is not markedly different between biface production and biface resharpening, as noted in the debitage data.

Since this was not what we had expected, we began exploring our experimental data to determine what might account for the ridge count drop at use-life event 5. One immediate pattern discovered was that the type of hammer used during the replication experiments gradually changed from hard hammer percussion to soft hammer percussion as the bifaces were progressively retouched. Other studies also suggest that hammer type and density can be important for flake removal (Andrefsky 2007; Cotterell and Kamminga 1987; Dibble 1995; Hayden and Hutchings 1989). Figure 4.12 charts our experimentally derived use-life events against the relative proportion of hard and soft hammer percussion used to remove flakes. The first three events are primarily hard hammer percussion; this changes to approximately 42% during event 4 and down to 2% during event 5, and then it goes back up to close to 30% during event 6. The steep drop in hard hammer percussion from events 3 through 5 and the subsequent rise at event 6 mirrors the ridge count pattern, and suggests to us that the ridge count index is sensitive to the type of hammer used in biface production and resharpening technology.

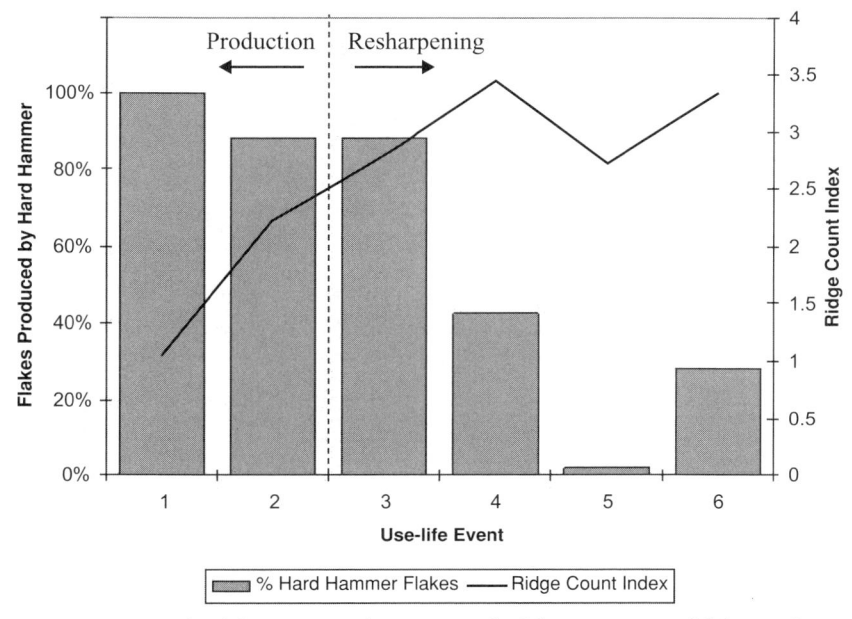

FIGURE 4.13. Graph of the average ridge count and of the percentage of flakes made by hard hammer percussion.

To explore this relationship further, we plotted the ridge count index and hammer type along with the use-life events (Figure 4.13). The ridge count index for use life events 4–6 is indeed similar to the relative percentages of hard hammer percussion. However, it is also apparent that the ridge count index is sensitive to previous flake removals on the biface. For instance, use-life events 1–3 have high values for hard hammer percussion, yet the ridge count index shows a steady increase from less than 1.0 to over 3.3. Essentially, the ridge count index is increasing as the original nodule is being progressively worked, even though the there is minimal change in the percussion technology.

However, we also feel that the ridge count index is related to the existing flake removal pattern on the biface and not solely associated with the type of percussion technology used. For example, Biface 2 in event 5 and Biface 3 in event 3 both have steep drops in the average ridge count (see Figure 4.11). During these particular times of the experiment, these bifaces had irregular flaws in the material that had to be removed in order to continue to use the biface for usage and resharpening episodes. In doing so, a large portion of the biface surface was removed, including the previous flake ridges, which may have greatly affected the number of flake ridges for particular analysis grids.

SUMMARY AND DISCUSSION

Although incomplete at this point, our analysis shows some interesting trends and potential avenues for further exploration with regard to retouch on bifaces. First, it appears that retouch on bifaces may not be the same as retouch on flake tools. Bifaces are retouched throughout the reduction sequence, even during the production phase. The biface core has to be reduced in a fashion where the edge is continually being modified or retouched. Thus, traditional measures linking retouch amount to curation amount may not be effective for bifaces, because they may have a high retouch score without ever having been used, as illustrated with the application of Clarkson's (2002) index of invasiveness.

Second, overall flake removal amount may be a good indicator of the use life events for bifaces. For instance, our experiment showed that flake removal patterns of biface surfaces tended to increase as the biface was progressively used and resharpened. However, flake removal amount is also sensitive to changes in hammer type. As hammer types change, so does the relative proportion of flake shapes and sizes, which influences the flake removal pattern found on the biface, i.e., raw material flaws or "problem areas." The hammer type used, soft hammer versus hard hammer, is an idiosyncratic choice that is not a constant. Depending on the skill and technique of the flintknapper, different types of hammers will be used to address or reduce the objective piece into the desired form. The goal of the various flintknappers may be the same but the technique/method will vary from person to person and possibly from stone tool to stone tool (even when the same type of stone tool is being made). This means that flake removal patterns on bifacial surfaces may be effective for interpreting reduction only if hammer type is held constant or can be accounted for in some other way.

Finally, it is also apparent from our data that flake removal amount is not sensitive to changes in biface production events vs. biface resharpening events. Even though these events are clearly visible with debitage characteristics, they are not evident from the surface of bifaces, because the flake removal pattern of bifacial surfaces is produced by a series of multiple technological factors. As previously discussed, these can include flintknapper skill and technique, raw material quality, hammer type, and reduction strategy.

In summary, retouch indices created for flake tools may not be suitable for understanding curation strategies for bifaces. As noted in several other papers in this volume, retouch is particular to different tool types (Andrefsky; Eren and Prendergast; Quinn et al.) and to different tool functions (MacDonald). Retouch does not always equate to tool curation. Retouch is a technique used to shape a tool within the context of tool production, use, and resharpening. All of these contexts must be considered in attempting to quantify tool curation.

REFERENCES CITED

Andrefsky, William, Jr. 2005. *Lithics: Macroscopic Approaches to Analysis*. Second edition. Cambridge University Press, Cambridge.
    2006. Experimental and Archaeological Verification of an Index of Retouch for Hafted Bifaces. *American Antiquity* 71:743–58.
    2007. The Application and Misapplication of Mass Analysis in Lithic Debitage Studies. *Journal of Archaeological Science* 34:392–402.
Bamforth, Douglas B. 1986. Technological Efficiency and Tool Curation. *American Antiquity*, 51:38–50.
Barton, C. Michael. 1988. *Lithic Variability and Middle Paleolithic Behavior*. International Series 408. British Archaeological Reports, Oxford.
Binford, Lewis R. 1973. Interassemblage Variability: The Mousterian and the "Functional" Argument. In *The Explanation of Cultural Change: Models in Prehistory*, edited by C. Renfrew, pp. 227–54. Duckworth, London.
    1979. Organization and Formation Processes: Looking at Curated Technologies. *Journal of Anthropological Research* 35:255–73.
Blades, Brooke S. 2003. End Scraper Reduction and Hunter Gatherer Mobility. *American Antiquity* 68:141–56.
Callahan, Errett. 1979. The Basics of Biface Knapping in the Eastern Fluted Point Tradition: A Manual for Flintknappers and Lithic Analysts. *Archaeology of Eastern North America* 7(1):1–180.

Clarkson, Chris. 2002. An Index of Invasiveness for the Measurement of Unifacial and Bifacial Retouch: A Theoretical, Experimental, and Archaeological Verification. *Journal of Archaeological Science* 29:65–75.

Cotterell, Brian, and Johann Kamminga. 1987. The Formation of Flakes. *American Antiquity* 2:675–708.

Davis, Zachary J., and John J. Shea. 1998. Quantifying Lithic Curation: An Experimental Test of Dibble and Pelcin's Original Flake-Tool Mass Predictor. *Journal of Archaeological Science* 25:603–10.

Dibble, Harold L. 1995. Middle Paleolithic Scraper Reduction: Background, Clarification, and Review of Evidence to Date. *Journal of Archaeological Method and Theory* 2:299–368.

———. 1997. Platform Variability and Flake Morphology: A Comparison of Experimental and Archaeological Data and Implications for Interpreting Prehistoric Lithic Technological Strategies. *Lithic Technology* 22:150–70.

Eren, Metin L., Manuel Dominguez-Rodrigo, Steven L. Kuhn, Daniel S. Adler, Ian Le, and Ofer Bar-Yosef. 2005. Defining and Measuring Reduction in Unifacial Stone Tools. *Journal of Archaeological Science* 32:1190–1201.

Hayden, Brian, and W. Karl Hutchings. 1989. Whither the Billet Flake? In *Experiments in Lithic Technology*, edited by D. S. Amick and R. P. Mauldin. International Series 528, pp. 235–58. British Archaeological Reports, Oxford.

Hiscock, Peter, and Chris Clarkson. 2005. Experimental Evaluation of Kuhn's Geometric Index of Reduction and the Flat-Flake Problem. *Journal of Archaeological Science* 32:1015–22.

Hoffman, C. Marshall. 1985. Projectile Point Maintenance and Typology: Assessment with Factor Analysis and Canonical Correlation. In *For Concordance in Archaeological Analysis: Bridging Data Structure, Quantitative Technique, and Theory*, edited by C. Carr, pp. 566–612. Westport Press, Kansas City.

Kuhn, Steven L. 1990. A Geometric Index of Reduction for Unifacial Stone Tools. *Journal of Archaeological Science* 17:585–93.

Nelson, Margaret C. 1991. The Study of Technological Organization. In *Archaeological Method and Theory*, volume 3, edited by M. B. Schiffer, pp. 57–100. University of Arizona Press, Tucson.

Shott, Michael J. 1989. On Tool-Class Use Lives and the Formation of Archaeological Assemblages. *American Antiquity* 54:9–30.

———. 1996. An Exegesis of the Curation Concept. *Journal of Archaeological Science* 27:653–63.

Whitaker, John C. 1994. *Flintknapping: Making and Understanding Stone Tools*. University of Texas Press, Austin.

Wilson, Jennifer Keeling, and William Andrefsky, Jr. 2006. The Debitage of Bifacial Technology: An Application of Experimental Data to the Archeological Record. Paper presented at the Fifty-Ninth Annual Northwest Anthropological Conference, March 29–April 1, Seattle, Washington.

5   PETER HISCOCK AND CHRIS CLARKSON

THE CONSTRUCTION OF MORPHOLOGICAL DIVERSITY: A STUDY OF MOUSTERIAN IMPLEMENT RETOUCHING AT COMBE GRENAL

Abstract

In this chapter we present a study of flake retouching on one level of the Combe Grenal, located in the Black Perigord of France. We use the results to reflect on existing explanations of Middle Paleolithic tool production and diversity. Our evidence indicates the nonstaged and multilinear character of implement production and the apparent importance of blank form in influencing the pattern of retouch distribution and intensity. This inference implies that models of the implement classes, as stages of reduction, are not a viable depiction of the retouching technology represented in Layer 21. Instead, our reconstruction of scraper retouching demonstrated that each of Bordes's implement types has multiple histories of retouching. Some implements received little retouch, whereas others were intensively retouched; retouch sometimes changed a specimen to such an extent that the type into which it was classified was altered, whereas other specimens remained typologically stable even though they received additional retouch. The possibility that different specimens belonging to a type had different histories is a reason that typological groups may make poor analytical units for many technological questions.

We acknowledge and appreciate the permission to examine the collection granted by Dr. J.-J. Cleyet-Merle, the Director of the Musèe National de Prèhistoire des Eyzies. We thank Dr. Cleyet-Merle and the Musèe National de Prèhistoire des Eyzies for providing their facilities for the prolonged duration of our project. For assistance and discussions we thank Alain Turq, Andre Morala, and Jean-Philippe Faivre. This research was funded by an Australian Research Council Discovery Grant (DP0451472 – A Reappraisal of Western European Mousterian Tools Australian Perspectives).

INTRODUCTION

Questions of artifact reduction have been central to a number of high-level debates in Paleolithic archaeology. In particular, there have been extensive discussions about whether the traditional practice of analyzing retouched flakes by classifying them into a number of normative categories, called implement types, is valid or problematic, and whether those types represent tools of distinctly different designs or arbitrary divisions in sets of morphologically variable objects. As the traditional "building blocks" for interpretations of Paleolithic life, inferences about these issues have underpinned the different explanations for the Mousterian facies and the opposing claims about whether Neanderthals conceived of a large number of tool designs or not (see Binford 1973; Binford and Binford 1966; Bordes 1972; Dibble 1984, 1988a; Dibble and Rolland 1992; Mellars 1996; Rolland and Dibble 1990). These debates about the nature of economy, technology, and cognitive states in ancient hominids are both significant and exciting, but they rest on the accuracy and clarity of depictions of artefact patterning and the meaning of morphological and technological diversity. Although much has been written on the characterization of retouched flake variability in Paleolithic assemblages, many aspects of the archaeological patterns remain unresolved.

In this chapter we present a study of flake retouching on one level of the famous Combe Grenal site, located in the Black Perigord of France. We use the results to reflect on existing explanations of Middle Paleolithic tool production and diversity. We note and focus on two different aspects of published explanations. One is the primacy of retouch intensity in models explaining morphological diversity, with some researchers arguing it to be the sole significant factor forming typological variation, whereas other researchers argue that intensity of retouch is one of many factors creating variation and that others are often more significant. We discuss these different interpretations below, in the context of theories about the production of Quina scrapers.

A second aspect of published discussions of Middle Paleolithic typology is ideas on how the intensity of retouching is related to different implement types. Many researchers have described the relationship as being one of two schemes, which we shall call a single divergent scheme and a multiple parallel scheme. By the term "single divergent scheme" we refer to ideas of retouching that depict the creation of typological variability as the diversification of a single reduction pathway, where increased retouching produces new implement types, either in a series of stages along the pathway and/or in the form of some branching off the main stem (Figure 5.1a). "Branching" describes a process in which one typological category gives rise to two or more typologically distinct categories. By the term "multiple parallel scheme" we refer to propositions that posit multiple parallel reduction pathways, with or without some exchange of specimens between pathways (Figure 5.1b). We note that whereas much of the existing modeling of Mousterian flake retouching conforms to one of these two formats, there is also the possibility of a third, previously little unexplored, scheme that combines features of both single divergent and multiple parallel schemes. We will call this a "parallel branching scheme," and in such an interpretation there are multiple parallel pathways, some or all of which also produce morphological diversity through branching (Figure 5.1c). Although no researcher believes that every flake in an assemblage or region followed exactly the same reduction pattern, normative sequences of retouching represented in these schemes are a device that has been commonly used to depict reduction processes and develop predictions for quantitative testing.

Although we have no dispute with the premise of the "reduction hypothesis," which suggests that the extent of retouching had a significant effect on the morphology of artefacts found in many assemblages, there is no obligation on researchers to accept models of prehistoric reduction that posit that retouch intensity was the only factor creating typological diversity, that reduction-related morphological change proceeded on only a single branching pathway, or that all specimens belonging to an implement type necessarily represent only a single stage of retouching and a single retouching process. In the following analysis of the contribution of retouch intensity to typological differences, we demonstrate that the construction of morphological diversity in Mousterian assemblages can be complex: variation in implements reflects factors other than retouch intensity, retouched flakes with the same typological form have diverse reduction histories,

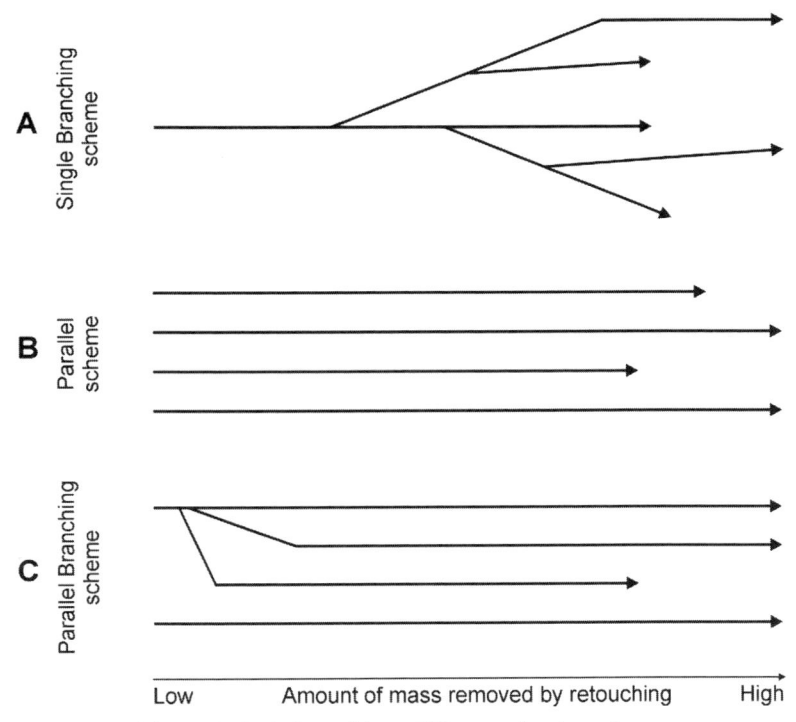

FIGURE 5.1. Schematic depiction of three different reduction schemes.

and reduction-related morphological changes sometimes follow multiple pathways rather than a single branching pathway. We provide a framework for this demonstration by examining some of the literature that deals with the relationship between different types of Quina scrapers and the extent of retouching that they have undergone.

THE QUESTION OF QUINA SCRAPERS AND REDUCTION

In English-language literature, one of the most well-known models for scraper reduction, and an example of the single branching scheme of implement variation for the Mousterian, was proposed by Harold Dibble (1984, 1987a, 1987b, 1988a, 1988b, 1995). Dibble hypothesized the transformation of scrapers from one typological class to another as they received additional reduction. His analysis examined

| A) Single Scraper | B) Double Scraper | C) Convergent Scraper | D) Transverse Scraper |

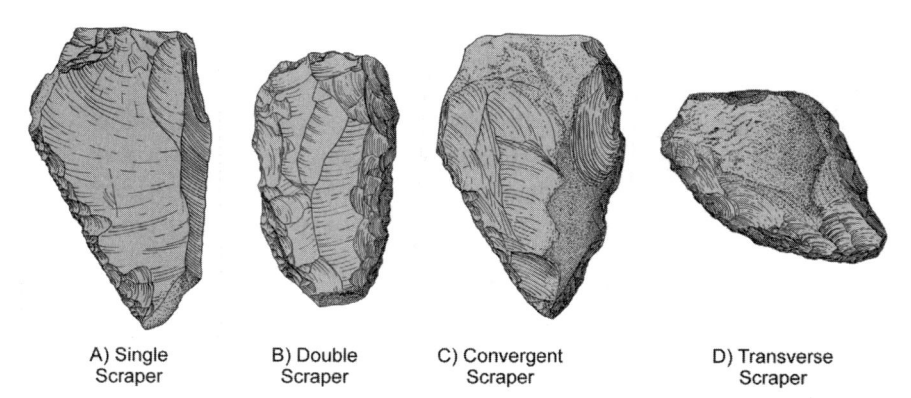

FIGURE 5.2. Examples of specimens classified into each of the four scraper classes: (A) single scraper, (B) double scraper, (C) convergent scraper, and (D) transverse scraper. All specimens are from Combe Grenal, Layer 21.

the proposition that the extent of reduction was the key factor causing differences between four implement classes with retouch onto their dorsal surface: (1) single-edged side scraper with retouch on one lateral margin (Bordes types 9–11), (2) double scrapers with two separate retouched edges (Bordes types 12–17), (3) convergent scrapers which have two retouched edges that touch (Bordes types 8, 18–21), and (4) transverse scrapers that have retouch across the distal end of the flake (Bordes types 22–24). Examples of these classes are provided in Figure 5.2. Dibble hypothesized that these four implement classes were the result of different amounts of reduction, in which all specimens began as single scrapers and with additional retouching were transformed either into transverse scrapers or alternatively into double and eventually convergent scrapers. This model, schematically shown in Figure 5.3, positions each of these four classes along a continuum of greater or lesser amounts of retouch, and reveals the proposition that there were two branches on which individual scrapers could travel from the same starting point. Dibble (1988a: 49; 1995: 319) suggested that those individual single scrapers that were further retouched were worked either at the distal end to become transverse forms or on the second lateral margin to become double/convergent implements. He suggested that the sequence that any individual specimen followed may have depended on the shape of the flake blank, with short/broad flakes being worked into transverse scrapers whereas

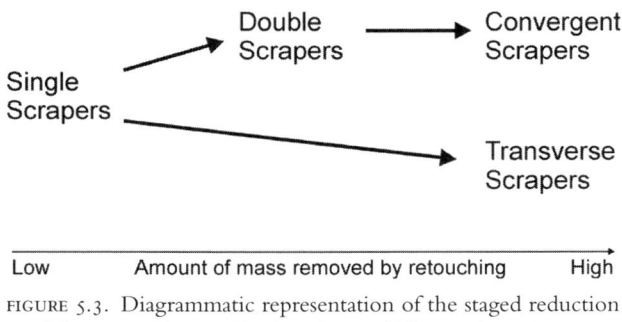

FIGURE 5.3. Diagrammatic representation of the staged reduction model proposed by Dibble (based on Dibble 1987b: 115).

longer, narrow flakes were retouched laterally to become double and convergent scrapers. However, he argued that much of the variation between implement classes, and specifically the morphology used to classify specimens, was a product of differences in the level of retouching and the length of time they had been used and maintained: single scrapers had undergone little retouching, whereas both transverse and convergent scrapers were more intensively retouched (Dibble 1995: 319). In a series of papers he has argued that the smaller average size of convergent and transverse scrapers, both in absolute terms and relative to their platform size, is evidence that this model was correct for Quina assemblages from southwest France and elsewhere. This conclusion has two significant implications for interpretations of Mousterian variability.

The first implication is that although he believed morphological variation in Mousterian retouched flakes took the form of a continuum created by differing extents of edge resharpening, he argued that traditional typology was valuable, because implement classes represented coherent stages in the continuum of retouch (Dibble and Rolland 1992: 11). Dibble (1988a: 52) therefore concluded that his model reinforced the value of typology by revealing a strong accord between the extent of utilization and the Bordesian implement types. He argued that traditional implement types could be employed as a proxy for the extent of tool maintenance/resharpening in archaeological assemblages.

Consequently Dibble developed a series of arguments that the composition of Mousterian industries did not reflect mental constructions of Neanderthals and that implement classes and industrial differences were not designed or functionally dedicated, but instead that differences between the implement classes and industries reflected differences in the intensity of tool use and by implication the nature of land use. For instance, Dibble and Rolland (1992: 17) argued that the production of industries dominated by convergent or transverse scrapers was a consequence of economic practices that encouraged more intensive tool use, such as the intensive maintenance of tools during cold paleoclimatic phases in which there were long winter residence and patterns of settlement based on the interception of migratory herds, situations in which provisioning of stone for tools could have proved difficult. They contrasted this with the contexts of industries dominated by side scrapers (and denticulates), which they hypothesized resulted from less intensive tool and site use that occurred under milder climatic phases in which the Neanderthal economy was focused on the pursuit of dispersed, mobile game. The value of these kinds of interpretations depends on the veracity of the characterization of traditional implement types as comparable units primarily reflecting differences in the extent of tool resharpening.

A number of commentaries and further studies have followed the publication of Dibble's model, many supporting his argument for the value of traditional implement types for studies of the extent of implement reduction (e.g., Gordon 1993; Holdaway et al. 1996). However, some reconsideration of the factors involved in implement creation has been offered. The most potent is the proposition that the extent of retouching is not a function of edge maintenance alone but was often a reflection of the size and morphology of the flake to which retouch had been applied. For instance, Dibble (1991: 266), Gordon (1993: 211), and Holdaway et al. (1996) all argued that larger flakes typically had greater potential for edge resharpening, and consequently in extensively reduced assemblages those larger specimens received more retouching than smaller ones. One result of the continued reduction of larger specimens, but not smaller ones, is that extensively retouched flakes were sometimes larger when discarded than less extensively retouched ones made on smaller blanks (Dibble 1991). Although this proposition has been applied to notched types (e.g., Holdaway et al. 1996; Hiscock and Clarkson 2007), its implications for the interpretation of other implement types and for the value of typology as a measure of the extent of retouching have received less attention. One obvious implication is that the amount of retouching applied to a specimen cannot be judged by its size (Dibble 1991), a realization that encouraged the development and growth of several methods for measuring retouch intensity on Middle Paleolithic tools (see Dibble 1995; Hiscock and Clarkson 2005). However, the existence of a strong relationship between blank form and retouch has been argued to create problems for the interpretation of implement types as reduction stages.

For example, if retouching is a response to blank form and there is variation in the size and shape of blanks being retouched, an almost inevitable reality in most prehistoric contexts, then the amount of mass removed during retouching may vary substantially between specimens assigned to any implement type. This appears to be the case in the data presented by Dibble (1987b: 113) for the La Quina scrapers, which display extraordinarily high levels of intratype variability in reduction measures, such as flake area/platform area ratios that show coefficients of variation of 125% for single scrapers, 49% for double scrapers, 91% for convergent scrapers, and 182% for transverse scrapers. Although Dibble still found statistically significant differences between the means of these four implement classes, the measured variability probably reflects very great differences in the amount of retouching between specimens in a single implement class. In such circumstances, the value of conventional types as units measuring the extent of reduction may be questioned, and Hiscock (1994) argued that analysts would be better able to discuss differences in amounts of retouching if they focused on measuring the manufacture of individual specimens rather than merely the contrast between types.

Furthermore, Kuhn (1992) has argued that if blank form played a significant role in the position and amount of retouch on each flake, this would change how variation between implements could be explained. In such situations, typological composition is not solely, or even principally, affected by the intensity of tool use, and so industrial variation may not directly correspond to different patterns of settlement and mobility. Instead, Kuhn argues, the typological composition of an assemblage would reflect the size and shape of available blanks, which in turn would reflect the form and availability of raw material and the tactics of core reduction. Although raw material procurement and core reduction may also be linked to economic and settlement patterns, the connection with the abundance of each implement type would be remote and indistinct. Although Kuhn did not deny the proposition that intensity of retouch may be an indicator of settlement/mobility systems, he argued that types are not reliable indicators of intensity of retouch, and that archaeologists will require dedicated measurements of retouch intensity prior to developing inferences about land use from lithic artefacts.

Long-term archaeological research in southwest France has yielded much evidence for the complex articulation of core reduction systems and the patterns of retouched tools made on the flakes produced in those systems (e.g., Bisson 2001; Bourguignon 1997; Bourguignon et al. 2004; Thiébaut 2003; Verjux 1988; Verjux and Rousseau 1986), reinforcing the possibility that blank form may have an important role in the construction of morphological diversity amongst Mousterian implements. Many discussions of blank-retouch relationships have posited a simple relationship between flake elongation and the position of retouch, suggesting that long flakes were often retouched on their lateral margins, whereas short, wide flakes were often worked at the distal end (e.g., Bordes 1961: 806, 1968: 101; Mellars 1992; Turq 1989). A number of researchers have argued that the flake blanks on which single and transverse scrapers were made were very different, and that regular production and/or selection of flakes with particular characteristics was a significant factor in the formation of the typological composition of any assemblage (e.g., Turq 1989, 1992). As a consequence, Turq (1989) argued that there were clear morphological discontinuities in the form of single and transverse scrapers in the Dordogne, evidence that would not be conformable to Dibble's reduction hypothesis. The hypothesized connection of blank form and systems of core reduction has been also been argued to be evidence for deliberate and planned acts of selection/production (e.g., Boëda 1988; Turq 1989, 1992).

Some models of the way the morphology of flake blanks strongly influenced the nature and typological category of implements have hypothesized that complex interactions between multiple characteristics were responsible for the nature of retouching. An example is Alain Turq's proposal that scrapers in Quina industries reflected a regular pattern of blank selection and retouching. He suggested that

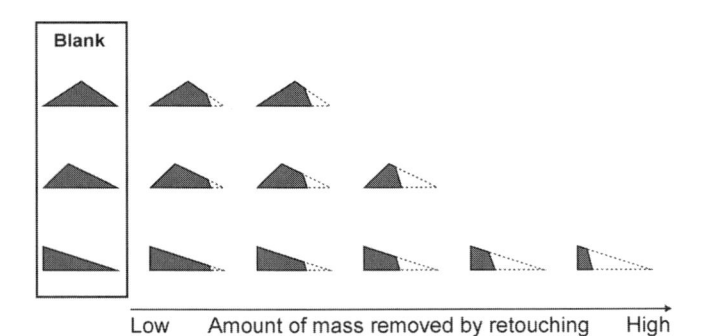

FIGURE 5.4. Schematic illustration of the relationship of blank cross section and extent of reduction for dorsally retouched Quina scrapers (after Turq 1992: Figure 6.2).

transverse scrapers, unlike single side scrapers, were made on flakes that were thick relative to their length and ventral surface area, a proposition that would account for differences between types in the relationship of platform and ventral areas, which Dibble (1984, 1987, 1995) had employed as evidence for different degrees of reduction. Furthermore, Turq argues that scrapers were typically made on flakes with asymmetrical cross sections and retouch was located on the flake margin furthest from the maximum thickness (Turq 1992: 75). In a diagram, presented here as Figure 5.4, Turq (1992: 77) implied that the potential for resharpening was related to the asymmetry of each blank selected for retouching, with symmetrical flakes having little mass removed before steep retouching came close to reaching the thickest part of the flake, wereas asymmetrical flakes could have considerably more mass removed through retouching before reaching the same state. This proposition linked variation in scraper morphology with blank form as well as with extent of reduction, implying that flake shape and blank selection were the proximate factors creating variation in both the location/orientation of retouch and the amount of mass removed by retouching on different specimens, and consequently the typological category into which each specimen was placed. Such a model not only contrasts with Dibble's in the emphasis given to flake form rather than extent of reduction, but also implies that there may be a great deal of difference in the extent of reduction of specimens with similar cross sections and placed in the same typological category.

Presented in this way, the distinctions between two different models of Quina scraper variability are clear. On the one hand, Dibble's reduction hypothesis, in the form of a single branching scheme, asserts that traditional typological categories represent different points/stages along a continuum of greater or lesser amounts of retouch, and the frequency of specimens in each type may therefore be used as a proxy for the intensity of reduction that an assemblage has undergone. From this perspective, intensity of reduction is the primary cause of typological variation, and although differences in flake blanks exist, their effect on typological variation is minimal. Consequently, typological diversity through time and space can be directly interpreted as a result of access to raw material and settlement/economic activities. On the other hand, what we might call the "blank–retouch interaction hypothesis" proposes that traditional typological categories represent complex patterns of morphological variation created by several factors, particularly differences in the distribution and intensity of retouch in response to blank form. Distinctions between conventional implement types may therefore have little coherent covariation with intensity of retouch and should not necessarily be treated as representing different points along a reduction continuum. This hypothesis implies that the frequency of specimens in each type may not be a reliable indicator of the intensity of retouching that an assemblage has undergone and that typological diversity through time and space is difficult to directly interpret in terms of settlement/economic activities. Instead, this hypothesis asserts that Borde's typology reflects morphological patterns created by a constellation of factors including blank form, material cost, and tool design, as well as amount of use-life/resharpening, and that the resulting typological patterns are not necessarily sensitive to variation in the intensity of retouch.

In this chapter we explore the applicability of these two opposing models to one Quina assemblage, recovered from Layer 21 in Combe Grenal. Although these models posit slightly different behavioral processes, and are consistent with different analytical practices and interpretations, they both invoke the extent of retouching as a mechanism constructing morphological variation; the two models differ in the way retouching is articulated to other technological and economic factors. We emphasize that there is no reason to expect that one hypothesis will inevitably be the most appropriate in all situations. It is possible for the reduction hypothesis to be correct for some assemblages and the blank–retouch interaction hypothesis to be correct for others. In this way, these opposing models are not competitors in a search for some ill-defined universal truth, but are actually expressions of the variable operation of multiple factors that may have created morphological variation in Mousterian implements. Consequently, our examination of these two models for Layer 21 at Combe Grenal is not a test of the general veracity of either model but actually an assessment of what kinds of processes were operating in the Neanderthal technological system at the time that the layer formed.

OUR APPROACH TO MEASURING THE EXTENT OF RETOUCHING

Our sample of artefacts for this chapter comes from the French Middle Paleolithic site of Combe Grenal, excavated by François Bordes (1972) and now held at the Musèe National de Prèhistoire des Eyzies. We measured a collection of dorsally retouched flakes from Layer 21, a level containing a representative Quina assemblage. Technological cores and unretouched flakes, broken specimens, and a small number of burins end scrapers, Mousterian points, and a truncated-faceted pieces were excluded from the analysis. For this paper our sample consists of 158 specimens representing each of the major typological categories: single scrapers ($N = 70$), double scrapers ($N = 23$), convergent scrapers ($N = 25$), and transverse scrapers ($N = 40$). The number of specimens in each of the Bordes type classes is listed in Table 5.1.

Our analysis of the reduction-relationship of these implement categories employs two measures of the position of retouching on each specimen and the amount of mass removed from each flake through retouching (Figure 5.5). The first is a version of Kuhn's index of unifacial reduction (GIUR), a measure we have experimentally verified (Hiscock and Clarkson 2005, 2007). Our experiments showed that scar height ratios, taken at multiple points around a retouched flake, yields an average GIUR value that has a nonlinear relationship with the mass removed by retouching (Hiscock and Clarkson 2005: 1019). Experimental retouching of flakes demonstrated that there was a strong log-linear relationship between the calculated Kuhn GIUR

Table 5.1. Sample of complete retouched flakes from Layer 21 used in analysis

Implement types	N
Single	
9. Single straight scraper	42
10. Single convex scraper	28
Double	
12. Double straight scraper	13
13. Double straight-convex scraper	6
15. Double convex scraper	3
16. Double concave scraper	1
Convergent	
8. Limace	2
18. Straight convergent scraper	2
21. Dejete scraper	21
Transverse	
22. Straight transverse scraper	6
23. Convex transverse scraper	33
24. Concave transverse scraper	1

and the percentage of original flake weight lost (Figure 5.6). This relationship appears to hold irrespective of whether retouching is applied to the lateral or distal margin (Hiscock and Clarkson 2005) or to one or more than one edge (Hiscock and Clarkson 2007). For instance, when we experimentally retouched flakes on one lateral margin, producing items similar to single side scrapers, there was a strong positive relationship between the index value and the mass removed by retouching ($r = 0.933, r^2 = 0.871$). When we experimentally retouched flakes on two lateral margins, the Kuhn GIUR was still strongly and significantly correlated with the proportion of mass lost from each flake ($r = 0.88, r^2 = 0.778$). We have argued elsewhere that although variations in the GIUR/mass-lost relationship occurred as a consequence of differences in the shape and size of flake blanks, a strong relationship exists for most flakes that are dorsally retouched, and consequently we take the Kuhn GIUR to be a reliable measure of the extent of dorsal, unifacial retouch in most instances, including the specimens discussed

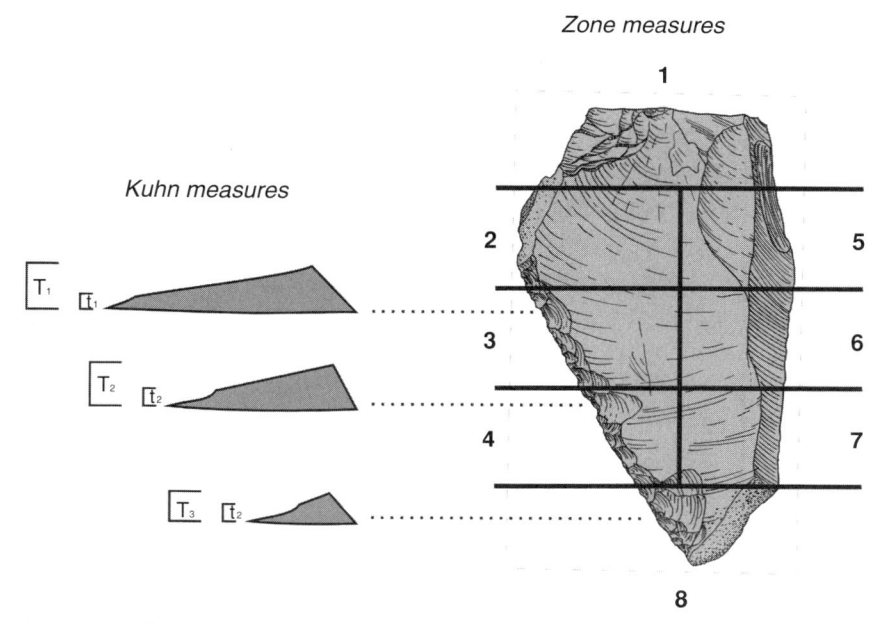

FIGURE 5.5. Illustration of the measurements of reduction used in this chapter: multiple values of Kuhn's (1990) unifacial reduction index and a count of the number of zones that have been retouched.

in this chapter, irrespective of the nature of the flake blank (Hiscock and Clarkson 2005: 1022). Furthermore, the high coefficient of determination (r^2) allows us to use the regression line and 95% confidence intervals shown in Figure 5.6 to estimate the approximate amount of mass removed during retouching.

A second measure of retouching, involving a record of the distribution of retouch on the margins of each flake, provides an indication of the lateral expansion of retouching around the specimen. This measure complements the Kuhn GIUR, which measures how far retouch has penetrated into the centre of a flake (Hiscock and Attenbrow 2005: 59). This "retouched zone index" was obtained by observing which of eight zones, illustrated in Figure 5.5, were retouched. The zones were defined in terms of five equal divisions of the percussion length, but with the left and right margins being separated to create eight locations (proximal, distal, three zones on the right margin, and three on the left). The face on which scars occurred was not relevant for

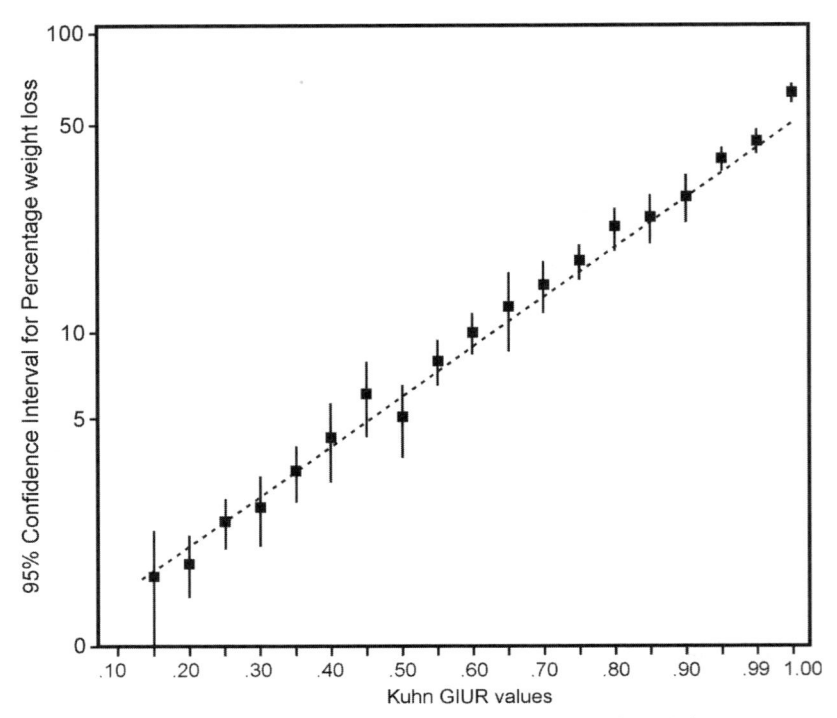

FIGURE 5.6. Relationship between Kuhn GIUR (as 0.05 intervals) and the percentage of original flake mass lost through retouching (shown with 95% confidence intervals) in the experimental dataset (Hiscock and Clarkson 2005). Broken line is the regression line ($r = 0.933$, $r^2 = 0.871$) published by Hiscock and Clarkson (2005).

this measure, giving retouched flakes values between 1 and 8 zones. This recording system not only was used to measure the amount of retouch around the flake margin, but also served as a way to compare the location of retouch on different specimens.

Other measures of flake retouching, such as the invasiveness index of Clarkson (2002) or the surface area/platform thickness ratio of Holdaway et al. (1996), were considered to be of lesser value on the steeply, unifacially retouched flakes in our sample and are not presented here. Although we have previously expressed doubt about the sensitivity of Dibble's (1987) surface area/platform area ratio as a measure of reduction, we have calculated this below as a comparison to published data that have been used to discuss models of Quina retouch.

Table 5.2. Descriptive statistics for the Kuhn GIUR, retouched zone index, and surface area/platform area ratio

	Kuhn GIUR	Retouched zone index	Surface area/ platform area
Single ($N = 68$)	0.51 ± 0.21	4.52 ± 1.33	18.52 ± 42.31
	$0.17 - 1.00$	$1 - 8$	$1.0 - 297.4$
Double ($N = 25$)	0.61 ± 0.14	6.28 ± 1.21	11.49 ± 7.88
	$0.31 - 0.86$	$2 - 8$	$2.9 - 31.2$
Convergent ($N = 26$)	0.66 ± 0.14	6.08 ± 1.67	12.52 ± 24.11
	$0.39 - 0.91$	$4 - 8$	$1.2 - 105.8$
Transverse ($N = 37$)	0.65 ± 0.21	3.78 ± 1.62	9.11 ± 18.11
	$0.36 - 1.00$	$1 - 6$	$0.5 - 102.7$

Note: Top line is mean and standard deviation; bottom line is the minimum and maximum values.

THE EXTENT OF RETOUCHING AND IMPLICATIONS FOR REDUCTION

With these measurements of the amount of retouching, we are able to evaluate whether the different implement categories (single, double, convergent, and transverse scrapers) actually represent clusters of specimens that have been reduced to different extents, as hypothesized by Dibble. Descriptive statistics for the reduction indices in our sample, presented in Table 5.2, show a pattern somewhat similar to that reported by Dibble (1987b: 113) for the La Quina site, and which he used in support of his reduction model. For instance, the mean surface area/platform area values are higher for single scrapers than for double and convergent ones, and transverse scrapers display the smallest mean; with the means being very similar to those Dibble found at La Quina. This is support for the proposition that, on average, single scrapers were less reduced than the other three scraper categories. Average values for the Kuhn GIUR and retouched zone index could also be used to suggest that single scrapers were less reduced than double or convergent scrapers, giving support to the inference of a single-double-convergent sequence of scraper transformations in Layer 21. ANOVA treatment of our data reveals statistically significant differences between the implement classes in the Kuhn GIUR ($F = 3.224$, d.f. $= 4$, $p = .014$, with the index broken into five groups: 0.01 – 0.19, 0.20 – 0.39, 0.40 – 0.59, 0.60 – 0.79, and 0.8 – 1.0) and in the retouched zone index ($F = 3.365$, d.f. $= 7$, $p = .002$); but not in the surface area/platform area ratio ($F = 0.749$, d.f. $= 5$, $p = .588$). These statistics all indicate that there is patterned variation in retouching intensity between the four classes of implements, at least in terms of the central tendencies for the classes.

However, the relationship of transverse and single scrapers is not entirely consistent with the predictions of Dibble's reduction model. Differences in mean Kuhn GIUR alone ($t = 3.361$, d.f. $= 103$, $p = .001$) conform to the predictions of Dibble's (1987b) model, although the question of how to interpret the large variation in each class is discussed below. However, the retouched zone index indicates that transverse scrapers have significantly less extensively retouched margins than single scrapers ($t = -2.600$, d.f. $= 64.9$, $p = .012$), which is a finding not consistent with Dibble's model, in which the addition of distal retouch converted single scrapers into transverse ones. These statistics imply a difference between single and transverse scrapers in intensity and location of reduction, but not necessarily as sequential stages, as Dibble argued in the model of his single-transverse sequences.

These data suggest that there are differences between these implement classes in the average degree of reduction, but such differences do not, by themselves, constitute evidence of the transformation of specimens from one implement class to another. An examination of the variation found within each implement class reveals that the archaeological evidence from Layer 21 does not simply conform to Dibble's reduction model. Each of the implement classes displays high levels of variation in the reduction measures. In particular, single and transverse scrapers show large ranges of reduction indices. For example, on single scrapers the coefficient of variation for Kuhn GIUR is 41% and for the retouched zone index it is 29%, whereas transverse scrapers have a coefficient of variation for the Kuhn GIUR of 32% and for the retouched zone index of 40%. This indicates that each of those typological groupings contains specimens with very different levels of retouch. Using the Kuhn GIUR to estimate the proportion of original flake mass removed through retouching indicates that single scrapers lost 2–66% of their weight, double scrapers 3–30%, convergent scrapers 5–35%, and transverse scrapers 4–66% of blank weight. When intensity of reduction is expressed in this way, it is clear that Dibble's reduction models, both the single–double–convergent path and the single–transverse one, do not account for all of the specimens in Layer 21. For instance, some single scrapers are extensively reduced; some are more than twice as reduced as any double or convergent scrapers. The existence of single scrapers with very high amounts of mass removed through retouching, and that were not converted into double or convergent forms, demonstrates that specimens typologically classified as single scrapers were not all "early-stage," with only little retouch. Conversely, the existence of double, convergent, and transverse scrapers with less than 5–10% of mass removed through retouching, representing the initial creation of the edge and perhaps one resharpening episode, demonstrates that such forms were not always more heavily retouched than single scrapers. Similarly, many transverse scrapers were not noticeably more reduced than many single scrapers, as might be expected if they were created at a later stage. However, other transverse scrapers have been extensively retouched, probably losing more than 50% of their original mass. This illustrates that the intensity of reduction within each implement class is highly variable. Further evidence for this within-class variation in retouching intensity, and its implications, is provided in the following sections.

SINGLE SCRAPERS

The striking characteristic of single scrapers in Layer 21, besides the strong pattern of retouch positioned on one lateral margin, is the great difference in the extent of reduction that different specimens had undergone. Some of the variation in the extent of retouching is displayed by the retouched zone index. The majority of single scrapers were retouched along much of one lateral margin, resulting in retouch scars in four or five zones (Figure 5.7). However, a few specimens had retouch restricted to a small portion of the lateral margin, only one or two zones; and some specimens also had small occurrences of retouch elsewhere on the flake-blank, in more than five zones. The distribution of retouch around the flake margin was clearly related to blank characteristics, such as edge angle, cross section, and distribution of cortex.

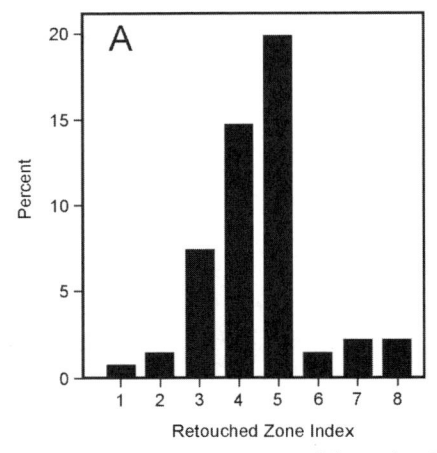

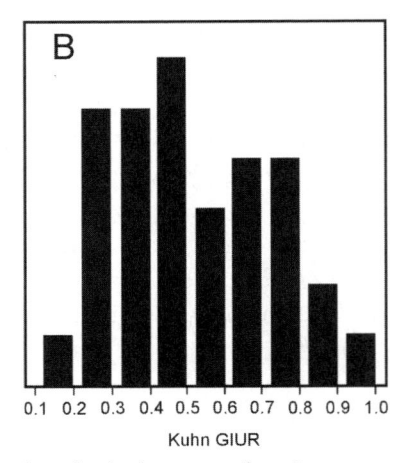

FIGURE 5.7. Histogram of the Kuhn GIUR values for single scrapers from Layer 21.

As a measure of another dimension of retouch intensity, the Kuhn GIUR also displays extreme variation. Figure 5.7 shows a histogram of the abundance of specimens with different levels of the Kuhn GIUR. Approximately 20% of single scrapers had a GIUR less than 0.3, equating to less than about 5% of the original flake mass removed by retouching. Most single scrapers had GIUR values of 0.3–0.8, representing about 5–20% mass loss. Some single scrapers, about 10%, had GIUR values greater than 0.8, representing retouch that probably removed from 30% to more than 60% of the original mass. Although conversion of GIUR values to mass lost through retouching in this way is only an estimate, it provides a behavioral expression of the large differences in retouch intensity that are evident on different single scrapers.

Differences in the extent of retouching on single scrapers are illustrated in Figure 5.8, which presents two specimens: one with a small amount of material removed by retouching and the other with a large amount. The first specimen is a long flake with a series of small retouch scars, mostly about 3 mm long, on three zones of the left lateral margin (Figure 5.8A). The Kuhn GIUR of 0.44 recorded for this specimen, in association with low unretouched edge angles of 20°–25° in the retouched zones, is consistent with less than 5–10% of the original flake mass being removed by retouching. The other specimen (Figure 5.8B) was the remnant of a wide, thick flake that has been extensively reduced through the removal of large flakes from along the entire right

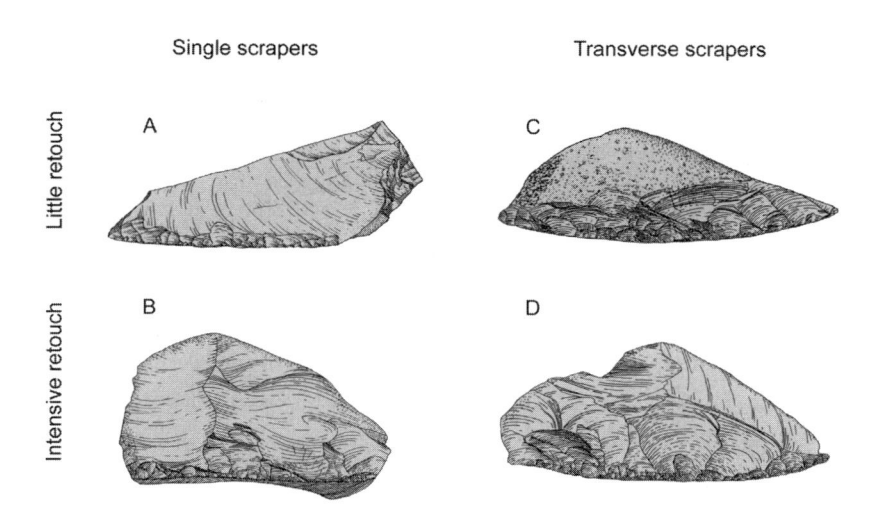

THE CONSTRUCTION OF MORPHOLOGICAL DIVERSITY 125

FIGURE 5.8. Examples of different levels of reduction on scrapers: A and B = single scrapers, C and D = transverse scrapers; A and C = little mass removed, B and D = extensive mass removed. A = single scraper with a Kuhn index of 0.44; B = single scraper with a GIUR index of 1.00; C = transverse scraper with a Kuhn index of 0.56; D = transverse scraper with a Kuhn index of 0.91.

lateral margin (retouch in five zones). This specimen has a GIUR of 1.00, with the retouch scars having removed the thickest part of the flake, a pattern consistent with the removal of approximately 45–65% or more of the original flake mass by retouching. Together these two illustrations exemplify the different levels of reduction present amongst single scrapers in Layer 21.

In conjunction with the statistics, these specimens demonstrate that some single scrapers in Layer 21 were minimally retouched whereas others were heavily retouched. The heavily retouched specimens, as indicated by the Kuhn GIUR, typically have retouch scars only on one lateral margin and had always been single scrapers throughout the retouching process. The evidence from such specimens shows that some single scrapers became very intensively retouched but that the level of reduction did not alter their typological status.

DOUBLE AND CONVERGENT SCRAPERS

Double and convergent scrapers are almost certainly made from single scrapers that had appropriate sizes and shapes, because one retouched margin must logically have been created before the other. The higher mean and minimum GIUR values for both classes, in comparison to those for single scrapers, are consistent with that interpretation. However, the evidence for Layer 21 does not conform to Dibble's proposed single–double–convergent sequence of type stages, for two reasons. The first has already been discussed: single scrapers were sometimes very intensively retouched and so specimens in that typological class do not always represent a stage of minimal reduction. This observation does not negate the conclusion that double/convergent scrapers were once single scrapers, but it refutes a stage-based model in which the former specimens are always highly retouched and the latter always minimally retouched.

Furthermore, for Layer 21, there appears to be little or no difference in the level of reduction of the double and convergent scrapers. There is no significant difference between these two classes for either the Kuhn GIUR ($t = -1.271$, d.f. $= 49$, $p = .210$) or the retouched zone index ($t = 0.496$, d.f. $= 49$, $p = .622$), and the ranges and distribution of values are comparable, evidence that all specimens in both groups show the same levels of retouch intensity. Because convergent scrapers in this assemblage are not more reduced than double scrapers, the notion of a double–convergent sequence of reduction is unlikely to be correct.

Instead, it seems likely that double and convergent scrapers are made on different flake blanks. A number of features of the blank are preserved on these retouched specimens and show statistically significant differences between the two classes. For example, the mean unretouched edge angles of double scrapers are significantly lower than those of convergent scrapers ($t = -2.138$, d.f. $= 49$, $p = .038$) and the mean thickness of double scrapers is also lower ($t = -3.218$, d.f. $= 49$, $p = .002$). These data raise the possibility that the differences in the relative positioning of retouched edges in double and convergent scrapers, leading them to be assigned to different types, primarily reflect dissimilarities in blank form rather than extent of retouch. Hence it is possible to conclude that many specimens classified as double and convergent scrapers were reworked single scrapers, but that many of the convergent scrapers are not more intensively retouched than double scrapers; the typological distinction largely reflects the influence of blank form.

TRANSVERSE SCRAPERS

Transverse scrapers also display large differences in the extent of retouching. Nearly 60% of transverse scrapers had a GIUR less than 0.6, probably indicating less than 10% mass lost through retouching; but 20% of specimens had values of 1.0 and probably had more than 40–50% of their initial mass removed. These differences can also be illustrated using specific implements as exemplars (Figure 5.8). For instance, Figure 5.8C shows a transverse scraper made on a primary decortication flake, which has had a series of small retouch scars in Zone 8, at the distal end. The Kuhn GIUR of 0.56 recorded for this specimen, in association with a low unretouched edge angle of 34° at the distal end, is consistent with less than 10% of the original flake mass being removed by retouching. In contrast, another transverse scraper shown in Figure 5.8D had a series of large flake scars at the distal end, with retouch removing the thickest part of the flake along one-half of the edge to give a GIUR of 0.91. This pattern is consistent with the removal of at least 30–35% of the original flake mass by retouching. These two illustrations exemplify the different levels of reduction present amongst transverse scrapers in this layer.

These large differences in retouch intensity between specimens classified as transverse scrapers may reflect blank form: specimens with GIUR of less than 0.6 have, on average, significantly smaller platform thickness ($t = 2.604$, d.f. $= 35$, $p = .014$), flake thickness ($t = 3.252$, d.f. $= 35$, $p = .003$), and lower unretouched edge angles ($t = 3.618$, d.f. $= 35$, $p = .001$). Reduction intensity was therefore probably connected to blank size and morphology, but despite the great variation in retouch between specimens in Layer 21, in response to different flake-blanks, retouch intensity did not alter the typological status of transverse scrapers. That inference is inconsistent with the notion that transverse scrapers were once single-side scrapers that had subsequently had additional retouch added to the distal end, and instead is evidence that many or all transverse scrapers were probably always typologically transverse forms.

This conclusion is reinforced by information about the distribution of retouch around the perimeters of flakes (Figure 5.9). Distribution of retouch around flake perimeters is not consistent with all transverse specimens originally being single scrapers. Nearly 50%

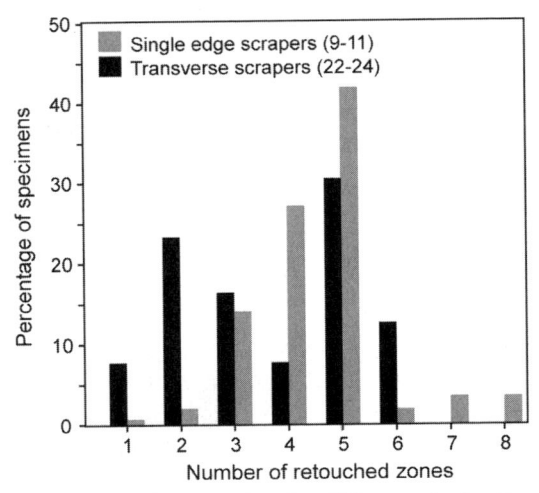

FIGURE 5.9. Histogram showing differences in the distribution of retouch on specimens classified as single scrapers and transverse scrapers.

of transverse scrapers have retouch only toward the distal end (<4 retouched zones). These specimens were never single scrapers, and we conclude that at least half the transverse scrapers began as transverse scrapers. Those with 4–6 retouched zones may once have been single-side scrapers that had retouch added to the distal end, or they may have begun as transverse scrapers that had retouch added to a lateral margin. Although it is possible that in Layer 21 Dibble's hypothesized transformation of single into transverse scrapers sometimes occurred, the evidence suggests that this was infrequent compared to the common process in transverse scrapers of beginning retouch at the distal end and continuing to retouch in that location.

The initiation and maintenance of restricted patterns of retouch, either at the distal end or on a margin, probably reflects the influence of blank form. Transverse and single scrapers were regularly made on different flake blanks. For example, flakes worked transversely are significantly thicker (3.021, d.f. $= 107$, $p = .003$) and with higher unretouched edge angles (2.420, d.f. $= 106$, $p = .017$) than those worked on only the lateral margin. Differences in flake shape and thickness are hypothesized to have been factors affecting the decision of knappers to begin working flakes laterally or distally.

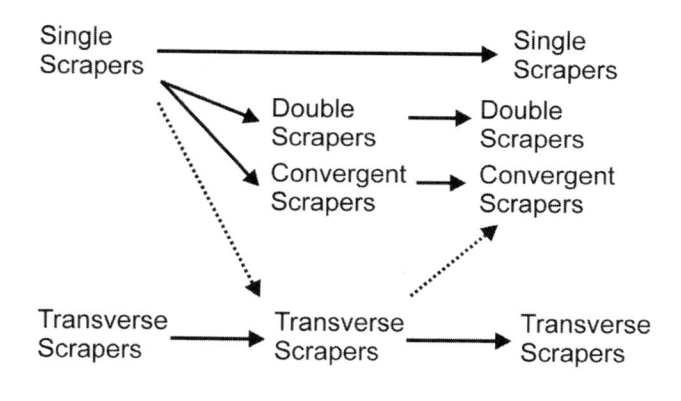

THE CONSTRUCTION OF MORPHOLOGICAL DIVERSITY 129

```
Single ─────────────────────→ Single
Scrapers                       Scrapers
       ↘   Double    ──→ Double
           Scrapers       Scrapers
       ↘   Convergent ──→ Convergent
           Scrapers       Scrapers
                              ↗
       ↘                   ↗
Transverse ──→ Transverse ──→ Transverse
Scrapers       Scrapers       Scrapers

|─────────────────────────────────────→
Low    Amount of mass removed by retouching    High
```

FIGURE 5.10. Illustration of the typological status and reduction history of flakes retouched to different degrees, using the same graphical conventions as Figure 5.2.

By definition, retouch on these specimens is concentrated at the distal end of each specimen, but often extends along some of one or both lateral margins, particularly in the production of the large curved retouched edges in convex transverse scrapers. Consequently, although the majority of specimens have retouch in two, three, or four retouch zones, some specimens have five or even six zones retouched. The creation of broad, transverse retouched edges involving retouch across several zones may also be conditioned by blank form. For instance, specimens with less than four retouched zones have, on the average, significantly smaller platform widths ($t = -2.056$, d.f. $= 36$, $p = .045$) and higher unretouched edge angles ($t = 2.392$, d.f. $= 31$, $p = .023$).

A RETOUCHING SCHEME FOR LAYER 21

This evidence already presented is consistent with a retouching scheme that is more elaborate and less stage-based in nature than the one proposed by Dibble (and shown in Figure 5.2). Our interpretation of the retouching processes creating the typological scraper groups in Layer 21 of Combe Grenal is a parallel branching scheme, graphically represented in Figure 5.10. Most frequently, single-side scrapers were retouched only on one margin for their entire history of production and maintenance. Some of those specimens were discarded after only a small amount of retouching, but others were very intensively retouched on the same margin but remained, in typological terms, single scrapers. Some single-side scrapers were retouched on additional margins to produce specimens classified as either double or convergent scrapers. Single scrapers were typically converted into double or convergent scrapers, but there is little evidence for double scrapers being reworked to form convergent ones. Double and convergent scrapers have comparable levels of retouch, and are not sequential stages of retouch; they represent alternative strategies applied to single scrapers with subtly different sizes and shapes. The choice of whether to continue retouching one margin or to begin working a second, and in the latter case to retouch parallel or converging edges, appears to be related to differences in blank form. The precise interaction of blank form and retouch intensity will be pursued in future publications.

Retouching of flakes to produce transverse scrapers appears to have been largely separate to patterns of lateral retouching leading to single, double, and convergent scrapers (see Figure 5.10). Some transformations of single-side scrapers to transverse scrapers, or the reworking of transverse scrapers into double/convergent scrapers, probably occurred, but in Layer 21 this was infrequent. Our interpretation of the evidence is that the majority of transverse scrapers were distally retouched throughout their "life span" and they had never been single-side scrapers. Transverse scrapers therefore principally represent the result of a parallel technological pattern that is separate from, and constitutes an alternative to, the retouching strategy that created single-side scrapers.

IMPLICATIONS FOR THE INTERPRETATION AND ANALYSIS OF MOUSTERIAN VARIABILITY

These interpretations of scraper retouching in Layer 21 of Combe Grenal carry a number of implications. Evidence presented indicates the nonstaged and multilinear character of implement production and the apparent importance of blank form in influencing the pattern of retouch distribution and intensity. This inference implies that Dibble's model of the implement classes as stages of reduction is not a viable depiction of the retouching technology represented in Layer 21, whereas our reconstruction of retouching processes conforms to many of the propositions contained in the blank–retouch interaction hypothesis.

Our reconstruction of scraper retouching in Layer 21 identified evidence that each of Bordes's implement types examined here may have multiple histories of retouching. Some implements received little retouch, whereas others were intensively retouched; retouch sometimes changed a specimen to such an extent that the type into which it was classified was altered, whereas other specimens remained typologically stable even though they received additional retouch; and so on. The possibility that different specimens belonging to a type had different histories is a reason that typological groups may make poor analytical units for many technological questions. In particular, the large variation in retouch intensity observed among different specimens classified in a single type demonstrates that the Bordes typology is not a reliable system for measuring retouch intensity. In assemblages such as this, Bordesian types tend to record the pattern and character of retouch preserved on flakes at the time they were discarded, but intensity of retouch cannot be accurately inferred from the type classification alone. Consequently, studies of spatial and temporal changes in retouch intensity will be more reliable when made on the basis of dedicated and experimentally verified systems of measurement, such as the Kuhn GIUR.

As discussed above, there have been extended discussions about the existence and meaning of an indistinct correspondence in southwest France between environmental conditions and the lithic "industry" that was in place, examining the proposition that different industries reflect different levels of retouch and core reduction intensity as knappers adjusted their technological practices to suit the prevailing economic conditions (e.g., Dibble 1984; Rolland 1981; Rolland and Dibble 1990). However, the coarse relationship between these phenomena has not convinced critical commentators of the reality of a functional link that would be capable of explaining industrial variation directly in terms of a connection between climate and intensity of tool use (e.g., Mellars 1996:343). The analysis we have presented in this chapter will reopen debate about this question by revealing that the typological practices employed to identify different industries and interpret them in terms of retouch intensity need not be accurate or sensitive to variations in the level of tool production and maintenance. Consequently it is possible that patterns of typological change, largely created by other factors, have partly obscured the strong connection between retouch intensity and ecological–economic contexts, or alternatively that there is little correspondence between reduction intensity and climatic phases and that typological patterns signal the operation of some other cultural process. For this reason, instead of pursuing difficult-to-interpret typological descriptions, we prefer to directly examine questions of retouch intensity in our future research through the use of nontypological techniques dedicated to measuring intensity of retouch/reduction.

Finally, this analysis revealed that the traditional implement typology is indeed a complex product of multiple processes and not principally a signal of differing levels of retouch, even though retouch intensity is undoubtedly one of a number of factors creating morphological variation between specimens. Our conclusion that at Combe Grenal, and perhaps for many Mousterian assemblages, there is a strong interaction between blank and the nature of retouch also invites consideration of broad questions about the interpretation of implement patterns. For several decades, a number of debates about how to understand the variation and regularity of retouched flakes in Mousterian assemblages were polarized between Dibble's model that morphology indicated only the intensity of tool resharpening and the more traditional hypothesis that regular patterns of implement shape represent knapping according to a fixed design or "mental template." Although our inferred retouching scheme for Layer 21 implies that sequential transformation of retouched flakes from one implement type to another, as demanded by the single branching scheme, was rare, and that instead there were multiple, albeit branching, pathways of reduction, this need not imply that conventionally defined implement types represent different preconceived tool designs. Demonstrating that retouching processes need not involve typological conversions does not indicate that typological stability during reduction entailed a specific design. In the instance of Combe Grenal the strong connection of retouch location, form, and extent with blank form may provide a mechanism for regular and stable implement shapes over the reduction process, even if no well-defined, formal design was in place. Habitual application of production rules to blanks of different shapes may maintain stability in the appearance and location of a retouched edge during extended reduction. However, as Turq, Boëda, and others have proposed, such production rules connecting blank form and retouching process to produce regularity in implement form might be considered a kind of design system for Middle Paleolithic hominids. Debates on how we should think of goal-oriented behavior in the Mousterian, and indeed the nature of the technological processes that were involved and their articulation with economic and ecological contexts, still require exploration in the quest to understand the construction of morphological diversity in Middle Paleolithic implements. The evaluation of what constitutes meaningful and valuable units of measurement, and how they may or may not be connected to traditional implement types, is not resolved; on the contrary, this discussion is merely beginning.

REFERENCES CITED

Binford, Lewis R. 1973. Interassemblage Variability – The Mousterian and the "Functional" Argument. In *The Explanation of Culture Change*, edited by C. Renfrew, pp. 227–54. Duckworth, Surrey.

Binford, L. R., and S. R. Binford. 1966. A Preliminary Analysis of Functional Variability in the Mousterian of Levallois Facies. *American Antiquity* 68(2):238–95.

Bisson, M. S. 2001. Interview with a Neanderthal: An Experimental Approach for Reconstructing Scraper Production Rules, and Their Implications for Imposed Form in Middle Palaeolithic Tools. *Cambridge Archaeological Journal* 11(2):165–84.

Bordes, F. 1968. *The Old Stone Age*. McGraw-Hill, New York.

——— 1972. *A Tale of Two Caves*. Harper and Row, New York.

Bourguignon, L. 1997. *Le Moustérien de type Quina : Définition d'une nou-velle entité technique*. Thèse de Doctorat de l'Université de Paris X, Nanterre.

Bourguignon, L., J.-P. Faivre, and A. Turq. 2004. Ramification des chaînes opératoires: Une spécificité du Moustérien? *Paleo* 16:37–48.

Clarkson, Chris. 2002. An Index of Invasiveness for the Measurement of Unifacial and Bifacial Retouch: A Theoretical, Experimental and Archaeological Verification. *Journal of Archaeological Science* 29:65–75.

Debenath, A., and H. L. Dibble. 1994. *Handbook of Paleolithic Typology. Volume One. Lower and Middle Paleolithic of Europe*. University Museum, University of Pennsylvania, Philadelphia.

Dibble, Harold L. 1984. Interpreting Typological Variation of Middle Paleolithic Scrapers: Function, Style, or Sequence of Reduction? *Journal of Field Archaeology* 11:431–6.

1987a. Reduction Sequences in the Manufacture of Mousterian Implements of France. In *The Pleistocene Old World: Regional Perspectives*, edited by O. Soffer, pp. 33–45. Plenum Press, New York.

1987b. The Interpretation of Middle Paleolithic Scraper Morphology. *American Antiquity* 52(1):109–17.

1988a. The Interpretation of Middle Paleolithic Scraper Reduction Patterns. In *L'Homme de Néandertal, vol 4, La Technique*, Actes du Colloque International de Liége, L'Homme de Neandertal, pp. 49–58.

1988b. Typological Aspects of Reduction and Intensity of Utilization of Lithic Resources in the French Mousterian. In *Upper Pleistocene Prehistory of Western Eurasia*, edited by H. Dibble and A. Montet-White, pp. 181–94. University Museum, University of Pennsylvania, Philadelphia.

1991. Rebuttal to Close. *Journal of Field Archaeology* 18:264–9.

1995. Middle Paleolithic Scraper Reduction: Background, Clarification, and Review of Evidence to Date. *Journal of Archaeological Method and Theory* 2:299–368.

Dibble, Harold L., and Nicholas Rolland. 1992. On Assemblage Variability in the Middle Paleolithic of Western Europe: History, Perspectives, and a New Synthesis. In *The Middle Paleolithic: Adaptation, Behavior, and Variability*, edited by H. L. Dibble and P. Mellars, pp. 1–28. University Museum, University of Pennsylvania, Philadelphia.

Gordon, D. 1993. Mousterian Tool Selection, Reduction and Discard at Ghar, Israel. *Journal of Field Archaeology* 20:205–18.

Hiscock, Peter. 1994. The End of Points. In *Archaeology in the North*, edited by M. Sullivan, S. Brockwell, and A. Webb, pp. 72–83. North Australia Research Unit, Australian National University.

Hiscock, Peter, and Val Attenbrow. 2005. *Australia's Eastern Regional Sequence Revisited: Technology and Change at Capertee 3*. British Archaeological Reports. International Monograph Series. Archaeopress, Oxford.

Hiscock, Peter, and Chris Clarkson. 2005. Experimental Evaluation of Kuhn's Geometric Index of Reduction and the Flat-Flake Problem. *Journal of Archaeological Science* 32:1015–22.

2007. Retouched Notches at Combe Grenal (France) and the Reduction Hypothesis. *American Antiquity* 72:176–90.

Holdaway, S., S. McPherron, and B. Roth. 1996. Notched Tool Reuse and Raw Material Availability in French Middle Paleolithic Sites. *American Antiquity* 61:377–87.

Kuhn, Steven L. 1990. A Geometric Index of Reduction for Unifacial Stone Tools. *Journal of Archaeological Science* 17:585–93.

1992. Blank Morphology and Reduction as Determinants of Mousterian Scraper Morphology. *American Antiquity* 57:115–28.

Mellars, Paul. 1992. Technological Change in the Mousterian of Southwest France. In *The Middle Paleolithic: Adaptation, Behavior, and Variability*, edited by H. L. Dibble and P. Mellars, pp. 29–43. University Museum, University of Pennsylvania, Philadelphia.

1996. *The Neanderthal Legacy*. Princeton University Press, Princeton, NJ.

Rolland, Nicholas. 1981. The Interpretation of Middle Paleolithic Variability. *Man* 16:15–42.

Rolland, Nicholas, and Harold L. Dibble. 1990. A New Synthesis of Middle Paleolithic Assemblage Variability. *American Antiquity* 55:480–99.

Thiébaut, C. 2003. L'industrie lithique de la couche III du Roc de Marsal: Le probléme de l'attribution d'une série lithique au Moustérien à Denticulés. *Paléo* 15:141–68.

Turq, A. 1989. Approche technologique et économique du faciés Moustérien de type Quina: Etude préliminaire. *Bulletin de la Société Préhistorique Française* 86:244–56.

1992. Raw Material and Technological Studies of the Quina Mousterian in Perigord. In *The Middle Paleolithic: Adaptation, Behavior, and Variability*, edited by H. L. Dibble and P. Mellars, pp. 75–85. University Museum, University of Pennsylvania, Philadelphia.

Verjux, C. 1988. Les Denticules Mousteriens. In *L'Homme de Néandertal, Vol 4., La Technique*, Actes du Colloque International de Liége, L'Homme de Neandertal, pp. 197–204.

Verjux, C., and D.-D. Rousseau. 1986. La retouche Quina: Une mise au point. *Bulletin de la Société Préhistorique Française* 11–12:404–15.

6 BROOKE BLADES

REDUCTION AND RETOUCH AS INDEPENDENT MEASURES OF INTENSITY

Abstract
This paper presents the argument that common interpretations of "reduction intensity" in fact conflate two different and at times independent processes. Reduction intensity should be restricted to an analysis of technological stages of raw material reduction and blank production, an overall process commonly referred to as the reduction sequence. Retouch intensity, by contrast, reflects changes to finished blanks and technological remnants arising from and related to function and use. The importance of the distinction lies in the identification of separate processes that may reveal elements of mobility, settlement pattern, and social intensification among prehistoric populations. This paper proposes that intensity should be analyzed in a reduction/retouch matrix and presents examples of such analyses.

INTRODUCTION

The "reduction thesis" (Shott 2005) has become the most powerful framework for understanding the most durable material element of the prehistoric archaeological record. The framework has emerged from analysis of varied cultural and temporal contexts by numerous researchers. This research, however, shares a common recognition of

I wish to express my sincere appreciation to Bill Andrefsky for the dual invitations to participate in the 2006 SAA session in San Juan and to contribute to this volume, and for his comments on this paper. I also would like to thank my friend and colleague Jehanne Féblot-Augustins for reading and commenting on an earlier draft of this paper.

lithic reduction as a process of continual material removal that may profoundly affect the shape and size of any stone artifact.

Similarly to other papers in this volume (Andrefsky; Quinn et al.; Wilson and Andrefsky), this paper will argue that the common assessment of what is termed "reduction" intensity most frequently evaluates only one aspect of those effects: retouch or more generally utilization intensity, that is, changes arising from and related to function and use. The term reduction intensity will, in the context of this paper, be restricted to reflections of varying extents of technological reduction of raw material and blank production within the framework of the lithic reduction sequence.

This distinction involves more than an attempt at semantic precision. As will be discussed, reduction and retouch intensities are not dependent variables, because one reflects the degree of technological reduction and the other evaluates the extent to which the products of that reduction were retouched and/or utilized. It is certainly true that both measures of intensity may mirror the influences of the same natural or cultural phenomena, which include the following:

- access to raw material
- constraints related to group or individual movement (frequency and magnitude)
- mobility parameters within the overall settlement system
- specific components (i.e., site type) within the settlement system
- social dimensions (intensification, risk minimization, etc.)
- However, although dimensions of reduction and retouch may measure

Complementary influences, the important point to consider is that they should be assessed independently because they may vary independently.

Reduction intensity measures the complexity of technological activity along a continuum of lithic reduction. Measures of reduction intensity evaluate the extent to which a given lithic material is fully integrated into the technological structure. Clearly, each raw material may reflect a different degree of reduction intensity.

Retouch or utilization intensity measures the degree or extent of retouch or utilization on the raw material blanks that emerge from the technological action of reduction. Although all technological systems are oriented to the production of a specific outcome or outcomes, it must be recognized that virtually any lithic piece at any technological stage, even in the most technologically specialized systems, may be retouched or utilized at least in an expedient fashion.

MEASUREMENT OF INTENSITY

This paper advocates a separation of measures of evaluation rather than a radical change in the manner in which such evaluations are undertaken. However, simply incorporating the various measures of what is herein termed retouch intensity into a unified structure would be a daunting task. Varying scales of precision and extents of applicability exist, as recently reviewed by Shott (2005). Such means of evaluation include geometric measures (Andrefsky 2006; Clarkson 2002; Eren et al. 2005; Hiscock and Clarkson, this volume; Kuhn 1990; Quinn et al., this volume), retouch type or extent (Dibble 1987; Movius et al. 1968), and various measurements focusing on blank allometry or relative sizes of blank elements (Blades 2003; Dibble and Pelcin 1995; Grimes and Grimes 1985; Holdaway 1991).

A similar range of variability exists in the evaluation of reduction intensity. Dibble et al. (1995) have summarized a wide range of data to argue that as core reduction increases, the degree of core preparation and number of blanks per core increase (Bar-Yosef 1991; Marks 1988; Montet-White 1991; Munday 1977), whereas average core size, flake size, flake platform area, and cortex all decrease (Henry 1989; Marks et al. 1991; Newcomer 1971; Stahle and Dunn 1982). The use of certain specific measures should therefore monitor reduction intensity:

- blank-to-core ratio (or flake-to-biface ratio);
- core size;
- blank size;
- amount of cortical covering.

As commonly employed, these measures provide a means of comparing two or more assemblages – or two or more raw materials within the same assemblage. Although the measures themselves generate ratio or even interval data, the resulting intra-assemblage or interassemblage comparisons rely on ordinal (i.e., greater or lesser intensity) distinctions. The problem of assigning lithic reduction remnants to discrete stages or of evaluation along a reduction continuum has been examined by Bradbury and Carr (1999).

Other challenges to the measurement or quantification of reduction intensity arise when separate measures are ambiguous or contradictory, or are not universally accepted as accurate reflections of reduction (see Eren and Prendergast, this volume, for a detailed discussion of this issue). Magne (1985) argued that dorsal and platform scar counts could be effectively correlated with experimental reduction stages. Some have adopted the Magne stages (Carr 1994), but Mauldin and Amick (1989) also relied on experimental data to argue that dorsal scar counts are not reliable indicators of reduction stages. The utility of a comparison of flake size and cortical covering has some experimental support (Mauldin and Amick 1989), but the use of cortical covering alone (Bradbury and Carr 1995; Mauldin and Amick 1989) or of flake size (Magne 1985) has been criticized.

These various complications and contradictions do not arise from the intensity dichotomy proposed herein, but do complicate the measurement of intensity. It is important to recognize, however, that these contradictions at least in part reflect realities of prehistoric technological and utilization activities. It is expected that different lithic remnants and products will reflect different rates of retouch, utilization, or consumption. Repeated occupations within differing settlement system structures, whether occurring in palimpsest rock shelter contexts or in near-surface plow-disturbed zones, may generate confused and apparently contradictory indications of intensity. The comparisons presented below proceed with an awareness of these concerns and are offered as theoretical possibilities despite being derived from archaeological examples.

COMPARISONS OF REDUCTION AND RETOUCH

Reduction and retouch intensities may be compared in an X–Y matrix, with categories 1 through 3 roughly equating with low to high intensities. One of the advantages of such a presentation is its flexible nature. For example, a given assemblage may be categorized by its constituent raw material types or by specific elements such as cores, bifaces, scrapers, and flakes. Various assemblages within a specific site or sites within a region may be summarized, but with a consequent loss of resolution. The latter case, however, does provide an opportunity to evaluate regional components of settlement systems, as will be illustrated below.

Conversion of intensity evaluations to a numerical scale may be undertaken in variety of ways. For example, the conversion of the Magne dorsal scar counts to the reduction intensity scale might proceed as follows:

0 to 1 scar = 63% of intact flakes = 0.63 × 1 = 0.63
2 scars = 27% of intact flakes = 0.27 × 2 = 0.54
3 + scars = 10% of intact flakes = 0.10 × 3 = 0.30
reduction intensity sum = 1.47 (on scale from 1 to 3)

Alternatively, it may be preferable to express the conversion in terms of central tendency (mean) and dispersion (standard deviation). The comparison of mean values alone may mask important differences in the distributions of flakes within the scar categories that would be reflected by a measure of dispersion.

This value would, of course, represent only one of several potential means of evaluating reduction intensity within a given assemblage. Reduction intensity values may be developed for cores and other technological pieces using a combination of criteria, including but not limited to size/weight comparisons, quantity of negative scars, evidence of platform creation/rejuvenation, and measures of volume. Retouch intensity may be assessed according to the various measures discussed earlier in this paper. It is anticipated that researchers would adopt those measures of reduction and retouch that were most clearly reflected in the lithic assemblages under consideration.

The application of reduction/retouch comparisons will be explored herein through four scenarios derived from archaeological data. The placements of the various data – technological stages, raw materials, site assemblages – are based on analyses of specific sites or combinations of sites, except for Example 4. However, the numerical scale values assigned to these data are not derived from measurement standards such as that based on the Magne scar count. The examples therefore represent logically derived but still theoretical presentations.

Pennsylvania Jasper Quarry

[Figure: scatter plot with Retouch Intensity (x-axis, 1–3) vs Reduction Intensity (y-axis, 1–3). Points: "preform blanks" at approximately (1, 2.5); "discarded tool (local material)" at approximately (2, 2.5); "discarded tool (distant material)" at approximately (2.5, 2.5); "cores, flakes" at approximately (1, 1.2).]

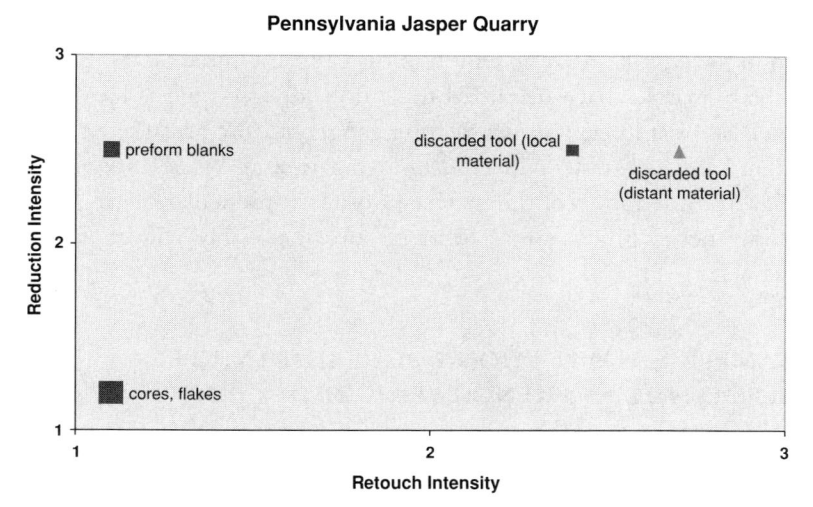

FIGURE 6.1. A retouch intensity model for a quarry site in Pennsylvania.

EXAMPLE 1: TECHNOLOGICAL SETTING (PENNSYLVANIA JASPER QUARRY)

The quarry component manifests low intensities of reduction and retouch (Figure 6.1). Ratios of flakes or blanks to cores are low (due to larger numbers of cores), cores are large or at least variable in size (i.e., are not exhausted), and sizes of waste flakes are large. However, other dimensions of intensity for the locally quarried material are possible. The production of "finished" blanks or smaller prepared cores may be indicated, reflecting higher reduction intensity but still low utilization intensity.

It is also possible to find a very different pattern reflected in components that are more heavily retouched or utilized. Flakes and points that have been retouched to the point of exhaustion have been found at quarries. Such pieces in Pennsylvania may be made of locally available chert or of materials such as argillite from more distant sources. In the case of the distant materials, the transport of retouched pieces or blanks that were later retouched – resulting in high reduction and retouch intensities – would be implied. The presence of heavily retouched and discarded pieces of locally available materials may suggest cyclical patterns of movement away from and back to the quarry locus. Stewart

(2003: 7) has argued for a temporal distinction in such quarry utilization in the Middle Atlantic region of the eastern United States. Earlier Archaic quarries have discarded tools from diverse quarry locations, which Stewart interprets as reflecting geographically broad settlement systems. By contrast, a given later Archaic and Woodland quarry yields discarded tools of the material from that particular quarry, suggesting more cyclical use of lithic resources, possibly within smaller territories.

EXAMPLE 2: HABITATION LOCUS (EARLY UPPER PALEOLITHIC IN FRENCH PÉRIGORD)

These loci were not quarry sites but often were located within a few kilometers of available chert materials. The dominant local material – dark Senonian chert – reflected a low blank-to-core ratio in comparison to the other raw materials present (Figure 6.2). Blank sizes were variable and comparatively few tools exist relative to numerous flakes. Retouch intensities were variable. Another locally available material – coarser quartzite – was present as larger pieces, few of which were retouched.

Materials from distant sources (i.e., more than 30 km in this instance) were present in limited quantities, with consistently high reduction intensities but variable retouch intensity. A few highly reduced cores of Bergerac chert (30 km west) were found. The material is present primarily as retouched blade blanks, although the intensity was not necessarily greater than found on the local Senonian blades. Fumel chert (40 km south) was not found in the form of cores but as blanks with limited retouch, implying that the more heavily retouched pieces were probably either transported elsewhere or not recovered in the excavations. Charente chert (100 km west) was present as retouched pieces and some debitage.

The general pattern indicated is one of increasing intensity of reduction – but not necessarily retouch – with greater distance from the source, a classic "down-the-line" distribution that essentially reflects a direct relationship between reduction and retouch or utilization intensities. Much variation exists in the lithic utilization patterns during the Upper Paleolithic in Europe, as Féblot-Augustins (1997, 1999) has thoroughly documented. The reduction/retouch

Habitation (early Upper Paleolithic, French Perigord)

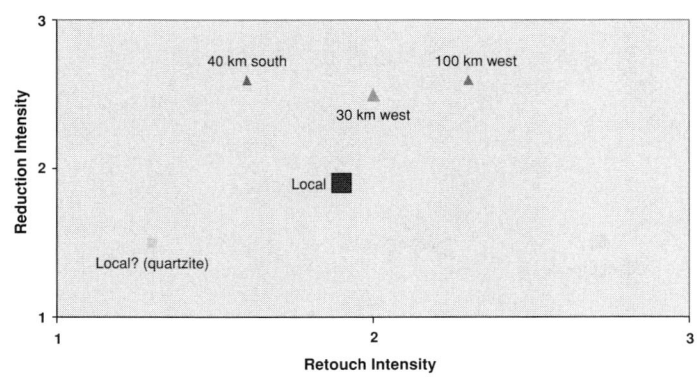

FIGURE 6.2. A retouch intensity model for an Upper Paleolithic habitation site.

comparisons should be useful in illustrating directional distributions in which lithics from greater distances appear in larger quantities and perhaps in different technological forms than those from sources closer to specific site loci. Such directional distributions are often correlated with networks of social intensification.

EXAMPLE 3: SETTLEMENT PATTERN (LATER PREHISTORIC, SOUTHWEST PENNSYLVANIA AND WEST VIRGINIA)

The reduction/retouch matrix provides a means of categorizing and summarizing the various spatial components that modern researchers regard as constituting a settlement pattern. Transect survey and excavations at the base of Chestnut Ridge revealed aspects of such a settlement pattern for the later Archaic and Woodland periods in southwestern Pennsylvania and northern West Virginia (Figure 6.3). The survey revealed three site "types":

- loci focused on secondary reduction of local black chert (high flake:biface ratios): 36Fa426/T3, 46Mg103 and 104;
- small seasonal? base camps with varied raw materials, more bifaces, hearths, and storage features: 36Fa 411, 418 Woodland, 418 lower, and 426/T2;
- numerous very small loci with few or no bifaces and small numbers of flakes, suggesting special purpose hunting and biface repair stations.

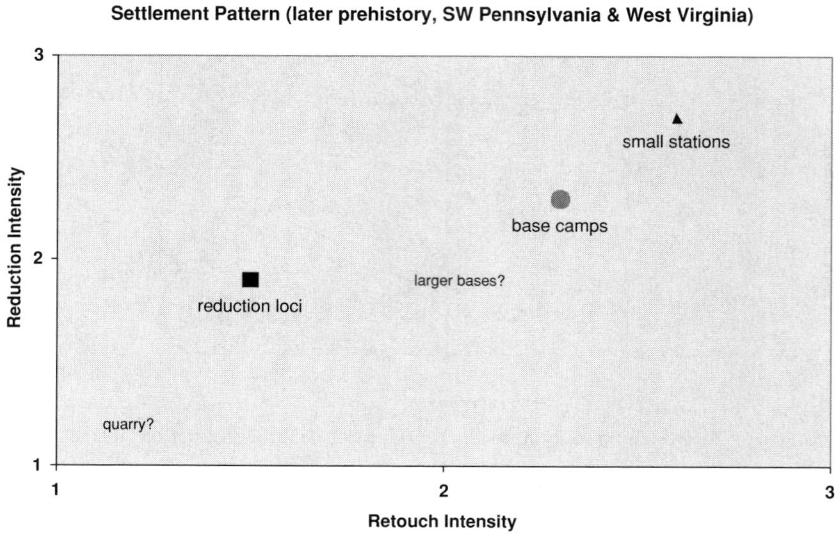

Settlement Pattern (later prehistory, SW Pennsylvania & West Virginia)

[Scatter plot with x-axis "Retouch Intensity" (1–3) and y-axis "Reduction Intensity" (1–3). Points plotted: small stations (~2.7, 2.7); base camps (~2.3, 2.3); larger bases? (~2, 1.8); reduction loci (~1.5, 1.8); quarry? (~1.3, 1.2).]

FIGURE 6.3. A retouch intensity model for a late prehistoric settlement pattern in the eastern United States.

The matrix comparison of these sites reveals more variability in retouch intensity than in reduction intensity. Clearly, lower reduction intensity loci (quarries) are missing elements. Larger base camps, particularly during the later Woodland, were probably located elsewhere, possibly along the Monongahela River to the west of Chestnut Ridge. The overall pattern has been interpreted as a logistical one in this highly seasonal environment, but such an attribution of course raises the question of what sort of reduction and retouch structures would be expected along the forager/collector continuum.

EXAMPLE 4: FORAGER/COLLECTOR (BASED IN PART ON CARR 1994)

Numerous attempts to create a framework for utilizing the Binford forager/collector continuum to interpret archaeological lithic assemblages exist in the literature. For example, Carr (1994) has proposed specific raw material and reduction expectations for forager residences, collector residences, and collector field camps (Figure 6.4):

- Forager residence: 50% local, 50% distant material; curated technology with emphasis on maintenance; distant material bifaces as cores

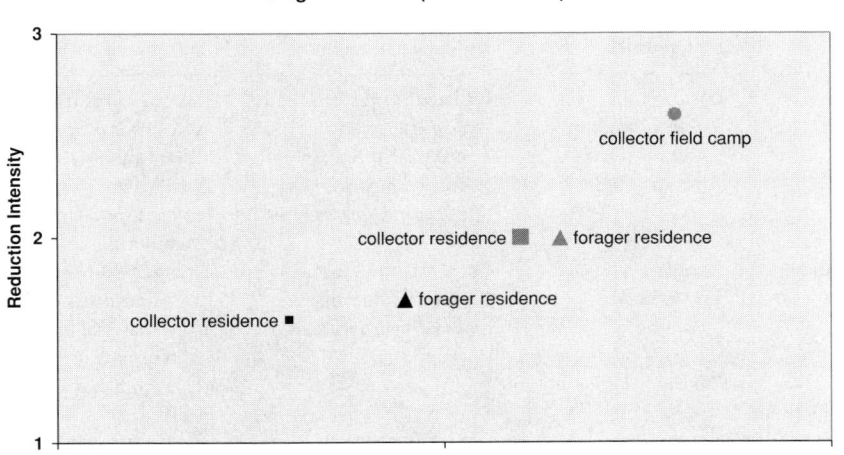

Forager/Collector (after Carr 1994)

[Chart: Reduction Intensity (y-axis, 1–3) vs Retouch Intensity (x-axis, 1–3), showing:
- collector field camp (~2.6, 2.6)
- collector residence (~2.0, 1.7)
- forager residence (~2.0, 2.1)
- forager residence (~1.6, 1.7)
- collector residence (~1.5, 1.2)]

FIGURE 6.4. A retouch intensity model for generalized foragers and collectors.

or tools; local materials as expedient tools and replacements for distant ones.
- Collector residence: 25% local, 75% distant material; curated but reliable technology; local materials used almost entirely expediently; task groups have transported distant materials as bifacial cores and reliable tools.
- Collector field camp: distant material almost exclusively (because focused on specific task, unable to exploit local materials); technology also curated and reliable; lithic debris with broken tools and some evidence tool resharpening (since reliable tools made at times other than use).

The purpose of the graphic depiction in Figure 6.4 is not to critique the thoughtful model put forth by Carr, an evaluation of which Carr himself undertakes (1994). Rather, the intention is to demonstrate what such a model might look like when evaluated from the standpoints of reduction and retouch intensities. The expedient utilization of local materials would suggest a relatively low reduction input and variable retouch/utilization intensities. Distant materials would probably reveal greater retouch intensity, particularly when used as reliable tools by "collectors." However, the transport of bifaces as cores would imply lower reduction intensity than would be found in an assemblage of transported finished tools.

Table 6.1. Reduction/retouch intensity interpretations

	Low retouch	Moderate retouch	High retouch
High reduction	Local blanks with few or no cores, distant blanks transported	Utilized blanks, finished points (varied materials), few or no cores *occupation locus with varied procurement*	Local points reshaped (retouch flakes only?), distant materials (if present) are heavily utilized *specialized procurement*
Moderate reduction	*Secondary reduction at/near quarry*	Local (at times distant) material dominance *varied components base camp*	*Shorter term base camp* at greater distance from material source
Low reduction	*Primary reduction quarry*		Early stage flakes heavily utilized

Collector assemblages may have somewhat lower retouch intensities than those deposited by foragers due to greater mobility frequencies of the latter, but such general categorizations are – and should be – controversial. The special-purpose collector camp had high-intensity distant material debris from tool use and repair. It is certainly possible, however, to envision camps with similar intensities of reduction and utilization of locally available materials in both residential and collector mobility orientations. One of the primary distinctions in the original configuration drawn by Binford (1980) lay in the greater diversity of special-purpose loci created by collectors from a centralized location.

CONCLUSIONS

A very preliminary summary of signatures and interpretations derived from the reduction/retouch matrix is presented in Table 6.1. The matrix comparisons serve to isolate those outcomes or manifestations that would be unexpected, such as pieces with high retouch intensity in assemblages with overall low intensity of reduction. The reduction/retouch matrix is particularly useful for interpreting various patterns of technological organization reflected in differing raw materials.

Comparisons of settlement pattern data (i.e., assemblage or site-level data) are more homogeneous than intra-assemblage comparisons of raw materials or specific technological categories, but do facilitate evaluation of various forager/collector models and in particular serve to highlight those components of settlement systems that are expected or anticipated but have not been identified.

The examples offered above revealed that direct relationships between reduction and retouch intensities are often indicated, which reflects the realities of prehistoric lithic technology and functional behaviors. However, deviations from direct variations exist, which may indicate an incomplete archaeological record but also the technological and behavioral flexibility that has frequently been observed in ethnographic settings and that should be expected in prehistoric adaptations.

REFERENCES CITED

Andrefsky, William, Jr. 2006. Experimental and Archaeological Verification of an Index of Retouch for Hafted Bifaces. *American Antiquity* 71:743–57.

Bar-Yosef, O. 1991. Raw Material Exploitation in the Levantine Epi-Paleolithic. In *Raw Material Economies among Prehistoric Hunter-Gatherers*, edited by A. Montet-White and S. Holen, pp. 357–97. University of Kansas Publications in Anthropology No. 19, Lawrence.

Binford, L. 1980. Willow Smoke and Dogs' Tails: Hunter–Gatherer Settlement Systems and Archaeological Site Formation. *American Antiquity* 45:4–20.

Blades, B. 2003. End Scraper Reduction and Hunter–Gatherer Mobility. *American Antiquity* 68:141–56.

Bradbury, A., and P. Carr. 1995. Flake Typologies and Alternative Approaches: An Experimental Assessment. *Lithic Technology* 20:100–115.

——— 1999. Examining Stage and Continuum Models of Flake Debris Analysis: An Experimental Approach. *Journal of Archaeological Science* 26:105–16.

Carr, P. 1994. Technological Organization and Prehistoric Hunter–Gatherer Mobility: Examination of the Hayes Site. In *The Organization of North American Prehistoric Chipped Stone Tool Technologies*, edited by P. Carr, pp. 35–44. International Monographs in Prehistory. University of Michigan Press, Ann Arbor.

Clarkson, C. 2002. An Index of Invasiveness for the Measurement of Unifacial and Bifacial Retouch: A Theoretical, Experimental, and Archaeological Verification. *Journal of Archaeological Science* 29:65–75.

Dibble, H. 1987. The Interpretation of Middle Paleolithic Scraper Morphology. *American Antiquity* 52:109–17.

Dibble, H., and A. Pelcin 1995. The Effect of Hammer Mass and Velocity on Flake Weight, *Journal of Archaeological Science* 22:429–39.

Dibble, H., B. Roth, and M. Lenoir. 1995. The Use of Raw Materials at Combe–Capelle Bas. In *The Middle Paleolithic Site of Combe–Capelle Bas (France)*, edited by H. Dibble, B. Roth, and M. Lenoir, pp. 259–87. University Museum Press, Philadelphia.

Eren, Metin I., Manual Dominguez-Rodrigo, Steven L. Kuhn, Daniel S. Adler, Ian Le, and Ofer Bar-Yosef. 2005. Defining and Measuring Reduction in Unifacial Stone Tools. *Journal of Archaeological Science* 32:1190–1206.

Féblot-Augustins, J. 1997. *La circulation des matières premières au Paléolithique* (two volumes). ERAUL 75, Liège.

——— 1999. La mobilité des groupes paléolithiques. *Bulletins et Mémoires de la Société d'Anthropologie de Paris*, n.s., 11:219–60.

Grimes, J., and B. Grimes. 1985. Flakeshavers: Morphometric, Functional, and Life-Cycle Analyses of a Paleoindian Unifacial Tool Class. *Archaeology of Eastern North America* 13:35–57.

Henry, D. 1989. Correlations between Reduction Strategies and Settlement Patterns. In *Alternative Approaches to Lithic Analysis*, edited by D. Henry and G. Odell, pp. 139–55. Archaeological Papers of the American Anthropological Association 1, Washington, DC.

Holdaway, S. 1991. *Resharpening Reduction and Lithic Assemblage Variability across the Middle to Upper Paleolithic Transition*. Ph.D. dissertation, University of Pennsylvania. University Microfilms, Ann Arbor.

Kuhn, S. 1990. A Geometric Index of Reduction for Unifacial Stone Tools. *Journal of Archaeological Science* 17:583–93.

Magne, M. P. R. 1985. *Lithics and Livelihood: Stone Tool Technologies of Central and Southern Interior British Columbia*. Archaeological Survey of Canada Paper 13. National Museums of Canada, Ottawa.

Marks, A. 1988. The Curation of Stone Tools during the Upper Pleistocene: A View from the Central Negev, Israel. In *Upper Pleistocene Prehistory of Western Eurasia*, edited by H. Dibble and A. Montet-White, pp. 87–94. University Museum Press, Philadelphia.

Marks, A., J. Shokler, and J. Zilhão. 1991 Raw Material Usage in the Paleolithic. The Effect of Local Availability on Selection and Economy. In *Raw Material Economies among Prehistoric Hunter–Gatherers*, edited by A. Montet-White and S. Holen, pp. 127–39. University of Kansas Publications in Anthropology No. 19, Lawrence.

Mauldin, R., and D. Amick. 1989. Investigating Patterning in Debitage from Experimental Bifacial Core Reduction. In *Experiments in Lithic Technology*, edited by D. Amick and R. Mauldin, pp. 67–88. British Archaeological Reports International Series 528, Oxford.

Montet-White, A. 1991. Lithic Acquisition, Settlements, and Territory in the Epigravettian of Central Europe. In *Raw Material Economies among*

Prehistoric Hunter–Gatherers, edited by A. Montet-White and S. Holen, pp. 205–20. University of Kansas Publications in Anthropology No. 19, Lawrence.

Movius, H. L., Jr., N. David, H. Bricker, and R. B. Clay. 1968. *The Analysis of Certain Major Classes of Upper Paleolithic Tools*. American School of Prehistoric Research Bulletin 26. Peabody Museum, Harvard University, Cambridge.

Munday, F. 1977. Intersite Variability in the Mousterian Occupation of the Avdat/Aqev Area. In *Prehistory and Paleoenvironments in the Central Negev, Israel, Vol. 1. The Avdat/Aqev Area, Part 1*, edited by A. Marks, pp. 113–40. Southern Methodist University Press, Dallas.

Newcomer, M. 1971. Some Quantitative Experiments in Handaxe Manufacture. *World Archaeology* 3:85–94.

Shott, M. 2005. The Reduction Thesis and Its Discontents: Overview of the Volume. In *Lithics "Down Under": Australian Perspectives on Lithic Reduction, Use and Classification*, edited by C. Clarkson and L. Lamb, pp. 109–25. British Archaeological Reports International Series 1408, Oxford.

Stahle, D., and J. Dunn. 1982. An Analysis and Application of the Size Distribution of Waste Flakes from the Manufacture of Bifacial Stone Tools. *World Archaeology* 14:84–97.

Stewart, R. M. 2003. A Regional Perspective on Early and Middle Woodland Prehistory in Pennsylvania. In *Foragers and Farmers of Early and Middle Woodland Periods in Pennsylvania*, edited by P. Raber and V. Cowan, pp. 1–33. Pennsylvania Historical and Museum Commission, Harrisburg.

7 COLIN PATRICK QUINN, WILLIAM ANDREFSKY, JR., IAN KUIJT, AND BILL FINLAYSON

PERFORATION WITH STONE TOOLS AND RETOUCH INTENSITY: A NEOLITHIC CASE STUDY

Abstract

A measure of retouch intensity, the EKCI, was devised based upon function and archaeological context. To arrive at the function of Pre-Pottery Neolithic A el-Khiam points from the Near East, controlled experiments were performed to determine the relative density of the contact material, which could affect use and retouch patterns. It was shown that el-Khiam points were likely used to pierce and scrape soft materials such as leather. The EKCI was then devised, measured, and tested. Experimental replication showed that the EKCI was an accurate measure of retouch intensity, and application of the EKCI to the lithic assemblage at Dhra' reaffirmed the EKCI's utility for analyzing PPNA archaeological assemblages. Although this curation index is effective for el-Khiam points, it may not be applicable to other hafted point types, which highlights the need for independently developed measures of retouch that account for the form, function, and context of the artifacts rather than attempting to generate universal measures of curation.

INTRODUCTION

Archaeological assemblages from the first farming villages in the Southern Levant have produced high-quality and large-quantity lithic data sets that Near Eastern archaeologists rely upon for interpreting the past. This vast resource of prehistoric knowledge has remained relatively untapped as a source of understanding individual decision-making in prehistoric lithic technology, especially from the perspective of artifact life histories and retouch intensity. Before archaeologists can begin to debate life cycles and retouch patterns of lithic artifacts, however, we must first develop the means of quantifying change in artifact morphology and assemblage characteristics that are directly linked to individual decision-making processes. In this study we provide a preliminary exploration of the practice of lithic curation in Near Eastern Neolithic assemblages, assessing retouch on perforating stone tools based on form, function, and archaeological context, which can be used to address issues of economic, social, and technological organization.

The concept of curation, defined as the ratio of realized to maximum utility of lithic artifacts (Shott 1996; Shott and Sillitoe 2005), has interpretive benefit for understanding lithic technological organization in the past. Assessing curation requires researchers to identify the intensity of use of lithic artifacts. Toward this goal, archaeologists over the past two decades have attempted to create ways of measuring retouch and applying those measurements to archaeological collections (Andrefsky 2006; Blades 2003; Clarkson 2002; Eren et al. 2005; Kuhn 1990). There are a few baseline assumptions upon which measures of curation are built. Curation is equated on a one-to-one basis with postproduction retouch. The key characteristic of postproduction retouch is morphological modification of the artifact. By quantifying the morphological change of an artifact, researchers gain a proxy measure of curation. Additionally, the measures of morphological change should be directly related to postproduction retouch. Every event of postproduction retouch will change the morphological characteristics of the artifact, even if in minor ways. It is the job of the researcher, therefore, to identify and define which morphological characteristics are changing and then develop a system of recording those characteristics. The resulting documentation system must be quantified in a way that equates increasing curation to increasing measurement values. With these two assumptions in mind, measures of retouch are important tools for discussing lithic curation intensity and become effective for interpreting archaeological assemblages only after results of experimental studies provide empirical patterning that validates their accuracy.

In building any measure of retouch, it is important to take into account manufacturing techniques, the form and function of the artifact, and retouch techniques. Indices that attempt to measure curation without taking these variables into account can lead to spurious interpretations based on overstepping the boundaries of the measurements (Andrefsky 2006; Davis and Shea 1998). Many archaeologists who are interested in measuring retouch have often tried to create general indices that can be applied to artifacts with varying forms, functions, and archaeological contexts. As a result, universal measures of retouch have been critiqued when they fail to work for a certain type of tool or a certain assemblage (see Eren and Prendergast, this volume; Hiscock and Clarkson, this volume; Wilson and Andrefsky, this volume). Variability in form, function, and archaeological context should be noted by archaeologists when they are developing retouch indices, as these variations often dictate the amount and type of retouch evident on tools. Contextually specific curation indices account for morphological change in lithic artifacts based upon select formal and functional requirements of the artifacts. Among other things, morphological changes in lithic artifacts are dictated by the form of the artifact, the way in which it is used, the contact material it is used upon, the temporal and spatial archaeological setting, and the site type, raw material availability, and retouch techniques. Different tool types are more effectively measured for curation with different indices based on the shape of the tool, how it is used, how it is resharpened, the type of site where it is used, and the context of that site in the larger spatial and temporal conditions of the region. Therefore, when attempting to quantify curation in the archaeological record, we must build contextually specific indices that actually measure the morphological change of the artifacts being studied.

This study is designed to quantify the retouch intensity on el-Khiam projectile points that are found in the Pre-Pottery Neolithic A period (PPNA) (11,500–10,500 cal. yr. B.P.) in the Southern Levant. Recent microwear studies performed by Sam Smith from the University of Reading suggested that these points functioned as perforators, though microwear patterns were inconclusive with regard to contact material (Smith 2005). In this study we test a variety of contact materials that may produce different wear and retouch patterns. Knowing how point morphology is affected by use and retouch technique is imperative in building measures of curation. Once attributes of macroscopic wear are defined and related to use, we develop a curation index to measure retouch on el-Khiam points. The new curation index, dubbed the el-Khiam curation index (EKCI), is later verified through controlled experiments and analysis of the archaeological assemblage from Dhra', Jordan.

BUILDING A CURATION INDEX: FORM, FUNCTION, AND CONTEXT

Materials and Methods

This study examines the lithic assemblage from the Pre-Pottery Neolithic A period site of Dhra' Jordan (occupied between approximately 11,500 and 11,200 cal. yr. B.P.) located in the Jordan Valley 5 km from the southeastern tip of the Dead Sea (Figure 7.1) (Finlayson et al. 2003; Goodale et al. 2002; Kuijt 1994, 2001; Kuijt and Finlayson 2001; Kuijt and Mahasneh 1995, 1998). From four field excavation seasons, over one million lithic artifacts were recovered at this early farming community, including over 800 el-Khiam points (Goodale and Smith 2001; Goodale et al. 2002). In this study, all of the complete and a nonrandom sample of the broken el-Khiam points from the 2004 field season were analyzed. These points come from numerous locations and contexts within the site and likely represent much of the variability in manufacture, use, and discard within the site.

In order to build a curation index, we first assessed the points with regard to form, function, and archaeological context. For many years, Near Eastern archaeologists assumed explicit functional attributes of stone tools based on their morphological characteristics. Due to their morphology, el-Khiam points have traditionally been classified as projectile points (e.g., Bar-Yosef and Gopher 1997). Although some el-Khiam points were undoubtedly used as projectile technology, the abundance of these points in the residential context of Dhra', Jordan, suggests that these points had an additional function. Recently, Smith (2005) has employed microwear studies to demonstrate that they were also used as perforators. Based on microwear patterns, it has been argued that these points were being used to drill beads (Goodale and Smith 2001; Smith 2005). Building on this previous research, we conducted a series of controlled experiments to test the efficiency of el-Khiam points as perforators on hard and soft contact materials

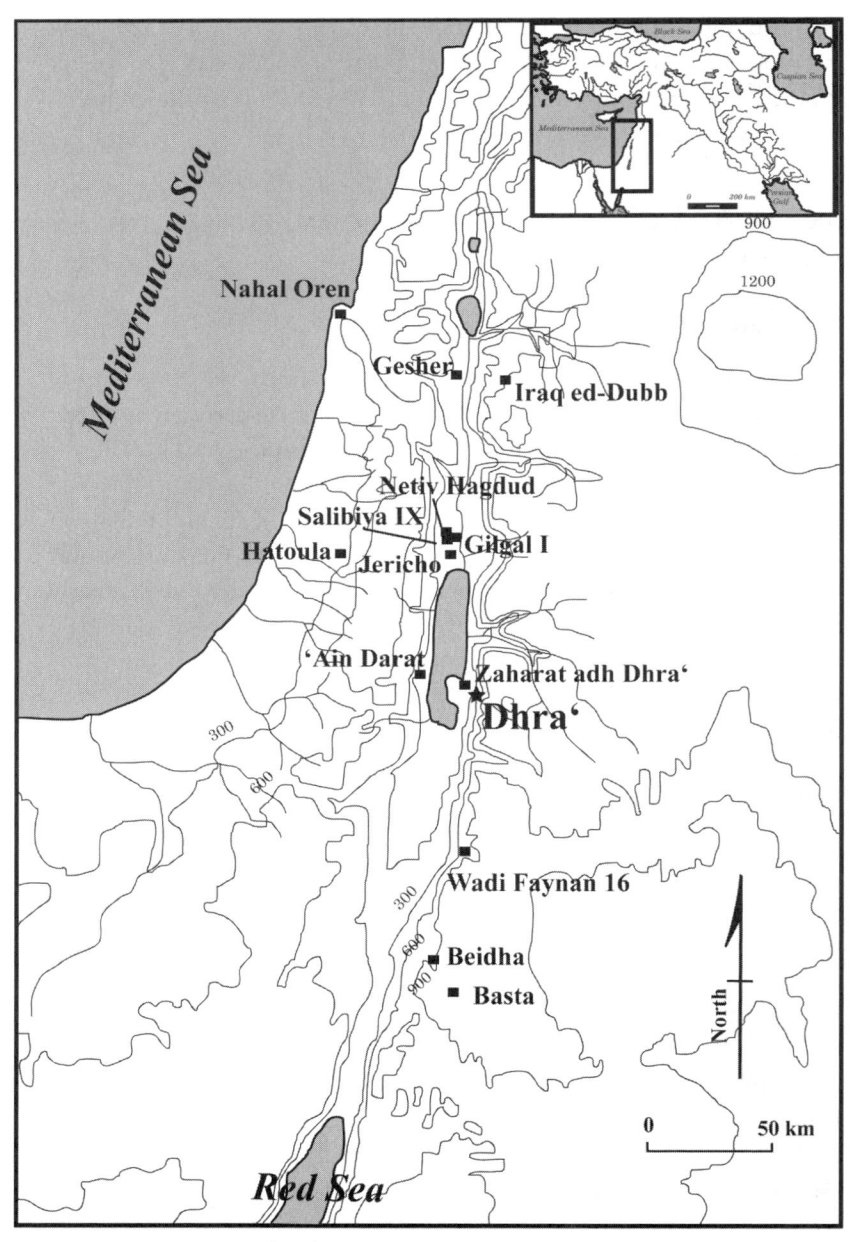

FIGURE 7.1. Location of study area.

Possible Uses of el-Khiam Points as Perforators

Drilling Hard Material **Puncturing Soft Material**

FIGURE 7.2. Illustration of possible uses of el-Khiam points.

(Figure 7.2). Our study uses four lines of macroscopic use-wear evidence, (1) a qualitative estimate of effectiveness in the task, (2) location of retouch, (3) breakage patterns, and (4) an index of point sharpness, to assess the effectiveness of el-Khiam points from Dhra' to perforate materials of various density and hardness.

The experiment began with the production of an el-Khiam point assemblage. First, blades were removed from a flint nodule using a soft hammer indentor made of antler. This nodule was taken from the same flint source, located 30 m off site, used by the prehistoric occupants of Dhra'. The blades that had a single dorsal arris, that were twice as long as they were wide, and that had margins roughly parallel to each other were selected for making el-Khiam points. An antler tine and a wooden anvil were then used to shape the blades into thirteen notched points. Finally, the el-Khiam points were hafted to shafts of willow and ocean spray wood using mastic and binding. These items replicated past technologies and binding materials available to PPNA peoples for creating a strong haft element. Twelve of the specimens were used in a drilling motion, with three points drilling each of the following

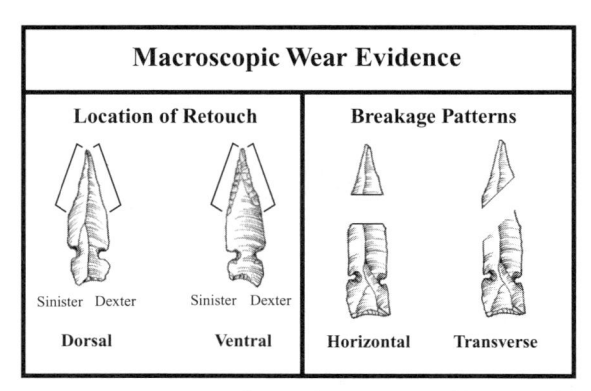

FIGURE 7.3. Location of retouch and types of breakage for el-Khiam points.

materials: limestone, malachite, willow, and alder. The points were used to bore holes into the materials using both a hand drill and a bow drill. The use-life of the points ended when either the point broke or the point became useless for the task of drilling. The points were subsequently photographed and data were recorded for several macroscopic use-wear attributes. The data were analyzed using several statistical techniques, as well as a new index for measuring point sharpness.

Our first assessment of point function and contact material was a qualitative measure of drilling effectiveness. It was hypothesized that if these points were being used to perforate hard material, they would be effective at drilling through hard material such as stone. If the points did not effectively drill holes in hard materials, then this would make it unlikely that PPNA peoples used el-Khiam points to drill stone. Likewise, if the points were effective when puncturing and scraping soft materials such as animal hides, then there is a possibility that PPNA peoples were using the points for this task. Estimating the effectiveness of el-Khiam points in performing perforating tasks, although important, is somewhat of a qualitative venture. Therefore, additional quantitative measures were taken into account to compare the assemblage of experimentally produced points to a nonrandom sample of points from the archaeological assemblage at Dhra'.

The second assessment measure examines the location of retouch. Four areas on the tips were examined (dorsal dexter, dorsal sinister, ventral dexter, ventral sinister) for evidence of flake removals (Figure 7.3). Manufacturing retouch on the el-Khiam points is almost universally isolated to one surface (either dorsal or ventral) per margin.

Sharpness Index

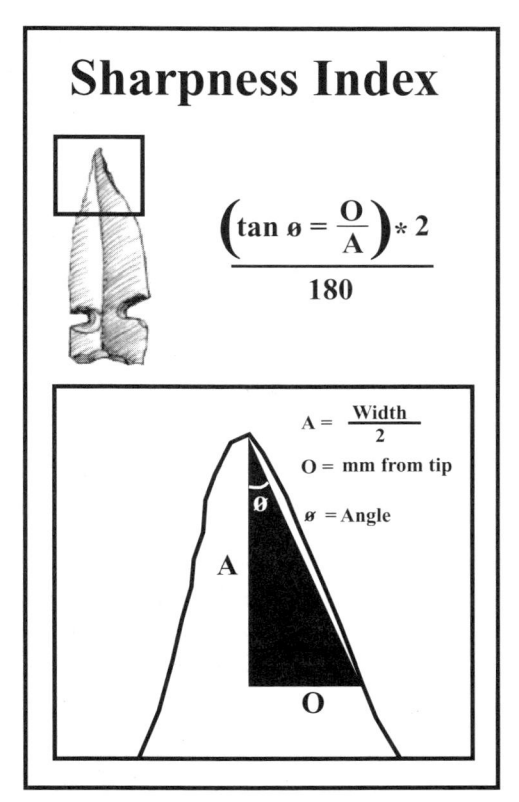

$$\frac{\left(\tan \theta = \frac{O}{A}\right) * 2}{180}$$

$A = \frac{\text{Width}}{2}$

O = mm from tip

θ = Angle

FIGURE 7.4. Calculation method for the sharpness index.

When there are flake scars on one or both of the two remaining tip locations, we assume that these flake removals were created by use rather than production. In an attempt to quantify use-related wear, we recorded the presence or absence of use-related flake removals for the experimental and archaeological collections.

Breakage patterns are also important for determining the function of the points. Variation in perforating actions, the properties of the contact material, and the application of force can cause the points to break in different ways. In this study, we look at two types of breakage patterns, horizontal and transverse (Figure 7.3), in both the experimental and archaeological collections, to see if the breakage patterns with experimental points used to perforate hard or soft materials replicate those from the archaeological collection.

The "sharpness index" (Figure 7.4) was developed in response to concerns raised by archaeologists about the accuracy of exterior edge angle measurements (cf. Andrefsky 2005) and was our fourth assessment measure. In order to avoid the possible pitfalls of measuring the exterior edge angle, this measure calculates the interior edge angle to determine the sharpness of a point. The interior edge angle is calculated at various locations on the points. First, intervals of 1 mm are taken from the tip of the point to 5 mm from the tip. Each millimeter, the width of the specimen is taken using a pair of digital calipers. In order to calculate the interior edge angle, the width at any given distance from the tip is divided in half. The given distance from the tip and one half of the width make up two sides of a right triangle, and using the Pythagorean Theorem, one-half of the interior angle can be calculated using this equation:

$\tan \phi =$ Opposite side (one half of the width)/ Adjacent side (distance from the tip).

This angle measure is doubled in order to determine the entire interior point angle (see Andrefsky 1986 for a similar calculation for flake curvature). In order to standardize the index from a range of 0 to 1, the interior angle is divided by 180 degrees (the maximum potential angle of the tip). Points that score high on the sharpness index will have the most acute interior angles, whereas the points that score lowest on the sharpness index will have interior angles that are high, with a maximum value of 180 degrees. The expectation is that the sharper the point, the more acute the interior angle, and conversely, the duller the point, the more obtuse the interior angle. The sharpness index, combined with efficiency, retouch location, and breakage patterns, provide the basis for evaluating the contact material of perforating el-Khiam points.

Results

For drilling the different materials, the most obvious qualitative assessment was the efficiency of the points. The el-Khiam points were able to bore holes in the willow, alder, and limestone with relative ease, whereas the malachite proved to be a more formidable material, but it was still possible to bore a hole. Although the flint was able to penetrate these materials, the perforations were much wider and shallower than those perforations observed on the ground stone beads at Dhra'. The

Perforation Attributes

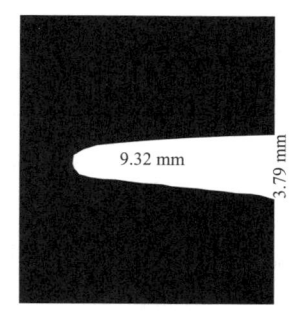

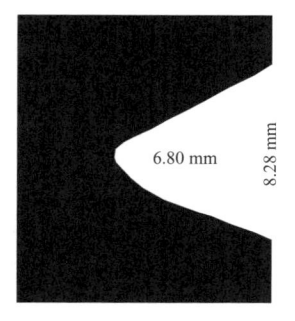

9.32 mm 3.79 mm

6.80 mm 8.28 mm

Dhra' Barrel-Elliptical Beads   Experimental Perforations

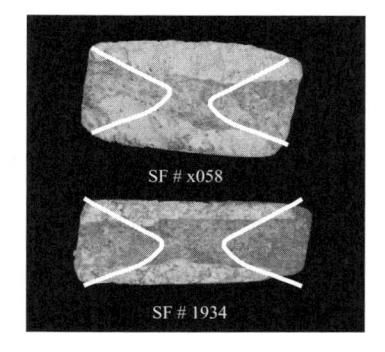

SF # x058

SF # 1934

Experimental Perforations overlying Dhra' Beads

FIGURE 7.5. Schematic illustration showing results of experimental and excavated mean perforation values.

perforations on the archaeological specimens are too deep and narrow for the flint points to have been used to drill them. The most telling evidence is the ratio of perforation depth to perforation width. To explore this further, we compared these results to the width and depth of perforations in beads from Dhra'. In the archaeological sample from Dhra', barrel-elliptical beads had a depth-to-width ratio of 9.32:3.79 (mm), whereas the experimental perforations in hard material had a reversed depth-to-width ratio of 6.80:8.28 (mm) (Figure 7.5).

Our experiments showed that the points were effective tools for puncturing holes in leather pieces of varying thickness and stiffness. Additionally, the points were effective in scraping activities due to the

Table 7.1. Retouch and breakage patterns on Dhra' and experimental assemblages.

	Use-related retouch		Breakage pattern	
	Present	Absent	Transverse	Horizontal
Hard material	11	1	4	0
Soft material	1	8	0	11
Dhra' sample	3	39	5	50

relatively steep edge angle that the manufacturing retouch created. As long as the points kept a sharp tip, it was an effective tool for puncturing leather. Importantly, the scraping edge continued to be efficient throughout the use-life of the tool. Unfortunately, comparisons between experimental perforations and archaeological perforations in leather are unachievable due to limited preservation of organic materials at Dhra'.

The damage patterns on el-Khiam points used to perforate hard materials are different from those found in the archaeological collection from Dhra' (Table 7.1). Experimental work shows that the Dhra' sample tends to conform more closely with retouch patterns produced from working soft materials as opposed to hard materials. The points at Dhra' rarely have use-related flake removals, whereas the experimental hard-material perforators have a high rate of use-related flake removals (Fisher's exact $p < .0001$). The points that were used to puncture leather exhibited very low rates of use-related flake removal, which is not surprising due to the physical properties of the soft material. Whereas the points used to drill hard material were very different from the archaeological assemblage at Dhra', the use-related retouch on the experimental leatherworking points and the Dhra' points was similar (Fisher's exact $p = 1$). These data suggest that the use-related damage that occurred on the archaeological points was not severe enough to produce use-related retouch. In sum, the data suggest that PPNA peoples were using the points on a material that would not produce severe use-wear.

Our study of experimentally produced breakage patterns indicates that the experimental points used to perforate stone and wood were different from the points found at Dhra' (Table 7.1). All four of the points that broke during drilling hard contact material had transverse fractures. This is significantly different from the Dhra' assemblage, where the breakage patterns of a random sample of 55 broken points were predominantly horizontal (Fisher's exact $p = .0003$). Again, this evidence undermines the hypothesis that the archaeological points from Dhra' were used to perforate stone or wood. On the other hand, the breakage patterns of tools used to perforate soft materials were not significantly different from the breakage patterns in the sample from Dhra' (Fisher's exact $p = .5802$). Although the breakage patterns do not support the use of el-Khiam points to perforate hard materials, the data do suggest that the points could have been used for perforating soft material, such as leather. In addition to variation based on contact material, the specific action that caused the fractures played a role in the breakage patterning within the experimental assemblage. The points that were used to drill stone and wood broke transversely more often, which may be a result of their use in hand drills and bow drills. The rotational torque placed on the points caused this type of fracture. The horizontal breaks in the leather-puncturing experimental assemblage were not a result of rotation, but rather of a failure in the point while being pressed straight into the material with little lateral rotation.

Finally, the sharpness of the used tip is important for determining contact material. The wood and stone drilling points all had significantly lower sharpness index values at each 1-mm interval from the tip than the archaeological specimens (Figure 7.6a). Additionally, there is little statistical difference between the sharpness of archaeological points and the points used to perforate leather at 1 and 5 mm from the tip. As a whole, the leatherworking points are distributed along with the points from Dhra' at the high end of the sharpness index, whereas the stone drilling points are distinctly duller (Figure 7.6b). The measure of sharpness using the interior angle did show that the archaeological samples were much sharper when they were discarded than the points used to drill stone, and were as sharp as the points used as leatherworking implements. This is important, as people were probably not inclined to resharpen their points immediately prior to discarding them. The damage sustained by drilling the two types of stone and two types of wood was visibly, and quantifiably, more severe than the use damage seen on the archaeological specimens. Likewise, the damage on the leatherworking points was visibly, and quantifiably, similar to the retouch patterns on the archaeological assemblage.

These data reveal that the el-Khiam points were not used to drill the ground stone beads or any other hard material at Dhra', and it is likely that these tools were not used to drill hard materials at other PPNA sites. The large number of points found at Dhra' (Kuijt 2001; Kuijt and Finlayson 2001) indicate that points were being used in multiple ways in addition to their possible utility as projectiles. The experimentally replicated assemblage of leatherworking el-Khiam points has produced data that are very similar to the archaeological collection in terms of retouch location (only manufacture retouch), breakage pattern (more horizontal breaks), and sharpness index (did not dull). The el-Khiam design is good for puncturing, with its sharp tip, as well as for scraping, due to the steep edge angle of the retouch. Additionally, the microwear analysis that suggested that the el-Khiam points were used as perforators noted the direction of striations (perpendicular to the edge and concentric around the tip) (Smith 2005), which could also be produced by rotating the point while perforating soft materials such as leather.

This study used controlled experiments to assess the effectiveness of el-Khiam points as perforating tools. The macroscopic wear, the efficiency of the tool in drilling, and the breakage patterns indicate that el-Khiam points were not used on hard materials such as stone and wood. The possible use of el-Khiam points as perforating implements on soft material such as leather, however, is supported by the experimental data generated in this study. All of this work on defining the function and contact material of tools is very important if archaeologists are to move past morphological classification systems and toward reconstructions of past behavior. The results of these experiments were useful in placing el-Khiam points within their functional context. The character and pattern of wear on the points conform to our expectations of tool use. Thus, it is now possible to derive and apply a measure of tool retouch that is consistent with the context of el-Khiam form and function.

MEASURING RETOUCH

Before settling on one curation index, we evaluated numerous existing measures, such as Kuhn's (1990) measure of retouch of unifacial stone tools and Clarkson's (2002) index of invasiveness for unifacial tools.

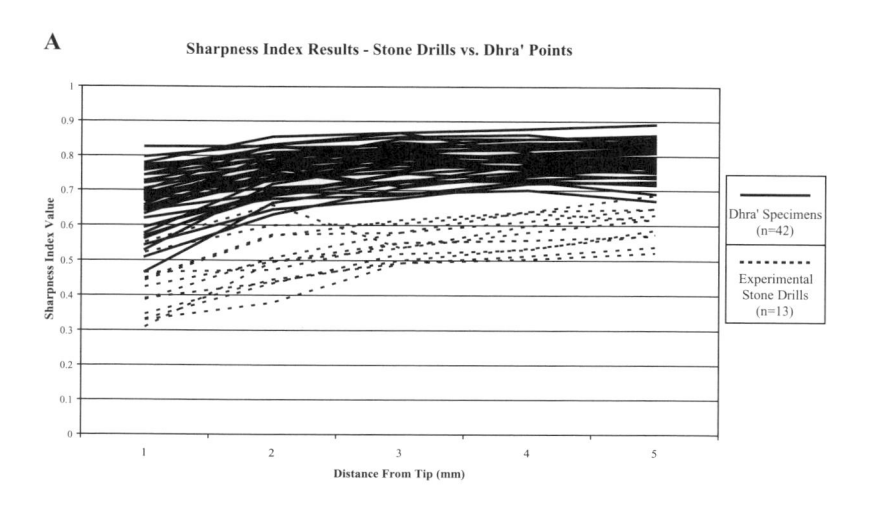

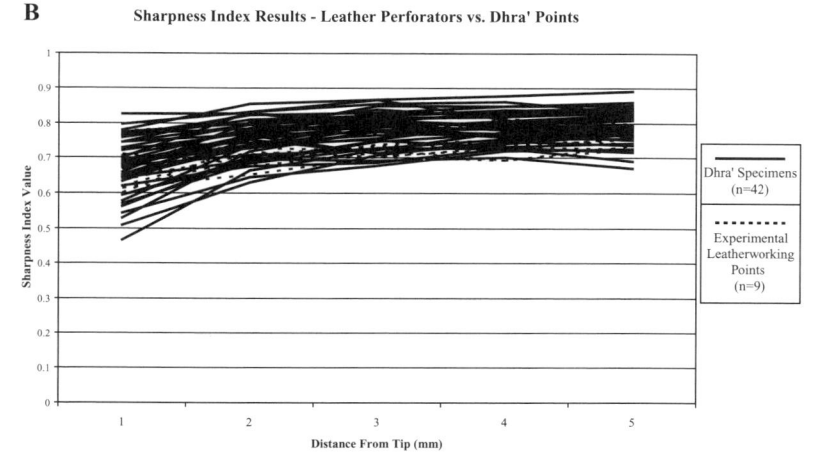

FIGURE 7.6. Sharpness index results comparing Dhra' points to experimental points for drilling stone (A) and for drilling leather (B).

These indices, however, were often inaccurate due to the form, function, and context of the points at Dhra' necessitating the development of a new index to record curation of el-Khiam points. Conceptually, the el-Khiam curation index (EKCI) estimates the original size of newly manufactured el-Khiam points based on an attribute that is preserved on manufactured points. Estimating original flake size is not a new concept in lithic analyses (Dibble 1998; Dibble and Pelcin 1995; Dibble and Whittaker 1981; Pelcin 1996, 1998; Shott et al. 2000).

Although there has been considerable debate about the accuracy of estimating original flake size (Davis and Shea 1998), most of the estimates of original flake size are based on platform characteristics, which can be difficult to measure and consistently replicate (Andrefsky 2005). The EKCI, however, utilizes blade thickness, which is preserved on the el-Khiam points even when they are heavily retouched.

The El-Khiam Curation Index

In order to estimate original blade size for the el-Khiam points, a nonrandom sample of fifty-eight pieces of unmodified debitage from the site of Dhra', Jordan, was analyzed to test predictability of blade length based on blade thickness. We selected blades, pieces of debitage that are twice as long as they are wide, from various contexts at the site to provide a representative sample of blades in the Dhra' lithic assemblage. These blades were chosen based on specific morphological characteristics (straight dorsal ridges, minimal blade curvature, and margins and distal ends with feathered terminations) that PPNA peoples likely used to select el-Khiam point preforms. First, the thickness of the blade was taken below the bulb of force, as this is often removed during el-Khiam point manufacturing. Second, the maximum length perpendicular to the striking platform was recorded for each blade. These attributes were then plotted against each other (Figure 7.7) and the best-fit linear regression line was calculated. The regression equation is

$$\text{Estimated Length} = 11.8 \times (\text{Thickness}) + 7.4.$$

This relationship between length and thickness is strong, and as a result, we can be confident that our estimates of original blade length are fairly accurate ($F = 84.569$, d.f. $= 1$, $p < .0005$, $r^2 = .602$). Although the relationship between the estimated original blade length and the actual length of points is a statistically viable way of measuring retouch ($F = 5.552$, d.f. $= 2$, $p = .011$), the practicality of this measure is questionable. Due to the fact that el-Khiam points were hafted, as seen with their notching elements, estimates of the potential usable portion of the points must not include the length of the blade that is covered by the haft element. As a result, the original blade

Estimating Blade Length Based on Blade Thickness at Dhra'

Estimated Length = (11.8 * Thickness) + 7.4
$r^2 = .602$
$p < .001$

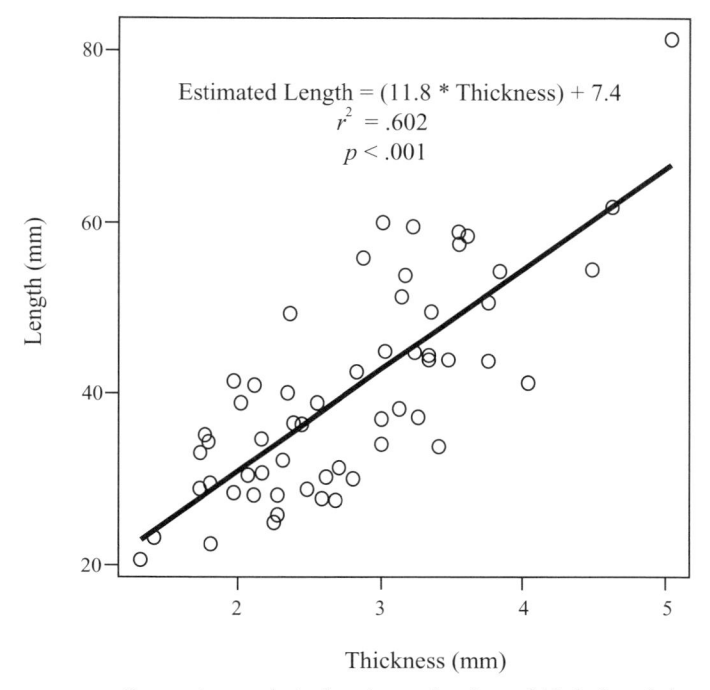

FIGURE 7.7. Regression analysis showing estimation of blade length based upon thickness values.

length estimation requires some tweaking to provide an estimate of the usable bit length. The el-Khiam curation index (EKCI) is calculated by quantifying the amount of bit that is lost through use and retouch (Figure 7.8). In this index, by subtracting the hafted portion of the blade (from the top corner of the notch to the base) from the estimated original blade length, a measure of bit length is devised:

Estimated Bit Length = Estimated Blade Length − Haft Element.

To calculate the amount of used bit length on a point, the maximum length of the point from the tip to the base is recorded (Figure 7.8a). As with the estimated bit length measurement, the haft element length (from the top corner of the notch to the base) is subtracted from the length of the point. The resulting number is the length of the bit that has not been removed:

Unused Bit Length = Total Length − Haft Element.

Because curation is the relationship of realized to maximum potential, the EKCI is the ratio of realized to maximum potential. To calculate this index, the unused bit length is subtracted from the estimated bit length, which is the length of the bit that has been removed by retouch (Figure 7.8b). This number is then divided by the maximum potential bit size, here represented as the estimated bit length:

EKCI = (Estimated Bit Length − Unused Bit Length)/
Estimated Bit Length.

The resulting number is the EKCI value (Figure 7.8c), ranging from a minimum amount of curation (0) to the maximum potential of the hafted el-Khiam point being realized (1). Due to possible slight errors in estimating the original blade length, some points can score in negative numbers based on this equation. In these rare cases, the value is rounded up to the lowest possible curation score of 0.

Experimental Verification of the El-Khiam Curation Index

To determine whether or not the EKCI actually quantifies retouch intensity, a number of experiments were conducted. A sample of el-Khiam points was initially manufactured and hafted to wooden handles. The EKCI for each point was measured prior to the points being used. Once these measurements were recorded, each point was used to perform a variety of leatherworking activities ranging from puncturing to scraping. Once the working edge of the points became dull or the tip of the bit snapped during use, the points were pressure-flaked to rejuvenate the edge as well as to resharpen the tip. After resharpening, the EKCI measurements were taken again. This process was repeated once more, giving a total of three EKCI measurements for each point, which accounts for two retouch events. A total of ten points were used to perform leatherworking experiments

The El-Khiam Curation Index (EKCI)

A — Estimated Bit Length, Estimated Blade Length, Haft Length

Estimated Blade Length − Haft Element = Estimated Bit Length

B — Unused Bit Length, Total Length, Haft Length

Total Length − Haft Element = Unused Bit Length

C — Material Removed by Retouch, Estimated Bit Length, Unused Bit Length

$$\frac{\text{(Estimated Bit Length} - \text{Unused Bit Length)}}{\text{Estimated Bit Length}} = \text{EKCI}$$

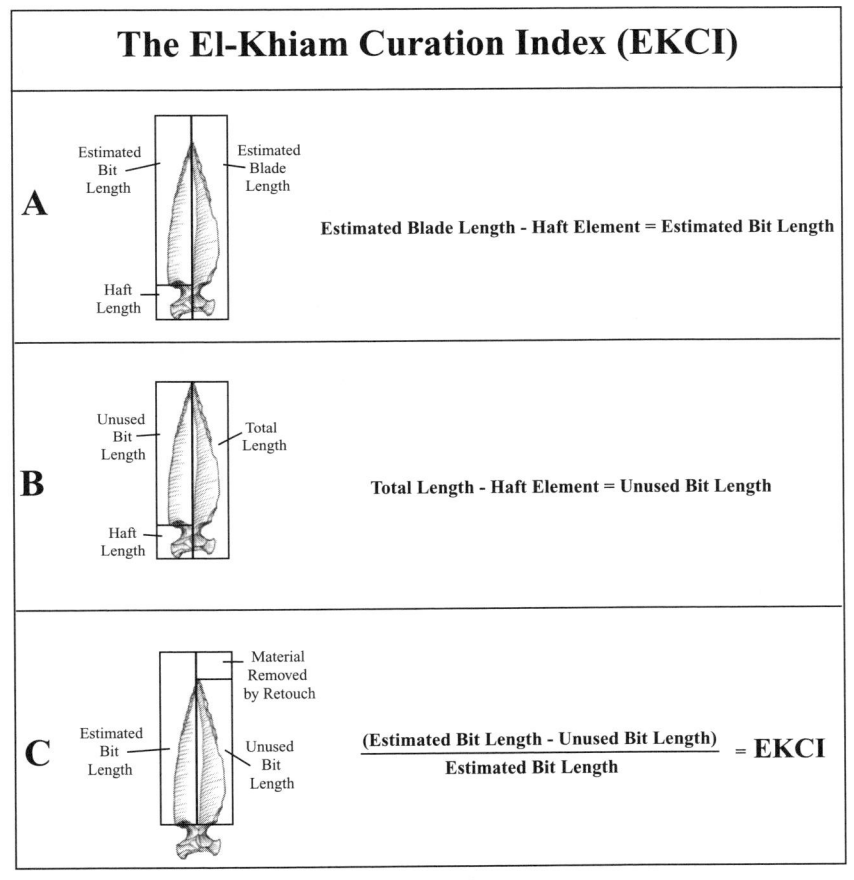

FIGURE 7.8. Schematic illustration showing method of calculating the el-Khiam index (EKCI).

in this fashion. Some of the experimental points were not resharpened as many times as others due to snapped bits that could not be rejuvenated.

The experimentally reduced assemblage had significant variation in the EKCI, with retouch values increasing with each subsequent stage of reduction ($F = 6.657$, d.f. $= 2$, $p = .005$). The EKCI value for each point increased each time it was retouched (Figure 7.9). Although none of the points have EKCI values that approach 1, this is to be expected from the nature of the measurements. For a score of 1, all of the bit must be removed, yet maintaining a tip is impossible once

Experimental Verification of the EKCI

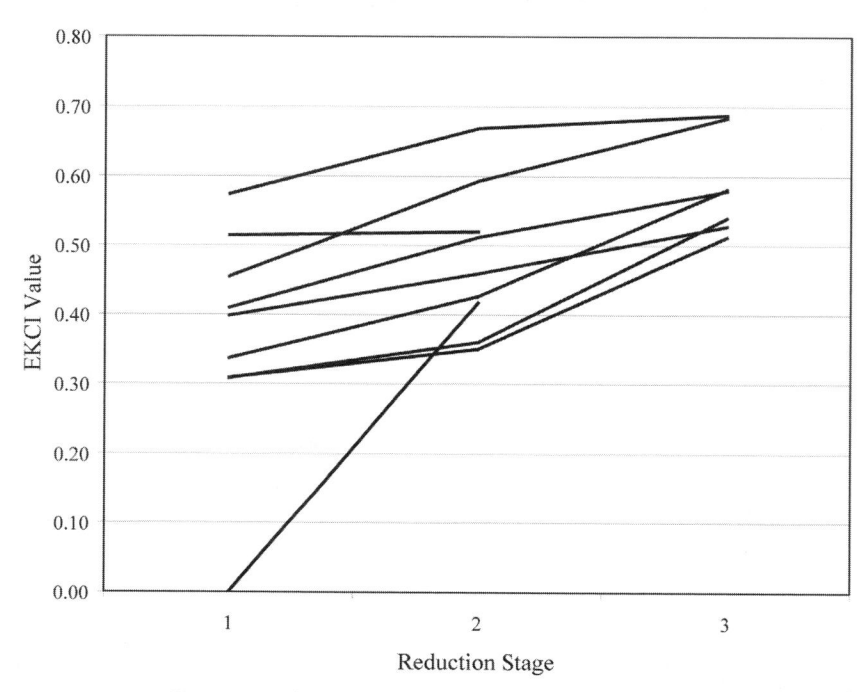

FIGURE 7.9. Experimental verification of the EKCI.

the available bit length is removed. As these data show, the EKCI is an effective measure of retouch intensity on el-Khiam points that are used for functions that produce minimal macroscopic use-wear, such as leatherworking.

RETOUCH AND THE DHRA' EL-KHIAM POINTS

A sample of el-Khiam points was taken from the 2004 excavation season at Dhra'. In all, forty-two complete el-Khiam points were included in the sample. The EKCI was calculated for each of these specimens (Table 7.2). Based upon the EKCI values, the archaeological points at Dhra' were discarded at various stages of their use-life, with unfinished points scoring low on the EKCI. Other points being discarded with nearly half of the usable bit length removed did appear intensively retouched (Figure 7.10). None of the discarded el-Khiam

Table 7.2. Raw data from experimental assemblage

Lab ID	Length	Thickness	Estimated blade length	Haft length	Bit length	Estimated bit length	EKCI value
4876	36.97	1.93	30.17	13.86	23.11	16.31	0.000
4303	33.41	1.67	27.11	9.75	23.66	17.36	0.000
4458	40.40	2.28	34.30	15.01	25.39	19.29	0.000
4467	48.09	2.76	39.97	9.16	38.93	30.81	0.000
4737	37.18	2.48	36.66	8.97	28.21	27.69	0.000
4736	31.16	2.02	31.24	8.13	23.03	23.11	0.003
4829	43.53	3.07	43.63	20.98	22.55	22.65	0.004
5068	31.45	2.11	32.30	10.19	21.26	22.11	0.038
4711	25.38	1.65	26.87	6.51	18.87	20.36	0.073
4446	39.29	2.99	42.68	14.04	25.25	28.64	0.118
4332	31.02	2.29	34.42	7.97	23.05	26.45	0.129
4369	29.81	2.29	34.42	7.25	22.56	27.17	0.170
4296	22.94	1.78	28.40	5.50	17.44	22.90	0.239
4452	32.46	2.73	39.61	11.75	20.71	27.86	0.257
4974	27.93	2.39	35.60	6.57	21.36	29.03	0.264
4554	32.71	2.71	39.38	14.85	17.86	24.53	0.272
5046	29.27	2.57	37.73	7.00	22.27	30.73	0.275
4625	30.69	2.67	38.91	10.16	20.53	28.75	0.286
5167	32.25	2.85	41.03	10.49	21.76	30.54	0.287
4747	28.85	2.53	37.25	8.10	20.75	29.15	0.288
4866	24.01	2.08	31.94	5.68	18.33	26.26	0.302
4386	40.05	3.86	52.95	11.09	28.96	41.86	0.308
4996	23.36	1.98	30.76	6.80	16.56	23.96	0.309
5125	26.29	2.36	35.25	6.90	19.39	28.35	0.316
4787	24.01	2.01	31.12	10.09	13.92	21.03	0.338
5143	18.88	1.64	26.75	4.56	14.32	22.19	0.355
4650	23.13	2.20	33.36	5.89	17.24	27.47	0.372
4456	19.41	1.77	28.29	4.53	14.88	23.76	0.374
4531	23.87	2.30	34.54	6.19	17.68	28.35	0.376
4523	27.15	2.70	39.26	7.18	19.97	32.08	0.377
4565	37.93	4.11	55.90	8.48	29.45	47.42	0.379
4951	32.70	3.44	47.99	8.21	24.49	39.78	0.384
4529	23.60	2.27	34.19	6.74	16.86	27.45	0.386
4325	25.86	2.53	37.25	9.54	16.32	27.71	0.411
4392	17.72	1.61	26.40	5.65	12.07	20.75	0.418
5194	29.05	2.61	38.20	16.49	12.56	21.71	0.421
4401	30.01	3.05	43.39	11.99	18.02	31.40	0.426
4877	18.45	1.82	28.88	5.94	12.51	22.94	0.455
4992	26.02	2.68	39.02	10.91	15.11	28.11	0.463
4433	18.03	1.84	29.11	5.40	12.63	23.71	0.467
5070	26.34	2.74	39.73	11.41	14.93	28.32	0.473
5035	18.68	1.76	28.17	8.21	10.47	19.96	0.475

points scored over .5 on the EKCI, suggesting that the points were not retouched as much as they might have been. Because the settlement of Dhra' is located within 30 m of a large flint source, the abundance of raw materials may have allowed the people at Dhra' to discard their points with usable bits remaining. The tasks for which the points were used, as fine tools associated with leatherworking, likely necessitated a sharp and narrow tip that would have been difficult to maintain when the bit became short.

It is important to note the differences between the EKCI values for the archaeological points and the experimentally produced points. The measurements for the experimental sample averaged .36, .48, and .59 at the first, second, and third retouch stages, respectively, whereas the Dhra' assemblage averaged .27 with no values over .47. The variation of these measurements seems to be attributable to the differences in skill of the researchers when compared with the skill of the PPNA el-Khiam point manufacturers in maximizing the amount of usable bit from a given blade during the primary manufacturing stage. Although the exact values from the archaeological points cannot be used to correspond with specific reduction events from the experiments, the EKCI values do accurately differentiate points throughout their use-life at Dhra'.

DISCUSSION

Among other things, this study shows that measures of curation may not be universally applied to all tool forms. Other researchers have noted this as well (Andrefsky 2006; Clarkson 2002; Eren and Prendergast, this volume; Wilson and Andrefsky, this volume). One universal truth in studies of retouch intensity, however, is the fact that all measurements of curation must conform to how the tool morphology changes through use and retouch, which is guided by the artifact's form, function, and context (MacDonald, this volume).

One consequence of researchers independently developing curation indices that are context-specific is the problem of comparing artifacts or assemblages with varying contexts. One simple way of doing this is to quantify our curation indices in a standardized way. We have followed work by Kuhn (1990), Clarkson (2002), and Andrefsky (2006; this volume), among others, that quantifies curation between

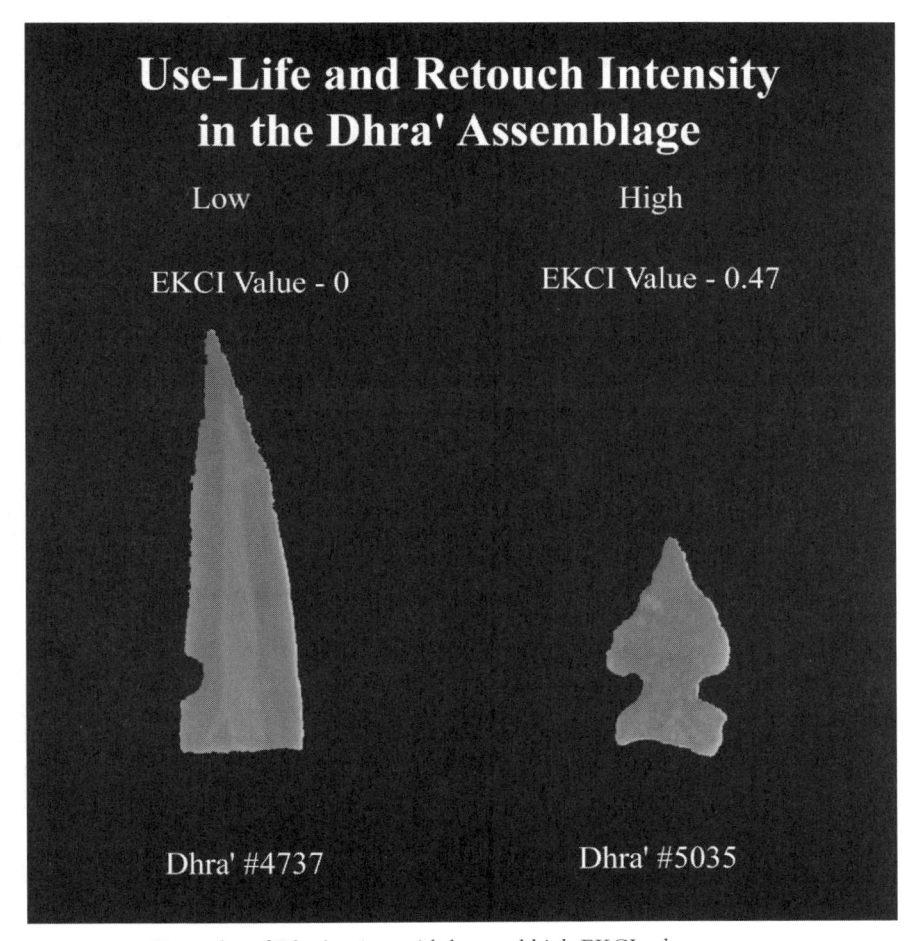

FIGURE 7.10. Examples of Dhra' points with low and high EKCI values.

the values of 0 and 1. We also realize that not all indices are created equal, and there may be distributional differences in the indices that do not fit a normal bell curve from 0 to 1. In these cases it is up to the author to explain the expected range of variation in the assemblage. For example, it is impossible for the EKCI to have a value of 1, as this would mean the entire bit was removed. El-Khiam points with their entire bit removed are not considered complete points, though values of 1 are possible with other curation indices (e.g., Andrefsky 2006). If authors are explicit with the expected ranges between 0 and 1 and how they correspond with low and high retouch intensity, we can start to compare retouch on artifacts that have little, if anything, to do with each other in terms of form, function, or context.

Lithic analysts can compare el-Khiam points that served as perforators at PPNA sites to side scrapers at Mousterian sites to hafted bifaces that were used as knives at sites in North America. Using a common language, ranging from low to high retouch intensity, researchers can then look to other variables to explain possible similarities and differences in retouch intensity, such as raw material availability, site type, and transport costs. The important thing to remember, however, is that each of those artifact types must have retouch intensity measured and tested with an independent index that is context-specific, rather than one index being used to measure all of them. Just because measures of curation must be developed for specific forms, functions, and contexts does not mean that we cannot compare retouch intensity on artifacts that vary in any of these attributes. This is a quantitative matter of scaling different measures from low to high so that such measures are comparable across different tool forms and different indices.

CONCLUSIONS

Based on the form, function, and archaeological context, we were able to devise a measure of retouch intensity, the EKCI, that provides a tool for researchers working in the Near East on PPNA assemblages. In order to better understand the function of el-Khiam points, controlled experiments were performed to determine relative density of the contact material, which could affect use and retouch patterns. It was shown that el-Khiam points were likely used to pierce and scrape soft materials such as leather. The EKCI was then devised, measured, and tested. Experimental replication showed that the EKCI was an accurate measure of retouch intensity, and application of the EKCI to the lithic assemblage at Dhra' reaffirmed the EKCI's utility for analyzing PPNA archaeological assemblages. This study has introduced a baseline technique with which future work can be compared using a standardized retouch intensity measurement. Although this curation index is effective for el-Khiam points, it may not be applicable to other hafted point types, which highlights the need for independently developed measures of retouch that accounts for form, function, and context of the artifacts rather than attempting to generate universal measures of curation.

REFERENCES CITED

Andrefsky, William, Jr. 1986. A Consideration of Blade and Flake Curvature. *Lithic Technology* 15(2):48–54.
  2005. *Lithics: Macroscopic Approaches to Analysis.* 2nd ed. Cambridge University Press, Cambridge.
  2006. Experimental and Archaeological Verification of an Index of Retouch for Hafted Bifaces. *American Antiquity* 71:743–58.
Bar-Yosef, Ofer, and Avi Gopher, eds. 1997. *An Early Neolithic Village in the Jordan Valley.* Part 1. *The Archaeology of Netiv Hagdud.* American School of Prehistoric Research, Peabody Museum, Harvard University Cambridge, MA.
Blades, Brooke S. 2003. End Scraper Reduction and Hunter–Gatherer Mobility. *American Antiquity* 68:141–56.
Clarkson, Chris. 2002. An Index of Invasiveness for the Measurement of Unifacial and Bifacial Retouch: A Theoretical, Experimental and Archaeological Verification. *Journal of Archaeological Science* 29:65–75.
Davis, Z. J., and J. J. Shea. 1998. Quantifying Lithic Curation: An Experimental Test of Dibble and Pelcin's Original Flake-Tool Mass Predictor. *Journal of Archaeological Science* 25:603–10.
Dibble, H. L. 1998. Comment on "Quantifying Lithic Curation: An Experimental Test of Dibble and Pelcin's Original Flake-Tool Mass Predictor," by Z. J. Davis and J. J. Shea. *Journal of Archaeological Science* 25:611–13.
Dibble, H. L., and A. W. Pelcin. 1995. The Effect of Hammer Mass and Velocity on Flake Mass. *Journal of Archaeological Science* 22:429–39.
Dibble, H. L., and J. C. Whittaker. 1981. New Experimental Evidence on the Relation between Percussion Flaking and Flake Variation. *Journal of Archaeological Science* 8:283–96.
Eren, Metin I., Manual Dominguez-Rodrigo, Steven L. Kuhn, Danial S. Adler, Ian Le, and Ofer Bar-Yosef. 2005. Defining and Measuring Reduction in Unifacial Stone Tools. *Journal of Archaeological Science* 32:1190–1206.
Finlayson, Bill, I. Kuijt, T. Arpin, M. Chesson, S. Dennis, N. Goodale, S. Kadowaki, L. Maher, S. Smith, M. Schurr, and J. McKay. 2003. Dhra' Excavation Project, 2002 Interim Report. *Levant* 35:1–38.
Goodale, Nathan, Ian Kuijt, and Bill Finlayson. 2002. The Chipped Stone Assemblage of Dhra', Jordan: Preliminary Results on Technology, Typology, and Intra-assemblage Variability. *Paleorient* 28(1): 115–30.

Goodale, Nathan, and Sam J. Smith. 2001. Pre-pottery Neolithic A Projectile Points at Dhra', Jordan: Preliminary Thoughts on Form, Function, and Site Interpretation. *Neo-Lithics* 2/01:1–5.

Kuhn, Steven L. 1990. A Geometric Index of Reduction for Unifacial Stone Tools. *Journal of Archaeological Science* 17:585–93.

Kuijt, Ian. 1994. Pre-pottery Neolithic A Settlement Variability: Evidence for Sociopolitical Developments in the Southern Levant. *Journal of Mediterranean Archaeology* 7(2):165–92.

——— 2001. Lithic Inter-assemblage Variability and Cultural–Historical Sequences: A Consideration of the Pre-pottery Neolithic A Period Occupation of Dhra', Jordan. *Paleorient* 27(1):107–26.

Kuijt, Ian, and Bill Finlayson. 2001. The 2001 Excavation Season at the Pre-pottery Neolithic A Period Settlement of Dhra' Jordan: Preliminary Results. *Neo-Lithics* 2/01:12–15.

Kuijt, Ian, and H. Mahasneh. 1995. Preliminary Excavation Results from Dhra' and 'Ain Waida. *American Journal of Archaeology* 99:508–11.

——— 1998. Dhra': An Early Neolithic Site in the Jordan Valley. *Journal of Field Archaeology* 25:153–61.

Pelcin, A. W. 1996. Controlled Experiments in the Production of Flake Attributes. Ph.D. thesis, University of Pennsylvania.

——— 1998. The Threshold Effect of Platform Width: A Reply to Davis and Shea. *Journal of Archaeological Science* 25:615–20.

Shott, Michael J. 1996. An Exegesis of the Curation Concept. *Journal of Anthropological Research* 52:259–80.

Shott, Michael J., Andrew P. Bradbury, Philip J. Carr, and George H. Odell. 2000. Flake Size from Platform Attributes: Predictive and Empirical Approaches. *Journal of Archaeological Science* 27:877–94.

Shott, Michael J., and Paul Sillitoe. 2005. Use Life and Curation in New Guinea Experimental Used Flakes. *Journal of Archaeological Science* 32:653–63.

Smith, Sam. 2005. *A Comparative Analysis of the Form and Function of Chipped Stone Artefacts from Wadi Faynan 16 and Dhra': Implications for the Description and Interpretation of Early Neolithic Chipped Stone Variability*. Ph.D. thesis, University of Reading.

8 CHERYL HARPER AND WILLIAM ANDREFSKY, JR.

EXPLORING THE DART AND ARROW DILEMMA: RETOUCH INDICES AS FUNCTIONAL DETERMINANTS

Abstract
Measuring retouch location and intensity on hafted bifaces is shown to be an effective technique for assessing artifact function. Unlike other areas of North America, where dart technology is replaced by arrow technology, Coalition Period occupations on the Pajarito Plateau of New Mexico contain both hafted biface forms used simultaneously. A stylistic analysis of dart points shows that hafted biface forms found in Coalition Period contexts were recycled from Middle and Late Archaic surface scatters. Furthermore, retouch location and intensity show that Coalition Period dart points were used for cutting and sawing activities and not as projectile technology.

INTRODUCTION

In the American Southwest, and throughout North America, dart-sized hafted bifaces identified as projectile points, normally associated with sites dating to the Paleoindian and Archaic time periods, are regularly found on sites dating to the past thousand years (cf. Kohler 2004; Turnbow 1997). Late period points were likely small and designed to be attached to the smaller arrow foreshaft. Although researchers have noted the presence of dart-sized points in settings where the bow and arrow were likely used, few have addressed the question of the context of manufacture or use of these larger hafted bifaces. In the Northern Rio Grande, the presence of Scottsbluff, Jay, Bajada, and other large dart points dating to the Late Paleoindian and Archaic in

Coalition and Classic period sites rarely elicits more than a description as a "curated" item or "heirloom," or as a knife replicating an older style. Are older styles of hafted bifaces being replicated by these later peoples as a component of a dart technology contemporaneous with the bow and arrow? Or are they being recycled and scavenged from Archaic sites to fulfill some functional or ritual requirements?

We contend that, rather than signaling the use of dart technology during the Ancestral Pueblo period, some large hafted bifaces recycled from Archaic sites served as cutting or sawing tools, fulfilling a need for Ancestral Pueblo people not filled by expedient flake tools. This study assesses technological variability on one of these hafted biface forms from the Pajarito Plateau in an attempt to explain the occurrence of this ancient dart form at later period Pueblo sites. Metric measurements of eighty-three large corner-notched projectile points, both complete and fragmentary, from Late Archaic and Ancestral Pueblo sites were compared to identify any differences related to function.

HAFTED BIFACES ON THE PAJARITO PLATEAU

The Pajarito Plateau in north central New Mexico (Figure 8.1) has a long and varied history of use, beginning approximately 10,000 years ago, evidenced by isolated finds of Paleoindian spear tips (Jolly 1970). This study focuses on the Late Archaic and Ancestral Pueblo periods. The Late Archaic period dates to between 800 B.C. and A.D. 600. During this period, prehistoric people followed a seasonal cycle of movement based upon the availability of plant resources within a number of elevation and vegetative zones. Archaic camps were loci of both hunting and gathering activities (Vierra and Foxx 2002). Dart points during this time period were produced as a result of very refined bifacial technology. Contracting stem and large corner-notched projectile points dominate the point assemblages, exhibiting beautiful bilateral symmetry, thin cross sections, and even flaking patterns. Those types have been identified across the North American West, and in the Southwest they are primarily associated with Late Archaic mobile foragers.

The Ancestral Pueblo period incorporates both the Northern Rio Grande Coalition and Classic Periods (Wendorf and Reed 1955). The

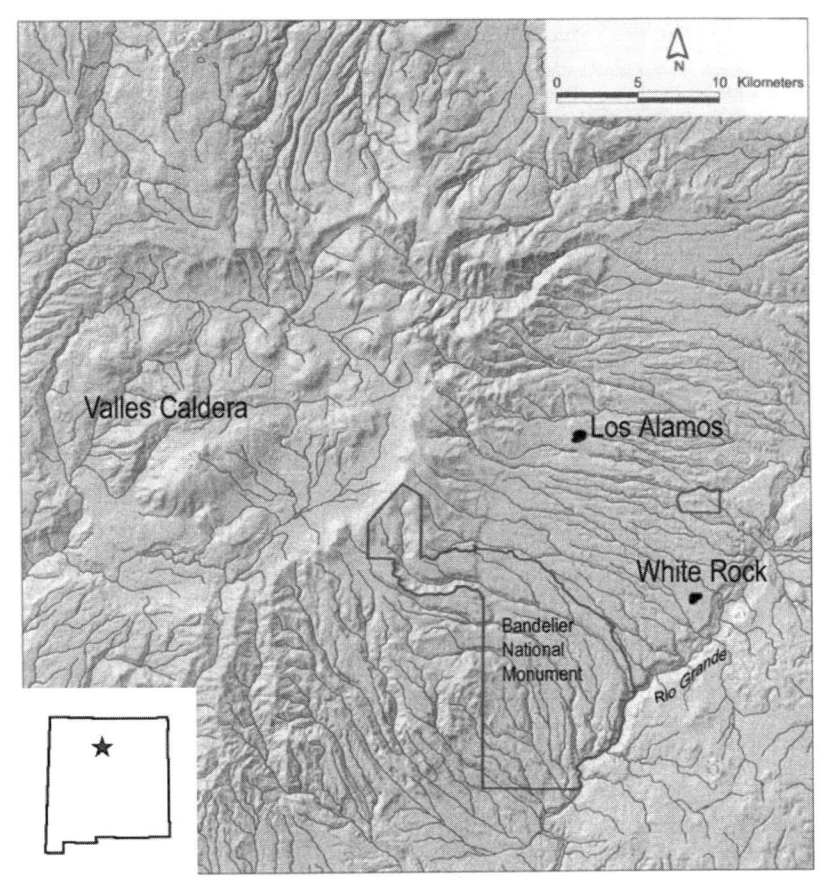

FIGURE 8.1. Project location, primarily within Bandelier National Monument (from Kohler 2004).

Coalition period (A.D. 1150 to 1325) is characterized by increasing sedentism and larger communities. Initially organized around individual households, homesteads made way for larger plaza pueblos as migrants, likely from the San Juan basin, moved onto the Pajarito Plateau (Kohler and Root 2004). Population peaked during the Late Coalition and began to decrease slightly during the early stages of the Classic period, dating to between A.D. 1325 and 1600. While population decreased, communities increased dramatically in size. Where there had once been dozens of small communities across the Pajarito Plateau during the Coalition, by the Late Classic most had been abandoned and the occupants moved into a handful of extremely large towns along the western bank of the Rio Grande (Kohler et al. 2004; Preurcel 1990). Hunters during this time period were utilizing the bow and arrow, and the associated projectile points were small side-notched forms, similar to those found throughout the West prior to Euro–American contact. They were created from small flake blanks, with the final shape often related to the shape of the original flake blank rather than a purposeful choice by the maker of the projectile tip.

In addition to small side-notched points, Coalition and Classic period sites also include a wide variety of larger points (Figure 8.2), which, upon first glance, would appear to be dart tips based upon stem width dimensions (Christenson 1986; Corliss 1972; Shott 1996). Only a few researchers have tried to identify the reasons behind the presence of these types of hafted bifaces in later contexts. Point types that would normally date to a period much earlier than the rest of the assemblage are normally listed as intrusive or as curated items, but rarely are these assumptions scientifically tested. Three possible theories to explain the presence of these points in later contexts have been identified in previous literature:

(1) The presence of large corner-notched points represents multiple occupation sites, with pueblos built on top of Archaic lithic scatters.
(2) Ancestral Pueblo people replicated the large corner-notched form for use as dart points or knives
(3) Ancestral Pueblo people collected old points when encountered to be reused as dart points or knives, or as items of ritual significance.

Because excavations in the region have found Archaic points in good stratigraphic contexts within Ancestral Pueblo sites associated with Coalition and Classic period deposits (e.g., Kohler and Root 2004; Turnbow 1997), the presence of older projectile point forms does not appear to represent multiple occupations.

Second, perhaps Ancestral Pueblo people, recognizing the efficiency of bifacial hafted tools, replicated the large bifacial corner-notched form for use as dart points or knives. If large hafted bifaces were being manufactured by Ancestral Pueblo people, then we would

FIGURE 8.2. Top row: small side-notched projectile points. Bottom row: large corner-notched projectile points.

expect that evidence of this manufacture would be found in the form of bifacial thinning flakes, bifacial tools, and bifacial cores. However, little evidence currently exists for a thriving biface technology during this time. Bifacial thinning flakes are extremely rare within the lithic database created from the Bandelier Archaeological Survey Project in the middle of the study area (Harro 1997). Only 1.4% of all debitage identified was described as originating from bifacial reduction (Head 1999). The amount of reduction taking place likely would not account for the number of Archaic points recovered in Ancestral Pueblo sites (Kohler et al. 2004). In addition, only 9% of all formal tools were identified as bifacial in form. Tyuonyi Annex, a small structure adjacent to the Classic Period pueblo of Tyuonyi at the bottom of Frijoles Canyon, excavated in 1988, had a much higher frequency of biface thinning flakes in its assemblage than any other Ancestral Pueblo site, at 6.6% of the total assemblage, and may represent a locus of projectile point manufacture; however, the rest of the assemblage is dominated by expedient flake tools and amorphous cores (Kohler et al. 2004). In regions such as Black Mesa, Arizona, researchers have found that, with increasing sedentism, bifacial technology tends to decrease, whereas expedient core technology increases. However, during both mobile and sedentary periods, neither bifaces nor expedient flake tools ever leave the assemblages (Parry and Christenson 1986). Tyuonyi Annex may represent specialist manufacture of projectile points, but the amount of reduction taking place at that site could not account for the number of Archaic style points recovered in Ancestral Pueblo sites (Kohler et al. 2004). Given these data, it is extremely unlikely that the large corner-notched projectile points were being created by Ancestral Pueblo people.

We contend that large corner-notched projectile points were picked up from old sites or as isolated artifacts and reused during later times. This type of reuse is not unknown in the ethnographic and archaeological records in the American Southwest. Recycled projectile points could have had some form of ideological meaning to Ancestral Pueblo people. Early and Middle Archaic points have been found as parts of ritualistic items such as shamans' wands (Thomas 1976) and pendants (Haury 1975) during these later time periods. Points were also collected by Pueblo people for use in medicine bundles.

Midwives would use older points as items of power, grinding them to help with labor (Hill 1938).

Evidence for use as a cutting or sawing tools, or as projectile points, can also be found in archaeological records. Use wear studies have shown that projectile points may have multiple uses as both hunting weapon tips and knives (Ahler 1971; Andrefsky, this volume; Truncer 1990). Excavation of a contact-period Jicarilla Apache campsite in Northeastern New Mexico unearthed a Large Corner-Notched hafted biface that had been reused (Gunnerson 1969). Although the majority of both surfaces was covered by a thick patina, retouch scars along both blade margins had patina removed along the edges, showing that these points were deposited on the ground surface long enough to gain a patina, and were reworked following collection by mobile Apaches (Gunnerson 1969). The late 19th-century Navajo believed that projectile points are physical representations of lightening, and would collect points from old sites for reuse (Hill and Lange 1982). In one source, physically creating an arrow point for use in hunting was taboo, and those that were actually made by Navajo were only used during rituals. Navajo men would collect projectile points any time they were found and then rehaft them onto new shafts (Hill 1938; Gunnerson 1959).

ANALYSIS

If the large corner-notched projectile points were being collected by Coalition and Classic period people and used as cutting tools, rather than ritual items, there should be some evidence of that use on the tools themselves. Large hafted bifaces were likely multipurpose tools, used as both projectile points and as knives (cf. Ahler 1971; Andrefsky 1997; Kay 1996; Truncer 1990) by Late Archaic people. Such tools are deposited into the archaeological record as a result of being lost when used as projectiles. Large points used as cutting tools may have been purposely discarded by Late Archaic people when they were worn. Ancestral Pueblo people would have needed only enough blade for a cutting tool, and could have then picked them up and recycled them for additional use. If the tools continued to be used for cutting or sawing actions, we would expect there to be a greater amount

Table 8.1. Retouched versus unretouched, large corner-notched and small side-notched ($\chi^2 = 13.2907$, d.f. $= 1$, $p = .0003$)*

	Retouched	Unretouched	Row total
Large corner-notched	26 (15)	47 (58)	73
Small side-notched	36 (47)	192 (181)	228
Column total	62	239	301

*Expected values in parentheses.

of retouch and wear on the reused points, representative of a longer use life.

The large corner-notched projectile points used in this analysis derive from surface contexts on the Pajarito Plateau in North-central New Mexico. Most were collected during the Bandelier Archaeological Survey Project, a large-scale five-year survey conducted from 1987 to 1992 at Bandelier National Monument (Powers and Van Zandt 1999). In addition, a smaller collection of points from surveys conducted at Los Alamos National Laboratory, north of Bandelier, was also included. Because of small sample size during the Classic Period, points from the Classic and Coalition periods were merged and analyzed as a single sample.

FUNCTION OF LARGE CORNER-NOTCHED POINTS

If hafted bifaces were being used as both projectile tips and cutting tools, we would expect to see evidence of resharpening of the blade edges in order for the tool to function efficiently in both realms. If, on the other hand, the tool only functioned as a projectile tip, there should be less retouch to resharpen the edges. In other words, if large corner-notched points were recycled as sawing and cutting tools during the Ancestral Pueblo period and not as dart points, whereas Small Side-Notched were used primarily as arrow points, we would expect to see a significant difference in how the large corner-notched points and small side-notched arrow points were maintained during this period. When the frequencies of retouched and unretouched points by point style are reviewed, this expectation appears to be met. Table 8.1 compares the frequency of retouch on small side-notched and large corner-notched hafted bifaces. Clearly the smaller arrow points were less often retouched than the large corner-notched points

($X^2 = 13.2907$, d.f. $= 1$, $p = .0003$). This suggests that large corner-notched points functioned in such a way that retouch was required during their use-life. On the other hand, retouch was not as frequently found on the edges of small side-notched points. We suggest that the small arrow point was used as the primary projectile weapon (but occasionally as a cutting tool) and the large corner-notched point was used primarily as a cutting and/or sawing tool.

However, if these points were picked up from Archaic sites by Ancestral Pueblo peoples, retouch seen on the large corner-notched points could originate from multifunctional use and subsequent retouch during the Archaic period just as easily as it could from the Ancestral Pueblo period. If there is a functional difference in large corner-notched points between the two periods, patterns in metric attributes of large corner-notched points from the Archaic and Ancestral Pueblo periods should differ significantly. If large corner-notched points from Ancestral Pueblo contexts were used solely as cutting tools and not as projectile points, because Ancestral Pueblo groups had adopted bow and arrow technology, we would expect to see differences in blade shape because of resharpening. During the Archaic, projectile points were multipurpose tools, but any retouch had to allow for continued use as a projectile point. The Ancestral Pueblo likely had no such constraints, and retouch would reflect only their use as cutting implements.

If, on the other hand, points were being collected for ritual purposes, there should be no difference in the morphology of large corner-notched points. If a point is included in a medicine bag or placed within a shrine, there should be no reason to retouch the edges, as it is not being used for cutting, slicing, or piercing. There should therefore be no difference in metric attributes related to ritual use.

A comparison of morphological characteristics of the two data sets (one from the Archaic and the other Ancestral Pueblo) relies on the assumption that prior to retouch, both data sets would be morphologically identical. If large corner-notched points were being retouched as a response to use in a cutting or sawing motion, then haft element attributes should not be affected. In order to achieve a sturdy haft, the hafting material would cover both the stem and lower portions of the shoulders, and would therefore not be retouched while

Blade Edge **Blade Tip**

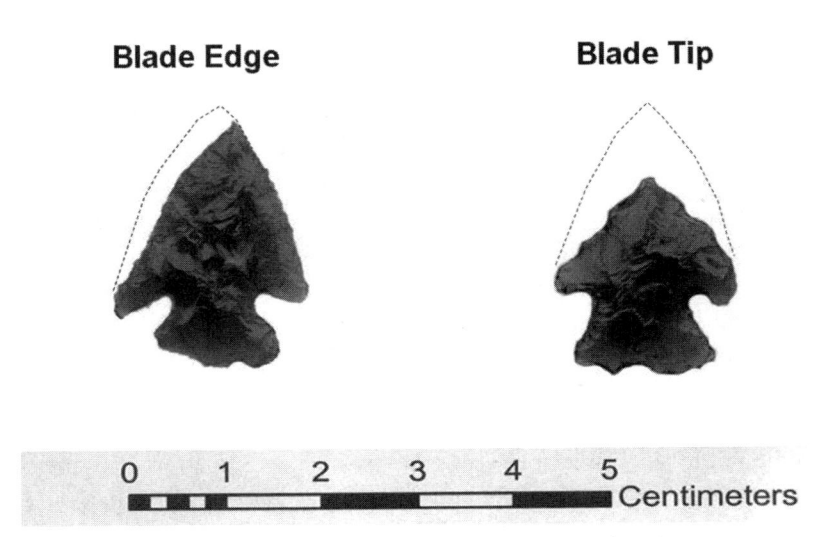

FIGURE 8.3. Types of retouch on large corner-notched projectile points.

still hafted. *T*-tests were run evaluating the differences in haft length ($p = .1620$), neck width ($p = .2106$), shoulder width ($p = .3926$), and basal width ($p = .7452$) of large corner-notched points from Archaic and Ancestral Pueblo settings, and found that there is no significant difference between the two data sets. Therefore, it is likely they come from the same population, and perhaps were manufactured during the Archaic period.

When hafted bifaces are used to cut or saw, the blade element often becomes narrower due to resharpening of the dulled edges, and the blade length often becomes shorter due to resharpening of tips that have been broken (Andrefsky 2006). Figure 8.3 shows examples of large corner-notched projectile points with evidence of resharpening. If our assumptions about large corner-notched points from Ancestral Pueblo sites are correct, both the blade lengths and the blade midpoint widths should be significantly shorter in Pueblo period sides than in Archaic ones. When the blade midpoint width and blade length values were compared, it was found that there was a significant difference between the two sets of data.

Differences in the blade midpoint width values, defined as the blade widths measured halfway between tip and tang, represent varying amounts of retouch placed on large corner-notched points. If they

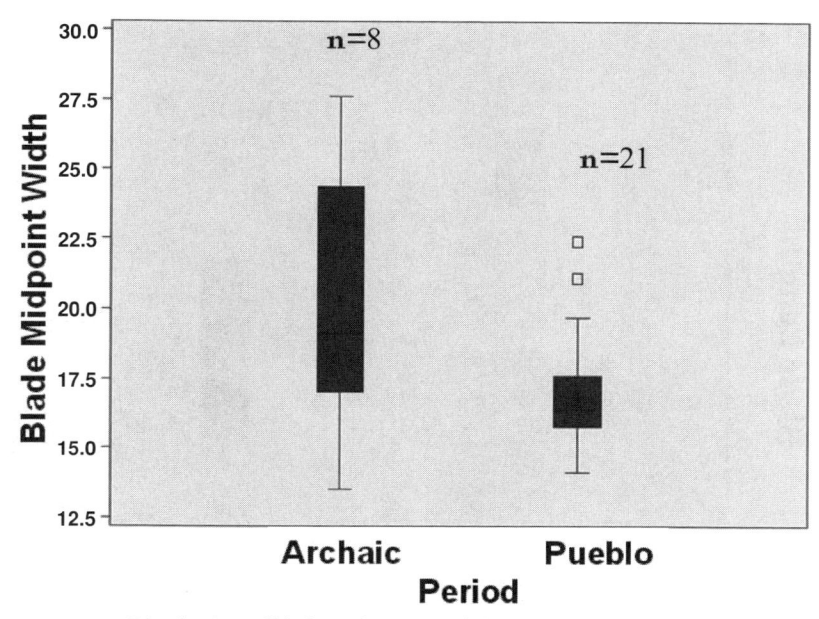

FIGURE 8.4. Distribution of blade midpoint width by period for large corner-notched points (t-test $F = 6.178$, d.f. $= 28$, $p = .0194$).

were being used more intensely as cutting utensils during the Ancestral Pueblo period, we would expect these tools to be retouched more often to resharpen their edges. This would result in narrower blade midpoint width values during the Ancestral Pueblo period than during the Archaic. When the blade midpoint width values were compared between the two periods, large corner-notched points collected from Ancestral Pueblo sites were significantly narrower at the blade midpoint than those collected from Archaic settings (Figure 8.4; $F = 6.178$, d.f. $= 28$, $p = .0194$). Retouch was not taking place low on the blade, as evidenced by comparable shoulder widths between periods, but was narrowing the blade where the most pressure would be placed in cutting motions.

If large corner-notched hafted bifaces are being used as knives rather than as projectile tips, the length of the blade should decrease in size as well. The blade length may decrease due to resharpening on one or both blade edges, along with the blade midpoint width. In addition, broken blades that would normally be discarded as unusable projectile tips during the Archaic might be resharpened into useable

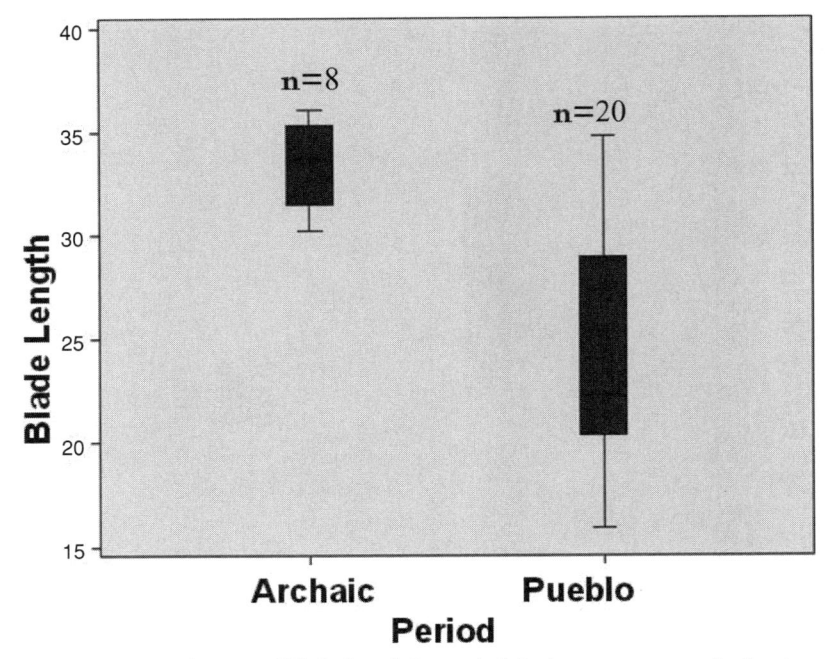

FIGURE 8.5. Distribution of blade length by period for large corner-notched points ($n = F = 6.780$, d.f. $= 27$, $p = .0150$).

cutting tools during the Ancestral Pueblo period. Blade length values between the two periods varied significantly, with blade lengths much shorter during the Ancestral Pueblo periods than those dating to the Archaic (Figure 8.5; $F = 6.780$, d.f. $= 27$, $p = .0150$). Archaic people discarded large corner-notched points once the blade was broken past a certain point. Ancestral Pueblo people may have found utility in these broken blades, resharpening them well below what was useful for Archaic people.

If large corner-notched projectile points were being used as knives, why would Ancestral Pueblo people have felt the need to collect them from old sites, when tool production was overwhelmingly expedient in nature? During the Ancestral Pueblo period, stone tool production and creation of arrow points were taking place within habitation sites (Pueblos), whereas nonhabitation sites such as field houses had lesser amounts of tool production (Head 1999). If large corner-notched points fulfilled the need for efficient stone tools in locations where tool

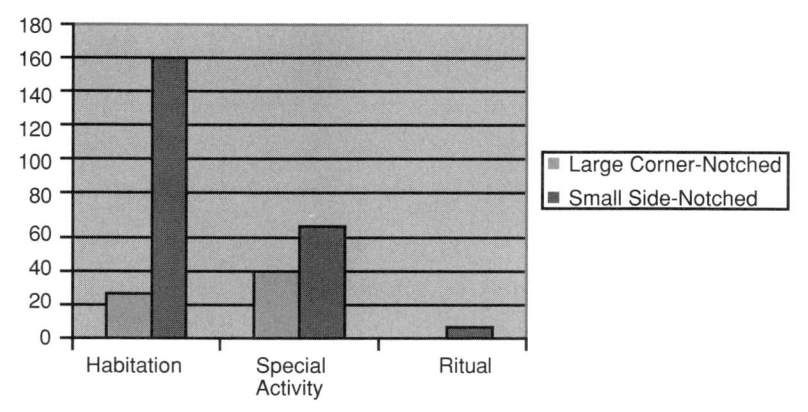

FIGURE 8.6. Frequency (by count) of projectile point styles by site type.

production was not taking place, then there should be a difference in the distribution of large corner-notched points and small side-notched points by site type. Although large corner-notched hafted bifaces were found throughout the Ancestral Pueblo Coalition and Classic periods, they were more frequently found at nonhabitation sites such as field houses, artifact scatters, and water control features than at longer-term habitation sites (Figure 8.6). The distribution of large corner-notched points from Coalition and Classic periods stands in contrast to that of small side-notched arrow points, which tend to be found more frequently in the habitation sites and ritual spaces such as kivas and shrines. A chi-squared test of the distribution of large corner-notched and small side-notched points comparing habitation and nonhabitation sites during the Ancestral Pueblo period found that the differences seen are statistically significant and are not a result of random sampling (Table 8.2; $\chi^2 = 32.017$, d.f. $=1$, $p < .0001$). Based upon this distribution, it is likely that large corner-notched and small side-notched hafted bifaces were being used in different manners associated with the different site activities or functions. Bifacial reduction was rarely taking place during the Ancestral Pueblo period, and only a few individuals may have had the skill or knowledge to create large bifacial tools. Large corner-notched points may have fulfilled the requirement of maintainable and reliable cutting tools at nonhabitation sites, where raw materials for expedient tool production would have been limited.

Table 8.2. Projectile point type by site type ($\chi^2 = 32.017$, d.f. = 1, $p < .0001$)

	Habitation	Special activity	Ritual	Row total
Large corner-notched	27	42	0	69
Small side-notched	158	63	5	226
Column total	185	105	5	295

DISCUSSION AND CONCLUSIONS

Expedient core technology appears to have been the cornerstone of the Ancestral Pueblo lithic technological organization (Arakawa 2000; Head 1999). Bifacial reduction was rarely taking place during the Coalition and Classic, and only a few individuals may have had the skill or knowledge to create large bifacial tools. The presence of large corner-notched projectile point tips does not fit with this picture of a lithic tool kit made up of expedient flake tools and small side-notched projectile points created from small flake blanks (Head 1999). As projectile points became smaller through time, they would have become less efficient as cutting tools (Christenson 1986, 1987; Fischer 1989).

Our analysis clearly shows that large corner-notched hafted bifaces were much more heavily retouched during the Coalition and Classic periods than during the Archaic periods. This was apparent from retouch expressed as a function of blade width and blade length. Such retouch has been linked to tool use as cutting or sawing utensils (Andrefsky 2006). We have also shown that these large hafted bifaces were more heavily retouched than the contemporaneously used small side-notched hafted bifaces, suggesting different functions for the tool forms during the Coalition and Classic periods. These trends suggest that Ancestral Pueblo people did not make large hafted bifaces to fulfill processing needs related to cutting and sawing. Instead, they used flake tools as knives for cutting and processing, and when available, they recycled and scavenged large hafted bifaces from Archaic sites to complete their cutting and processing requirements.

Others have shown that Ancestral Pueblo peoples were expedient tool makers and that even hafted biface forms such as small arrow points were made quickly. Our review of debitage suggests that large hafted bifaces were not produced on Ancestral Pueblo sites, and instead were probably recycled in completed form from Archaic contexts. Finally, because our study shows that these recycled dart points were used by Ancestral Pueblo people as cutting and sawing tools, we believe that dart and arrow projectile technology were not simultaneously in use during the Ancestral Pueblo times, as others have hypothesized.

REFERENCES CITED

Ahler, Stanley A. 1971. *Projectile Point Form and Function at Rodgers Rockshelter, Missouri*. Research Series No. 8. Missouri Archaeological Society, Columbia.

Andrefsky, William, Jr. 1997. Thoughts on Stone Tool Shape and Inferred Function. *Journal of Middle Atlantic Archaeology* 13:125–44.

———. 2006. Experimental and Archaeological Verification of an Index of Retouch for Hafted Bifaces. *American Antiquity* 71:743–58.

Arakawa, Fumi. 2000. Lithic Analysis of Yellow Jacket Pueblo as a Tool for Understanding and Visualizing Women's Roles in Procuring, Utilizing, and Making Stone Tools. M.A. thesis, University of Idaho, Moscow.

Christensen, Andrew L. 1986. Projectile Point Size and Projectile Aerodynamics: An Exploratory Study. *Plains Anthropologist* 31:109–28.

———. 1987. Projectile Points: Eight Millennia of Projectile Change on the Colorado Plateau. In *Prehistoric Stone Technology on Northern Black Mesa, Arizona*, edited by William J. Parry and Andrew L. Christenson, pp. 143–98. Occasional Paper No. 12. Southern Illinois University at Carbondale, Center for Archaeological Investigations.

Corliss, David W. 1972. *Neck Width of Projectile Points: An Index of Culture Continuity and Change*. Occasional Papers of the Idaho State University Museum, No. 29. Idaho State University Museum, Pocatello.

Fischer, A. 1989. Hunting with Flint-Tipped Arrows: Results and Experiences from Practical Experiments. In *The Mesolithic in Europe*, edited by C. Bonsall, pp. 29–39. John Donald, Edinburgh.

Gunnerson, Dolores. 1959. Tabu and Navajo Material Culture. *El Palacio* 64(1):1–9.

Gunnerson, James H. 1969. Apache Archaeology in Northeastern New Mexico. *American Antiquity* 34(1):23–39.

Harro, Douglas R. 1997. Patterns of Lithic Raw Material Procurement on the Pajarito Plateau, New Mexico. M.A. Thesis, Department of Anthropology, Washington State University, Pullman.

Haury, Emil W. 1975 (1950). *Ventana Cave*. University of Arizona Press, Tucson.

Head, Genevieve N. 1999. Lithic Artifacts. In *The Bandelier Archeological Survey, Volume II*, edited by Robert P. Powers and Janet D. Orcutt, pp. 469–549. Professional Paper No. 57. Intermountain Cultural Resources Management, Santa Fe, NM.

Hill, W. W. 1938. *The Agricultural and Hunting Methods of the Navajo Indians*. Yale University Publications in Anthropology 18, New Haven.

Hill, W. W., and Charles H. Lange. 1982. *An Ethnography of Santa Clara Pueblo, New Mexico*. University of New Mexico Press, Albuquerque.

Jolly, Fletcher, III. 1970. Fluted Points Reworked by Later Peoples. *Tennessee Archaeologist* 26(2):30–44.

Kay, Marvin. 1996. Microwear Analysis of Some Clovis and Experimental Chipped Stone Tools. In *Stone Tools: Theoretical Insights into Human Prehistory*, edited by George Odell, pp. 315–44. Plenum Press, New York.

Kohler, Timothy A. 2004. Introduction. In *Archaeology of Bandelier National Monument: Village Formation on the Pajarito Plateau, New Mexico*, edited Timothy A. Kohler, pp. 1–17. University of New Mexico Press, Albuquerque.

Kohler, Timothy A., Sarah Herr, and Matthew J. Root. 2004. The Rise and Fall of Towns on the Pajarito (A.D. 1375–1600). In *Archaeology of Bandelier National Monument: Village Formation on the Pajarito Plateau, New Mexico*, edited by Timothy A. Kohler, pp. 215–64. University of New Mexico Press, Albuquerque.

Kohler, Timothy A., and Matthew J. Root. 2004. The Late Coalition and Earliest Classic on the Pajarito Plateau (A.D. 1250–1375). In *Archaeology of Bandelier National Monument: Village Formation on the Pajarito Plateau, New Mexico*, edited by Timothy A. Kohler, pp. 173–214. University of New Mexico Press, Albuquerque.

Parry, William J., and Andrew L. Christenson. 1986. *Prehistoric Stone Technology on Northern Black Mesa, Arizona*. Occasional Paper of the Center for Archaeological Investigations, Southern Illinois University Press, Carbondale.

Powers, Robert P., and Tineke Van Zandt. 1999. An Introduction to Bandelier. In *The Bandelier Archaeological Survey, Volume 1*, edited by Robert P. Powers and Janet D. Orcutt, pp. 1–31. Intermountain Cultural Resources Management Professional Paper No. 57, Santa Fe.

Preucel, Robert W., Jr. 1990. *Seasonal Circulation and Dual Residence in the Pueblo Southwest: A Prehistoric Example from the Pajarito Plateau, New Mexico*. Garland Publishing, New York and London.

Shott, Michael J. 1996. Innovation and Selection in Prehistory: A Case Study from the American Bottom. In *Stone Tools: Theoretical Insights into Human Prehistory*, edited by George H. Odell, pp. 279–309. Plenum Press, New York.

Thomas, David Hurst. 1976. A Diegueño Shaman's Wand: An Object Lesson Illustrating the "Heirloom Hypothesis." *Journal of California Anthropology* 3(1):128–32.

Truncer, James J. 1990. Perkiomen Points: A Study in Variability. In *Experiments and Observations on the Terminal Archaic of the Middle Atlantic Region*, edited by R. W. Moeller, pp. 1–62. Archaeological Services, Bethlehem, CT.

Turnbow, Christopher A. 1997. Projectile Points as Chronological Indicators. In *OLE. Volume II: Artifacts*, edited by John C. Acklen, pp. 161–230. TRC Mariah Associates Inc., Albuquerque.

Vierra, Bradley J. and Teralene Foxx. 2002. Archaic Upland Resource Use: The View from the Pajarito Plateau, New Mexico. Paper presented at the 67th Annual meetings of the Society for American Archaeology, March, 2002.

Wendorf, Fred, and Erik K. Reed. 1955. An Alternative Reconstruction of Northern Rio Grande Prehistory. *El Palacio* 62:131–73.

PART THREE

NEW PERSPECTIVES ON LITHIC RAW MATERIAL AND TECHNOLOGY

9 WILLIAM ANDREFSKY, JR.

PROJECTILE POINT PROVISIONING STRATEGIES AND HUMAN LAND USE

Abstract

The classification of projectile points in North America often emphasizes the shape and size of the haft element and not of the blade element. Emphasis on the haft element in classification is an advisable strategy because the blade element morphology and size tend to change during the use life of the specimen. This is exactly why the characteristics of projectile point blade elements such as retouch amount, size, and shape are useful for inferring characteristics of technological organization. Variability in retouch amount and location on projectile point blade elements is shown to be directly associated with prehistoric hunter–gatherer land use patterns.

INTRODUCTION

Hafted biface provisioning strategies (production, consumption, discard) have been shown to be directly related to artifact function and

Field and laboratory components of this study were funded by grants from the College of Liberal Arts, Washington State University, and the U.S. Department of Interior, Bureau of Land Management. I am grateful for their efforts and assistance with this project. In particular I would like to thank Natalie Sudman, Diane Pritchard, and Richard Hanes for their ongoing support. I would also like to thank the Northwest Research Obsidian Studies Laboratory in Corvallis, Oregon for providing a grant to assess geochemical characterization of the hafted bifaces used in this study. I thank Lisa Centola and Eren Wallace for sampling the obsidian specimens. The staff of the University of Oregon, Museum of Natural History was generous with their time and help in characterizing and sharing information from the Paulina Lake site – thanks especially to Tom Connolly and Dennis Jenkins.

processing requirements for various tasks (Ahler 1971; Churchill 1993; Ellis 1997; Frison 1991; Hester and Green 1972; Odell and Cowan 1986; Tomka 2001; Truncer 1988). However, others have shown that hafted biface provisioning strategies are also directly related to human land-use practices and raw material availability (Andrefsky 1994, 2005; Daniel 2001; Flenniken and Wilke 1989; Greaves 1997; Hoffman 1985; Kelly 1988; Sassaman 1994; Tankersley 1994). This is particularly true of hunting and gathering populations, which often leave residential base camps for extended periods of time to acquire resources outside of a one- or two-day spatial range. Such tool makers and users must be equipped with an adequate supply of hafted bifaces while on the move or they must have the ability to resupply their tool kits while away. The known locations of lithic raw material sources within the hunter–gatherer circulation range will influence how the travelers provision their tool kits and ultimately consume and discard their tools. In this paper I explore the role of land-use practices on hafted biface provisioning strategies based upon the characteristics of hafted bifaces recovered from a hunter–gatherer residential site.

One of the challenges of linking human land-use practices to lithic technology is the difficulty of gathering independent data on the circulation range(s) of the aboriginal populations responsible for production of archaeological assemblages. If we knew where and particularly how far tool makers and users have circulated across the landscape we might be able to more confidently assess these circulation patterns in hafted biface provisioning tactics. Fortunately, XRF analysis of obsidian in hafted biface form provides accurate information on tool stone source locations. These source locations can be used as proxy data for circulation ranges and allow comparison of hafted biface retouch and provisioning trends based upon human land-use practices.

SITE CONTEXT AND OBSIDIAN SOURCES

This study examines hafted bifaces from a residential base camp in southeastern Oregon. The site contains obsidian from eleven known obsidian source locations and several unknown source locations as well. This site (Birch Creek, 35ML181) has a house pit village occupation and a pre–house pit occupation along the Owyhee River (Figure 9.1). The pre–house pit occupation is dated to between 5315 and 4865 B.P. (Beta 142362), and the house pit occupation ranges in age between

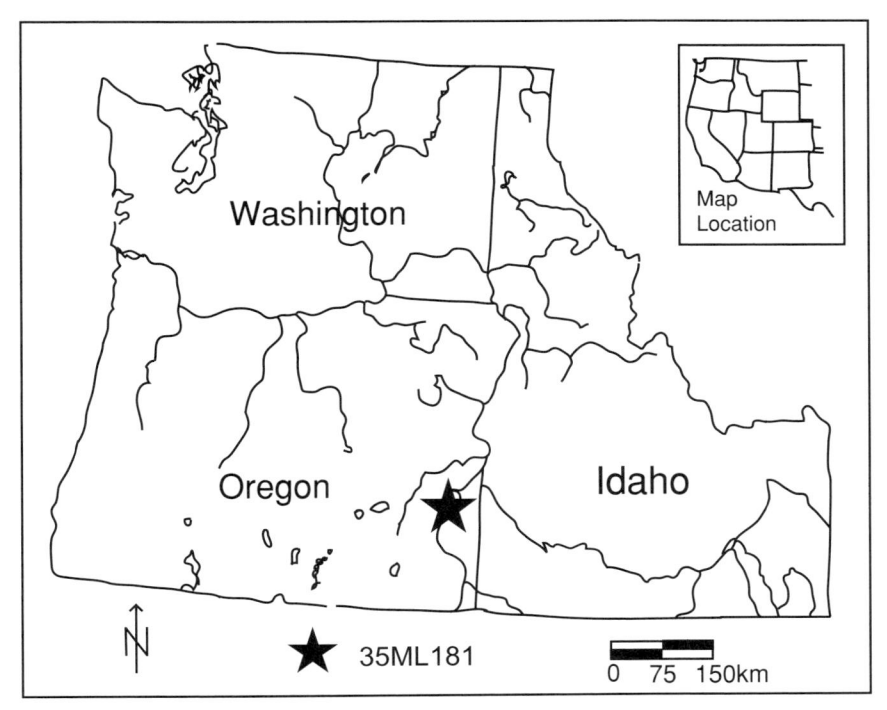

FIGURE 9.1. Birch Creek site location.

4030 and 2335 B.P. (Beta 130362, 130363, 165497). The house pit and the pre–house pit occupation had the same kinds of adaptive strategies, based upon stone tool composition, recovered faunal remains, and raw material acquisition (Andrefsky et al. 2003; Centola 2004; Wallace 2004). For approximately 3,000 years the site location was used (perhaps intermittently) as a winter season residence camp. During that course of time, not only were the same lithic raw material sources used, but those sources were used in exactly the same relative amounts (Wallace 2004). Additionally, site occupants during this course of time made the same relative amounts of chipped stone and ground stone tool types (Centola 2004; Cowan 2006). These patterns of stone tool use and lithic raw material selection suggest there was a great amount of continuity in hunter–gatherer adaptive strategies over a long period of time.

Over 200 hafted biface specimens were recovered from the two occupations at the Birch Creek site (Andrefsky et al. 2003). In an effort to understand hafted biface provisioning strategies, only the obsidian

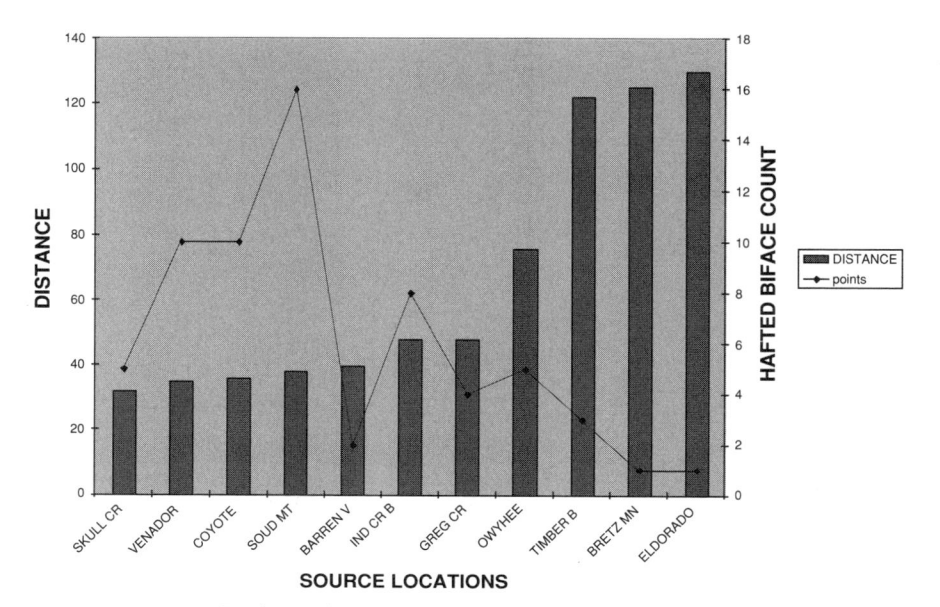

FIGURE 9.2. Birch Creek site hafted biface frequency and distances to obsidian source locations.

hafted bifaces that had been successfully linked to a known source location were analyzed in this study ($n = 52$). Chert and obsidian hafted bifaces without a known source location were not included in the study because they could not provide reliable information about circulation range. Figure 9.2 shows the frequency of hafted bifaces used in the study, along with the obsidian sources and the distance from each obsidian source to the residence camp. These data reveal some interesting patterns associated with regional circulation of aboriginal stone tool makers and users occupying the Birch Creek site. Notice that the closest obsidian source is 32 km away and the most distant source is 130 km away. Also notice that there is a fairly wide gap in obsidian source distances between the Gregory Creek and Owyhee sources. The Gregory Creek source is 48 km from the Birch Creek site and the Owyhee source is 76 km distant. In general, the obsidian sources less than 40 km from the Birch Creek site were used to make most of the hafted bifaces found at the site. The farther the obsidian source was from the residential site, the less often it was used to make hafted bifaces. Interestingly, 30–40 km is ethnographically within the normal daily one-way circulation range of Paiutes in the Great Basin

Table 9.1. Impact damage on hafted bifaces by obsidian source ($\chi^2 = 8.945$; d.f. = 1; $p < .0005$)

Obsidian source	Impact damage	
	Yes	No
Near	33	9
Distant	3	7

area (Fowler 1982; Kelly 1964). Distances greater than about 40 km are often included in the hunter–gatherer foraging ranges but are often greater than two days' travel time to go out and return to camp.

If we partition the obsidian sources into near and distant sources based upon the two-day travel range of Great Basin hunter – gatherers, we have a near group of obsidian sources between 32 and 48 km (Skull Creek, Venador, Coyote Wells, Sour Dough Mt., Barren Valley, Indian Creek Butte, Gregory Creek) and a distant group of obsidian sources between 76 and 130 km away from the site (Owyhee, Timber Butte, Bretz Mine, Eldorado).

HAFTED BIFACES AND IMPACT DAMAGE

Impact damage on hafted bifaces is potentially a very important information-laden characteristic. Not only does it suggest the function of the hafted biface as a projectile, but also it reveals information about the context in which the hafted biface was used. When archaeologists recover hafted bifaces with impact damage, it suggests that tool makers and users were in a situation that allowed them to discard the damaged tool – presumably to be replaced by another projectile tip. Under what contexts are impact-damaged hafted bifaces discarded and replaced? And what are the circumstances that lead some tool makers and users to resharpen hafted bifaces that have impact damage?

Table 9.1 shows an interesting and highly significant relationship between these two obsidian source distances and impact damage on hafted bifaces. Hafted bifaces from distant sources tend to have no impact damage and hafted bifaces from near sources tend to be associated with impact damage ($\chi^2 = 8.945$; d.f. = 1; $p < .0005$). I suggest that this pattern is not necessarily related to artifact function. In other words, distant obsidian and near obsidian made into hafted bifaces do not have different functional properties. I suggest that they are both used as projectile tips and as cutting tools. I also suggest that distant obsidian in hafted biface form shows significantly less impact damage because those specimens were made and used at locations greater than 1–2 days' distance from the residence, and when those specimens were damaged by impact fracture they were replaced and/or resharpened in the field. This stands in contrast to closer proximity obsidian in hafted biface form, which was not replaced or resharpened in the field, but instead was brought back to the residence and replaced as needed from obsidian gathered within a day's journey or cached directly at the residence site.

Aboriginal tool makers and users elected to retool at the residence site if they were within a day or two of the residence site when their hafted bifaces were damaged. However, if they were more than a day or two from the residence site, they elected to discard or resharpen their hafted bifaces while in the field. Of course, the discarded specimens would not be found at the residence location because they would have been discarded while away from the residence. However, the hafted bifaces with impact damage that could not be replaced in the field would potentially show signs of being resharpened or reworked.

If this hafted biface provisioning scenario is correct, we would expect to see hafted bifaces made from distant sources have more retouch evidence than those from near sources. To assess this expectation, I tabulated hafted biface data based upon the presence or absence of retouch on the blade element. Not all hafted bifaces in the study contained blade elements, because some were only base elements when discarded. Retouch was identified as present if the blade element showed twisted beveling, noticeably irregularly shaped lateral margins, or significantly shortened blade length from the mean blade length for that style (Andrefsky 2006; Ballenger 1998; Hoffman 1985; Nowell et al. 2003).

Figure 9.3 shows examples of hafted bifaces with configurations of retouch and also shows examples of impact-damaged specimens. Some specimens were snapped as a result of impact damage. The top specimens are examples of hafted bifaces that have resharpened blades with irregular margin profiles and with twisted beveling. The lower

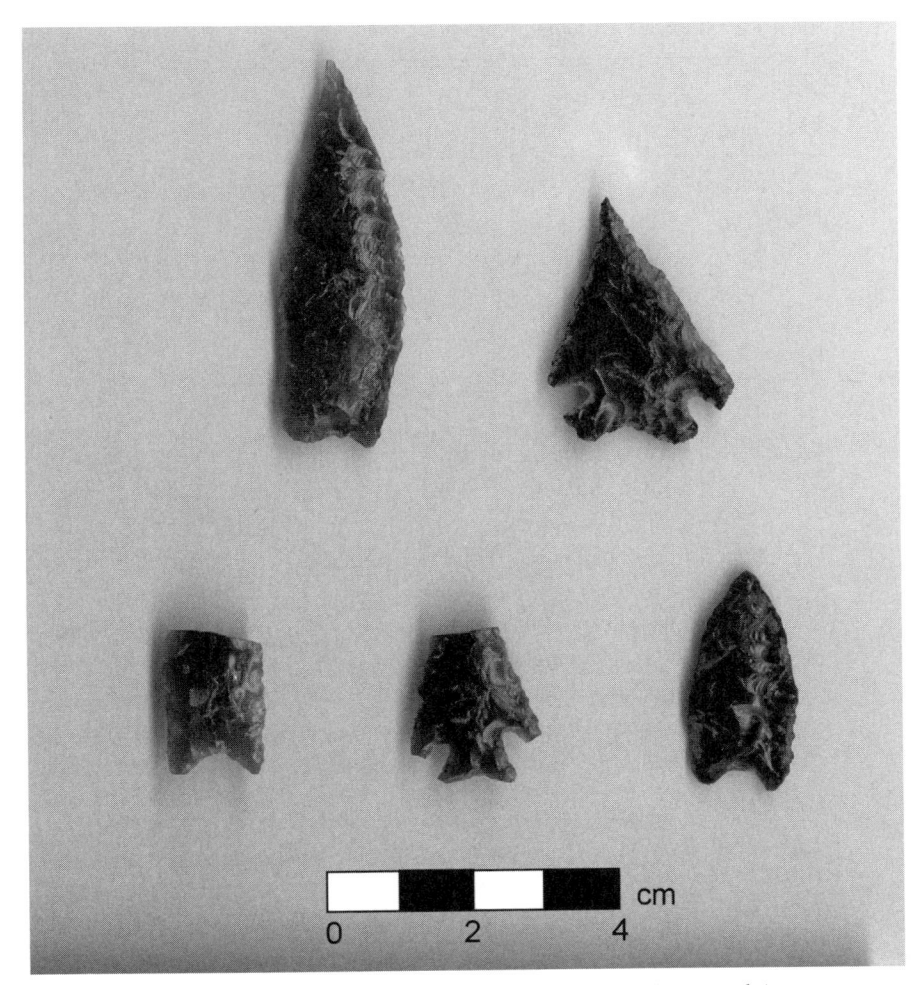

FIGURE 9.3. Examples of hafted bifaces with resharpening evidence and impact damage.

right specimen is an example of a hafted biface with impact damages that was resharpened to produce a usable tip after the impact damage.

Table 9.2 shows the frequency of retouch presence and absence for all hafted bifaces with blade elements and for hafted bifaces that show no impact damage. In both cases retouch is significantly associated with distance from obsidian source location ($\chi^2 = 10.602$; d.f. = 1; $p = .001$ and $\chi^2 = 9.679$; d.f. = 1; $p = .005$, respectively). Also note

Table 9.2. Retouch on hafted bifaces by obsidian source

Obsidian source	Retouch	
	Yes	No
All hafted bifaces ($\chi^2 = 10.602$; d.f. = 1; $p = .001$)		
Near	11	30
Distant	7	1
Hafted bifaces with no impact damage ($\chi^2 = 9.679$; d.f. = 1; $p = .005$)		
Near	2	7
Distant	7	0

that 100% of the distant hafted bifaces that showed no impact damage had evidence of resharpening. Only one of the distant hafted bifaces that had impact damage was not resharpened or discarded in the field. I suggest this specimen was kept because the impact damage caused its base to fracture, but there was enough of the specimen remaining that it could have been recycled in a pinch (Figure 9.4). Obviously it was discarded once the travelers reached the residence camp, the Birch Creek site. The other specimen is an example of a hafted biface base after impact damage. Specimens of this type were not included in the assessment of retouch because no blade remains for use as a cutting tool.

Certainly some hafted bifaces made of near obsidian sources are resharpened and brought back to residence camp. I think this relates to hafted bifaces as multifunctional tools. Not only are they used as projectile tips (as is evident with the impact damage), but they are also used as cutting tools (even the smaller ones). Given that these tools are multifunctional in character, it is not unreasonable to see resharpening on lateral margins. However, the extent to which hafted bifaces are resharpened, particularly those that break from impact damage, has much to do with how far the travelers are from their residence camp and their tool provisioning needs while away.

Another implication of this hafted biface provisioning strategy is that we would expect that hafted bifaces made at the residence camp would be manufactured from nearby sources of obsidian that was gathered as needed or cached at the site. Such production detail should be apparent from debitage and other retouched pieces. Table 9.3 lists all

FIGURE 9.4. Birch Creek site obsidian hafted bifaces with impact damage.

nonbifacial obsidian that was sourced at the residence camp. All of the production debris originates from the nearby source areas. No distant obsidian in debitage form was sourced from the site assemblage. This does not mean that there was none – only that of the sixty-one nonbifacial specimens sourced, all were from obsidian sources within a day or two of the residence site. This too supports the hafted biface provisioning scenario outlined above.

RESIDENCE CAMPS AND QUARRY CAMP LOCATIONS

The hafted biface provisioning strategy anticipates different kinds of hafted bifaces based upon distance and retouch patterns at residence camps versus quarry locations or camps away from the residence area. Unlike the hafted biface pattern of retouch and impact damage found

Table 9.3. Production debitage and tools by obsidian source

Obsidian source	Proximal flakes	Flake shatter	Cores and retouched flakes
Near	36	20	5
Distant	0	0	0

at the Birch Creek site, I would expect that quarry locations would contain few complete hafted bifaces, and those that were complete would have evidence of extensive resharpening because of use. I would also expect that quarry locations would have a relatively high frequency of hafted bifaces with impact damage that were discarded at the quarry during retooling efforts. Such a retooling scenario at quarry locations has been suggested in the archaeological literature before (Binford 1977; Gramly 1980; Hess 1999; Hester and Shafer 1987).

Unfortunately I do not have data collected from the raw materials source areas used by the inhabitants of the Birch Creek site. However, the Paulina Lake site in central Oregon is interpreted as a camp adjacent to a quarry area where aboriginal tool makers and users came to primarily replenish their tool kits (Connolly 1999), which is what I would also expect of quarry areas used by inhabitants of the Birch Creek site.

Figure 9.5 shows the frequency of hafted bifaces discarded at the Paulina Lake site from various source locations. This distribution is surprisingly similar to what was found at the Birch Creek site. In general, the closer obsidian sources were used to produce most of the hafted bifaces. Obsidian sources more than approximately 40 km from the site tended to have considerably fewer hafted bifaces represented in the collection.

However, the pattern of impact damage on hafted bifaces is considerably different at the two locations. The residence site at Birch Creek showed a significant relationship between sources and lack of impact damage. The quarry camp at Paulina Lake shows no such trend. Table 9.4 shows that almost all hafted bifaces at Paulina Lake have impact damage regardless of how far they are from the site. Note the very different impact damage pattern of hafted bifaces from the Birch Creek Site.

As predicted, residence camps and quarry camp locations have different hafted biface production, consumption, and discard patterns based upon foraging patterns and land-use strategies. These combined data suggest that hafted biface discard and use (including retouch) have much to do with the known availability of lithic raw materials as foraging hunters and gatherers make their way on short and long forays from their residence locations. These data show that hafted bifaces tend to have their use lives extended through retouch in contexts where

PROJECTILE POINT PROVISIONING STRATEGIES   205

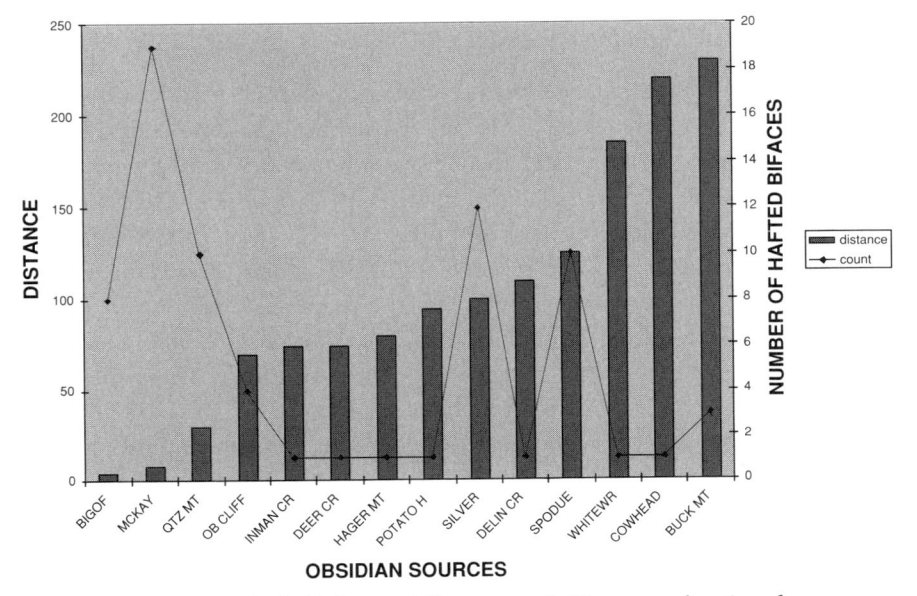

FIGURE 9.5. Numbers of hafted bifaces and distances to obsidian source locations from the Paulina Lake site.

foragers are some distance from their residence camps and are not able to retool at quarry locations. When foragers are closer to residence locations they tend not to retouch their hafted bifaces while in the field, but instead bring those damaged hafted bifaces to their residences for retooling.

Table 9.4. Impact damage on hafted bifaces by obsidian source

	Impact damage	
Obsidian source	Yes	No
Paulina Lake site ($\chi^2 = 0.264$; d.f. = 1; $p > .500$)		
Near	23	6
Distant	27	5
Birch Creek site ($\chi^2 = 8.945$; d.f. = 1; $p < .0005$)		
Near	33	9
Distant	3	7

SOURCE DISTANCES AND RETOUCH INTENSITY

If the provisioning scenario is correct, we should see one more pattern in the hafted biface data related to retouch. The hafted biface provisioning strategy described above suggests that at residence locations we should find hafted bifaces with progressively less retouch amount as obsidian sources get progressively closer to the residence location, and that more retouch should be found on specimens that are progressively farther away from the residence location. So far we have only examined retouch based upon presence or absence values as determined by blade irregularities and shortening. These measures do not assess progressive retouch values.

There are many techniques for measuring retouch amount on flake tools such as scrapers and knives (Blades 2003; Clarkson 2002; Davis and Shea 1998; Dibble 1997; Dibble and Pelcin 1995; Eren et al. 2005; Kuhn 1990; Morrow 1997; Shott et al. 2000). However, there are very few techniques available for measuring retouch amount on hafted bifaces (but see Hoffman 1985). The technique used to assess retouch amounts on hafted bifaces in this study has been explained and tested elsewhere (Andrefsky 2006), but it is worth briefly describing here to better understand hafted biface provisioning strategies at the Birch Creek.

The hafted biface retouch index (HRI) computes the overall amount of retouch along the lateral edges of the blade elements on hafted bifaces. In this case retouch is defined as secondary chipping along the edge that is found over the original or previous flake scars. In most cases secondary retouch is applied to the cutting edge in an effort to straighten the cutting surface or to renew the dulled margins. The HRI is measured only on the blade and not the haft element of the specimen. The blade element is partitioned into sixteen segments (eight on each face of the specimen). Each segment is assessed with a value based upon the appearance of edge resharpening within the segment. Segments that are dominated by flake scars originating from the bifacial edge and extending to the midline or beyond are given a value of zero. A value of zero is also given to those segments where the original flake scars do not extend to the midline, but instead meet flake scars that originate from the opposite margin. Essentially both cases represent original tool trimming without resharpening and are

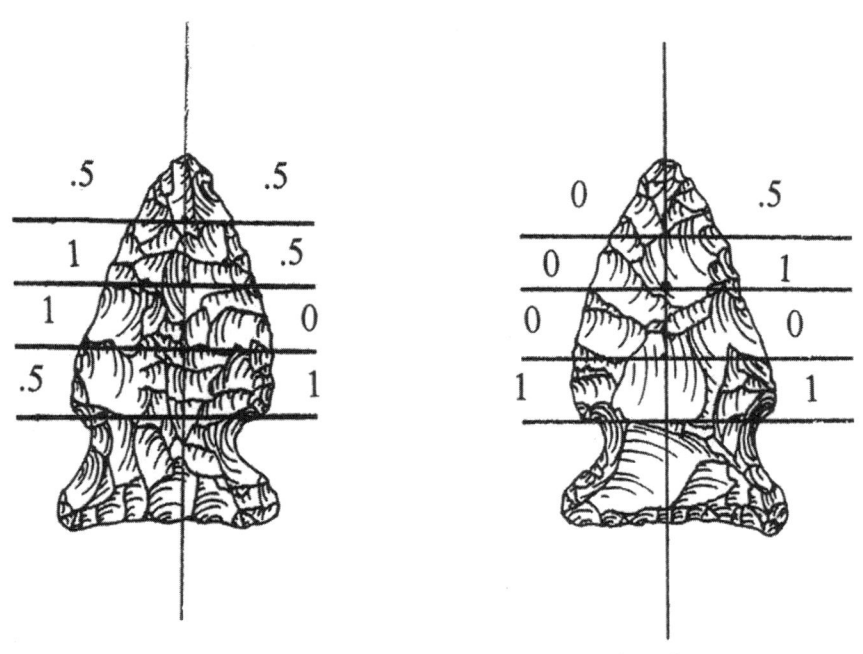

FIGURE 9.6. Example of a hafted biface with a calculated HRI value of 0.5312.

each given a value of zero. Segments where the entire edge contains resharpening flake scars, or flake scars that do not extend to the midline or to flake scars originating from the opposite lateral margin, are given a value of one. Segments that contain roughly equal amounts of retouch flake scars and flake scars that extend to the midline are given a value of 0.5. Both sides of the biface are assessed in this manner for a total of sixteen segments. The HRI is then calculated as the sum of all section scores divided by the total number of sections (sixteen). Figure 9.6 shows an example of a side-notched hafted biface with a HRI of 0.5312. In this case the total value of all segments is summed to 8.5. This value is divided by the total number of segments (16) to arrive at 0.5312 for the HRI. Because the blade element for hafted bifaces is partitioned into sixteen segments and each segment is scored with a standardized value (0, 0.5, or 1.0), all hafted bifaces can be compared to one another with the HRI regardless of the sizes of various blade elements. By dividing the total score of all segments by the number of segments, the HRI values are theoretically standardized from "0" (no retouch) to "1" (completely retouched).

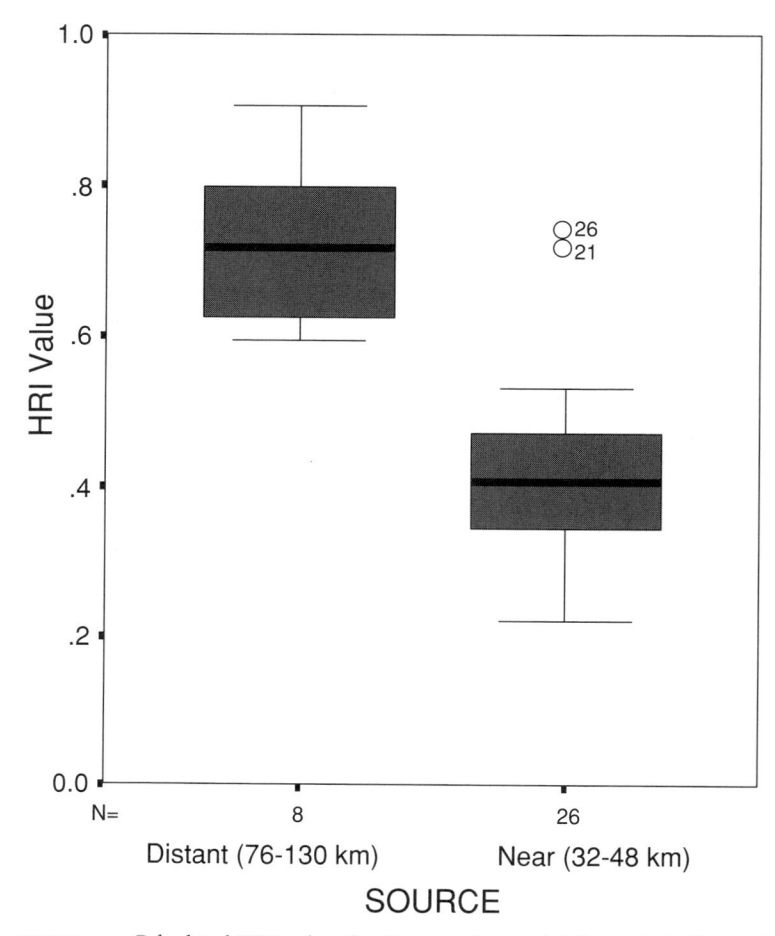

FIGURE 9.7. Calculated HRI values for distant and near obsidian in hafted biface form.

The HRI values for bifaces from near and distant sources were calculated and plotted on Figure 9.7. As expected, HRI values are generally lower for hafted bifaces made from relatively closer obsidian sources. This again supports the proposition that hafted bifaces used on foraging trips farther away from the residence camps will be drafted into service for longer use lives by resharpening the specimen (even those that have been damaged by impact). Shorter foraging trips (within 40 km) result in the damaged biface being brought back to the residence for discard and replacement by locally available obsidian. These bifaces tend not to be resharpened.

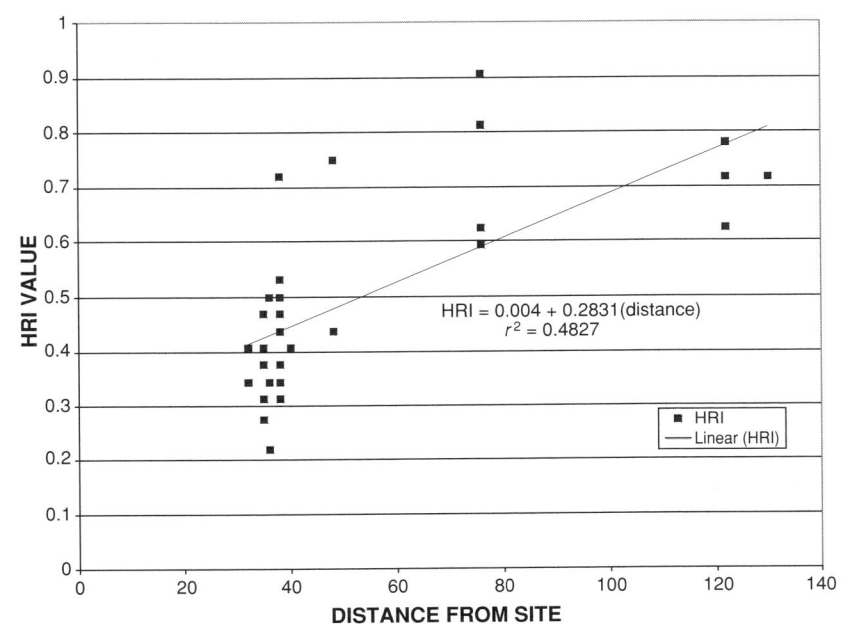

FIGURE 9.8. Calculated HRI values for obsidian at different distances from the Birch Creek site.

Also, given our provisioning scenario, we would expect to see a progressively greater amount of retouch based upon distance from sources. Figure 9.8 plots HRI values against distance from the Birch Creek site. With some dispersion around the linear regression, there is still a positive and significant association of retouch intensity and obsidian source locations ($F = 29.865$; d.f. $= 1$; $p < .0005$). Again, hafted bifaces made from distant sources tend to have greater resharpening values than hafted bifaces made from near sources.

DISCUSSION AND SUMMARY

Hafted biface configuration and use-life modification have been directly associated with artifact functional requirements. Hafted bifaces have been found in the archaeological record attached to arrow and dart shafts (Dixon et al. 2005; Elston 1986). This evidence, along with impact damage on specimens (Odell and Cowan 1986; Truncer 1990), use wear analysis (Greiser 1977; Kay 1996), and ethnographic analogy (Kelly and Fowler 1986; Witthoft 1968), has clearly linked hafted biface forms to functional properties of the artifact. However, in this study I have attempted to relate the production, consumption, and discard of hafted bifaces to issues outside of those specifically related to artifact function. I have shown that hafted bifaces are produced and consumed within a context of adaptive strategies, and it is these human organizational parameters that influence the final disposition and to a certain extent the final configuration of hafted bifaces.

Yes, impact damage on hafted bifaces suggests that the specimen was used as a projectile. Yes, marginal resharpening of hafted bifaces suggests that the specimen was used as a cutting tool. But why are some hafted bifaces discarded after being damaged by impact and others resharpened and reused? What are the conditions and contexts under which tool makers and users choose to resharpen a damaged hafted biface instead of replacing it? I have argued that human organizational strategies are critical for understanding hafted biface provisioning.

In this particular case, I have shown that lithic raw material proximity plays an important role in hafted biface retouch and/or discard. Foragers circulating in their resource range greater than two days distance will tend to retool if they opportunistically or intentionally encounter usable tool stone. This is evident at quarry camps such as Paulina Lake, where almost all hafted bifaces are damaged and discarded. If they do not encounter usable tool stone while on distant journeys, they will tend to resharpen and draft hafted bifaces into a longer service life. This is evident from the hafted biface assemblage at the Birch Creek site, where distant raw materials in hafted biface form show significantly more retouch and reconfiguration than hafted bifaces made from locally available raw materials.

The impact damage and resharpening trends evident from this study are not necessarily universal to all forager residence camps. The production, resharpening, and discard patterns of hafted bifaces from the Birch Creek site assemblage are unique to the Birch Creek site. Other sites with the same kinds of resource availability may show the same trends. However, artifact provisioning strategies are sensitive to human organizational contexts, and any particular site location may have a different context of use. It is important to remember that lithic artifact patterning, whether it is hafted biface impact damage or debitage size distributions, does not fit universal behavioral expectations. The reason we do not have a one-to-one fit between lithic artifact distributions and human behavior is that lithic technology is highly influenced by human organizational strategies, including raw material availability, aboriginal adaptive practices, and environmental constraints. This complicated association of human adaptive practices and lithic technological strategies is the very reason lithic artifacts are useful for interpreting aspects of human organizational strategies. In the case of the Birch Creek site, we can understand why some hafted bifaces were resharpened and discarded at the site when others were simply discarded upon breakage when we know something about the circulation range of Birch Creek occupants via source location studies of tool stone.

REFERENCES CITED

Ahler, Stanley A. 1971. *Projectile Point Form and Function at Roger's Shelter, Missouri*. College of Arts and Science, University of Missouri – Columbia and the Missouri Archaeological Society, Columbia, Missouri.

Andrefsky, William, Jr. 1994. Raw Material Availability and the Organization of Technology. *American Antiquity* 59:21–35.

2005. *Lithics: Macroscopic Approaches to Analysis.* 2nd ed. Cambridge University Press, Cambridge.

2006. Experimental and Archaeological Verification of an Index of Retouch for Hafted Bifaces. *American Antiquity* 71:743–58.

Andrefsky, William, Jr., Lisa Centola, Jason Cowan, and Erin Wallace, eds. 2003. *An Introduction to the Birch Creek Site: Six Seasons of Washington State University Archaeological Study*. Center for Northwest Anthropology. Contributions in Cultural Resource Management. No. 69. Washington State University, Pullman.

Ballenger, Jesse. 1998. The McKellips Site: Contributions to Dalton Occupation, Technology, and Mobility from Eastern Oklahoma. *Southeastern Archaeology* 17:158–65.

Binford, Lewis R. 1977. Forty-seven Trips. In *Stone Tools as Cultural Markers*, edited by R. S. V. Wright, pp. 24–36. Australian Institute of Aboriginal Studies, Canberra.

Blades, Brooke S. 2003. End Scraper Reduction and Hunter–Gatherer Mobility. *American Antiquity* 68:141–56.

Centola, Lisa. 2004. *Deconstructing Lithic Technology: A Study from the Birch Creek Site (35ML181), Southeastern Oregon*. M.A. thesis, Department of Anthropology, Washington State University, Pullman.

Churchill, Steven E. 1993. Weapon Technology, Prey Size Selection, and Hunting Methods in Modern Hunter–Gatherers: Implications for Hunting in the Paleolithic and Mesolithic. In *Hunting and Animal Exploitation in the Later Paleolithic and Mesolithic of Eurasia*, edited by Gail Larson Peterkin, Harvey M. Bricker, and Paul Mellars, pp. 11–24. Archaeological Papers of the American Anthropological Association, Number 4, Washington D.C.

Clarkson, Chris. 2002. An Index of Invasiveness for the Measurement of Unifacial and Bifacial Retouch: A Theoretical, Experimental and Archaeological Verification. *Journal of Archaeological Science* 29:65–75.

Connolly, Thomas J. 1999. *Newberry Crater: A Ten-Thousand Year Record of Human Occupation and Environmental Change in the Basin-Plateau Borderlands*. University of Utah Anthropological Papers, No. 121. University of Utah, Salt Lake City.

Cowan, Jason. 2006. *Grinding It Out: A Temporal Analysis of Ground Stone Assemblage Variation at the Birch Creek Site (35ML181) in Southeastern Oregon*. M.A. Thesis, Department of Anthropology, Washington State University, Pullman.

Daniel, I. Randolph, Jr. 2001. Stone Raw Material Availability and Early Archaic Settlement in the Southeastern United States. *American Antiquity* 66:237–66.

Davis, Z. J., and J. J. Shea. 1998. Quantifying Lithic Curation: An Experimental Test of Dibble and Pelcin's Original Flake-Tool Mass Predictor. *Journal of Archaeological Science* 25:603–10.

Dibble, Harold L. 1997. Platform Variability and Flake Morphology: A Comparison of Experimental and Archeological Data and Implications for Interpreting Prehistoric Lithic Technological Strategies. *Lithic Technology* 22:150–70.

Dibble, Harold L., and Andrew Pelcin. 1995. The Effect of Hammer Mass and Velocity on Flake Mass. *Journal of Archaeological Science* 22:429–39.

Dixon, E. James, William F. Manley, and Craig M. Lee. 2005. The Emerging Archaeology of Glaciers and Ice Patches: Examples from Alaska's Wrangell–St. Elias National Park and Preserve. *American Antiquity* 70:129–43.

Ellis, Christopher J. 1997. Factors Influencing the Use of Stone Projectile Tips: An Ethnographic Perspective. In *Projectile Technology*, edited by Heidi Knecht, pp. 37–78. Plenum Press, New York.

Elston, Robert G. 1986. Prehistory of the Western Area. In *Handbook of North American Indians*. Volume 11. *Great Basin*, edited by Warren L. D'Azevedo (volume editor), pp. 135–48. Smithsonian Institution Press, Washington, DC.

Eren, Metin I., Manual Dominguez-Rodrigo, Steven L. Kuhn, Daniel S. Adler, Ian Le, and Ofer Bar-Yosef. 2005. Defining and Measuring Reduction in Unifacial Stone Tools. *Journal of Archaeological Science* 32:1190–1206.

Flenniken, J. Jeffrey, and Philip J. Wilke. 1989. Typology, Technology, and Chronology of Great Basin Dart Points. *American Anthropologist* 91:149–58.

Fowler, Catherine S. 1982. Settlement Patterns and Subsistence Systems in the Great Basin: The Ethnographic Record. In *Man and Environment in the Great Basin*, edited by D. B. Madsen and J. F. O'Connell, pp. 121–38. Society for American Archaeology Press, Washington, DC.

Frison, George C. 1991. *Prehistoric Hunters of the High Plains*. 2nd ed. Academic Press, New York.

Gramly, R. Michael. 1980. Raw Material Source Areas and "Curated" Tool Assemblages. *American Antiquity* 45:823–33.

Greaves, Russel D. 1997. Hunting and Multifunctional Use of Bows and Arrows: Ethnoarchaeology of Technological Organization among Pume' Hunters of Venezuela. In *Projectile Technology*, edited by Heidi Knecht, pp. 287–320. Plenum Press, New York.

Greiser, Sally T. 1977. Micro-Analysis of Wear Patterns on Projectile Points and Knives from the Jurgens Site, Kersey, Colorado. *Plains Anthropologist* 22:107–16.

Hess, Sean C. 1999. *Rocks, Range, Renfrew: Using Distance–Decay Effects to Study Late Pre-Mazama Period Obsidian Acquisition and Mobility in Oregon and Washington*. Ph.D diss., Washington State University, Pullman.

Hester, Thomas R., and L. M. Green. 1972. Functional Analysis of Large Bifaces from San Saba County, Texas. *The Texas Journal of Science* 24:343–50.

Hester, Thomas R., and Harry J. Shafer. 1987. Observations on Ancient Maya Core Technology at Colha, Belize. In *The Organization of Core Technology*, edited by J. K. Johnson and C. A. Morrow, pp. 239–58. Westview Press, Boulder, CO.

Hoffman, C. Marshall. 1985. Projectile Point Maintenance and Typology: Assessment with Factor Analysis and Canonical Correlation. In *For Concordance in Archaeological Analysis: Bridging Data Structure, Quantitative Technique, and Theory*, edited by C. Carr, pp. 566–612. Westport Press, Kansas City.

Kay, Marvin. 1996. Microwear Analysis of Some Clovis and Experimental Chipped Stone Tools. In *Stone Tools: Theoretical Insights into Human Prehistory*, edited by George Odell, pp. 315–44. Plenum Press, New York.

Kelly, Isabel T. 1964. *Southern Paiute Ethnography*. University of Utah Anthropological Papers 69. University of Utah, Salt Lake City.

Kelly, Isabel T., and Catherine S. Fowler. 1986. Southern Paiute. In *Handbook of North American Indians*. Volume 11. *Great Basin*. Edited by Warren L. D'Azevedo (volume editor), pp. 368–97. Smithsonian Institution Press, Washington, DC.

Kelly, Robert L. 1988. The Three Sides of a Biface. *American Antiquity* 53:717–34.

Kuhn, Steven L. 1990. A Geometric Index of Reduction for Unifacial Stone Tools. *Journal of Archaeological Science* 17:585–93.

Morrow, Juliet. 1997. End Scraper Morphology and Use-Life: An Approach for Studying Paleoindian Lithic Technology and Mobility. *Lithic Technology* 22:70–85.

Nowell, April, Kyoungju Park, Dimitris Mutaxas, and Jinah Park. 2003. Deformation Modeling: A Methodology for the Analysis of Handaxe Morphology and Variability. In *Multiple Approaches to the Study of Bifacial Technologies*, edited by Marie Soressi and Harold L. Dibble, pp. 193–208. University of Pennsylvania Museum of Archaeology and Anthropology, Philadelphia.

Odell, George H., and Frank Cowan. 1986. Experiments with Spears and Arrows on Animal Targets. *Journal of Field Archaeology* 13(2):195–212.

Sassaman, Kenneth E. 1994. Changing Strategies of Biface Production in the South Carolina Coastal Plain. In *The Organization of North American Prehistoric Chipped Stone Tool Technologies*, ed. P. J. Carr, pp. 99–117. International Monographs in Prehistory: Archaeological Series 7. University of Michigan Press, Ann Arbor.

Shott, Michael J., Andrew P. Bradbury, Philip J. Carr, and George H. Odell. 2000. Flake Size from Platform Attributes: Predictive and Empirical Approaches. *Journal of Archaeological Science* 27:877–94.

Tankersley, Kenneth B. 2000. The Effects of Stone and Technology on Fluted-Point Morphometry. *American Antiquity* 59:498–509.

Tomka, Steve A. 2001. The Effect of Processing Requirements on Reduction Strategies and Tool Form: A New Perspective. In *Lithic Debitage: Context, Form, Meaning.* edited by Wm. Andrefsky, Jr., pp. 207–24. University of Utah Press, Salt Lake City.

Truncer, James J. 1988. Perkiomen Points: A Functional Analysis of a Terminal Archaic Point Type in the Middle Atlantic Region. *Journal of Middle Atlantic Archaeology* 4:61–70.

1990. Perkiomen Points: A Study in Variability. In *Experiments and Observations on the Terminal Archaic of the Middle Atlantic Region*, edited by R. W. Moeller, pp. 1–62. Archaeological Services, Bethlehem, CT.

Wallace, Erin. 2004. *Obsidian Projectile Points and Human Mobility around the Birch Creek Site (35ML181), Southeastern Oregon.* M.A. thesis, Department of Anthropology, Washington State University, Pullman.

Witthoft, John. 1968. Flint Arrowpoints from the Eskimo of Northwestern Alaska. *Expedition* 10:1–37.

10 DOUGLAS H. MACDONALD

THE ROLE OF LITHIC RAW MATERIAL AVAILABILITY AND QUALITY IN DETERMINING TOOL KIT SIZE, TOOL FUNCTION, AND DEGREE OF RETOUCH: A CASE STUDY FROM SKINK ROCKSHELTER (46NI445), WEST VIRGINIA

Abstract

Analysis of lithic artifact data from Skink Rockshelter (44NI445) in central West Virginia indicates that stone quality and availability were important in determining how Native Americans differentially utilized tools at the site. In turn, tool function influenced lithic raw material selection. Although local Kanawha chert was clearly preferred for projectile point and biface manufacture, nonlocal Upper Mercer chert was preferred for flake tool use. Skink Rockshelter lithic data do not support the original hypothesis of the paper, that expedient flake tool use would increase at the expense of curated tools in the Kanawha chert primary source area. Instead, individuals curated the comparatively high-quality Upper Mercer chert stone tools to the site and continued to use and retouch them, rather than replace them with tools produced from the inferior, but abundant, Kanawha chert. Curation of Upper Mercer flake tools, as well as projectile points, resulted in their markedly reduced sizes and higher hafted biface reduction index (HRI) measures compared to the local Kanawha chert tools.

I would like to thank Bill Andrefsky for inviting me to participate in the Society for American Archaeology symposium in San Juan, Puerto Rico, in 2006. The current paper is a revision of the paper presented at that symposium. Excavations at Skink Rockshelter were funded by Alex Energy, Inc., of Summersville, WV. GAI Consultants in Pittsburgh, PA was my employer during the Skink Rockshelter project and I owe them – especially Ben Resnick, Jon Lothrop, and Diane Landers – a debt of gratitude for their support and friendship between 1999 and 2006. Brent Shreckengost was field director during the excavations at Skink Rockshelter in the winter of 2002–3; the crew included Lisa Dugas, William Hill, Jon Boilegh, Damian Blanck,

INTRODUCTION

Although overlooked in most prior studies, retouch of stone tools is an important component of the wider technological organization strategy of mobile hunter–gatherers. Even though retouch is the main focus of this paper, the ultimate goal is to better understand the means by which hunter–gatherers achieved success in life with the help of stone. In that regard, retouching stone tools is but one facet of a broader risk-minimizing strategy for reducing the chance of failure and controlling future success, given uncertain future travel and subsistence realms.

This paper focuses on how lithic raw material availability and quality affect the degree of retouch on stone tools and overall toolkit size. The main hypotheses are that lithic raw material availability and quality were key factors in determining the size of the tool kit and the extent of tool retouch. In toolstone-rich settings, the size of curated tool kits and the degree of tool retouch may decrease in favor of expedient tool production using locally abundant lithic raw materials. In toolstone-deficient environments, increased curated toolkit size and increased tool retouch likely reduced the risk of tool depletion during forager travel. These hypotheses on retouch versus replacement decisions are tested utilizing stone tool data from Skink Rockshelter (46NI445), a multicomponent, stratified rockshelter in uplands of Nicholas County, West Virginia, as well as from assorted other case studies.

ORGANIZATION OF LITHIC TECHNOLOGY

The lithic technological organization literature is rife with examples showing the relationship between lithic raw material type and forager mobility and settlement patterns (Andrefsky 1994a; Bamforth 1986; Binford 1979; Nelson 1991). Andrefsky (1994b) showed that prehistoric Native Americans of the Columbia Plateau and elsewhere used local lithic raw materials when they were of high quality, workable morphology, and moderate to high availability. Archaeological sites

and Steve Brann. David L. Cremeens provided crucial insights into site formation and geomorphology during the interpretation of Skink Rockshelter soils and stratigraphy. I am also indebted to the University of Montana, Missoula, Department of Anthropology for providing resources during the completion of this paper.

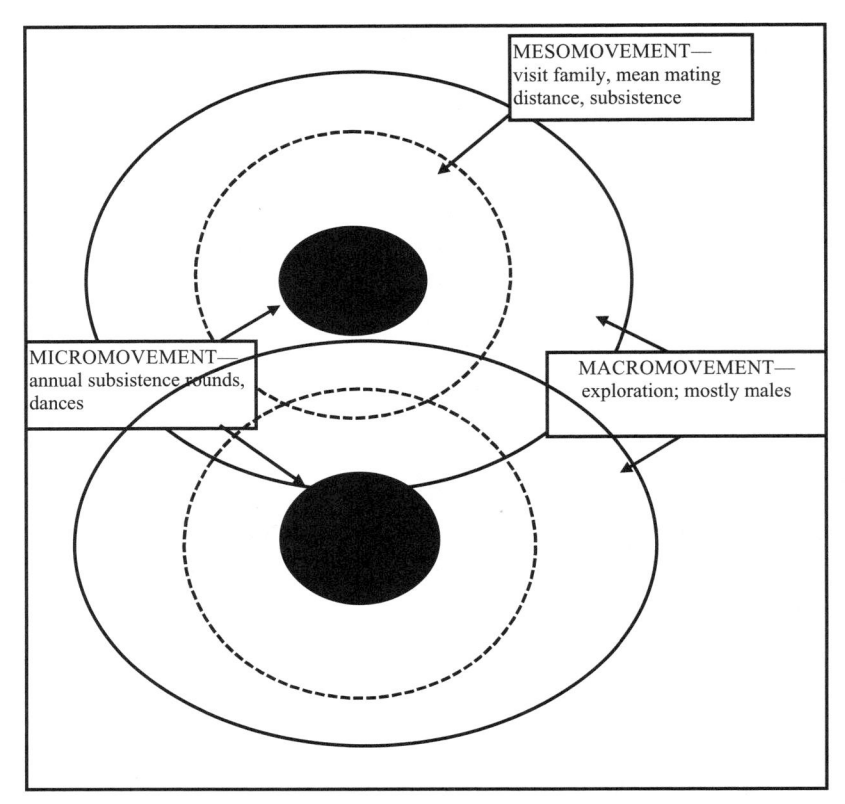

FIGURE 10.1. Schematic three-tier model of hunter–gatherer travel patterns.

in regions such as these invariably will contain high percentages of these local lithic raw materials. In the current paper, local lithic raw materials are defined as those occurring within bedrock or secondary deposits within 5–15 miles of a given archaeological site.

At the opposite end of the spectrum, if local lithic raw materials are scarce and/or of low quality, then foragers will curate higher-quality lithics with them in their travels. Archaeological sites in these areas will thus contain substantial amounts of these moderate- to high-quality lithic materials from 30 or more miles distant, considered here to be semilocal and nonlocal based on distance to source (Figure 10.1).

Applying previous models of forager mobility (Binford 1983; Mandryk 1993; Sampson 1988), I have suggested elsewhere that hunter–gatherers generally organize themselves within a three-tier

THE ROLE OF LITHIC RAW MATERIAL   219

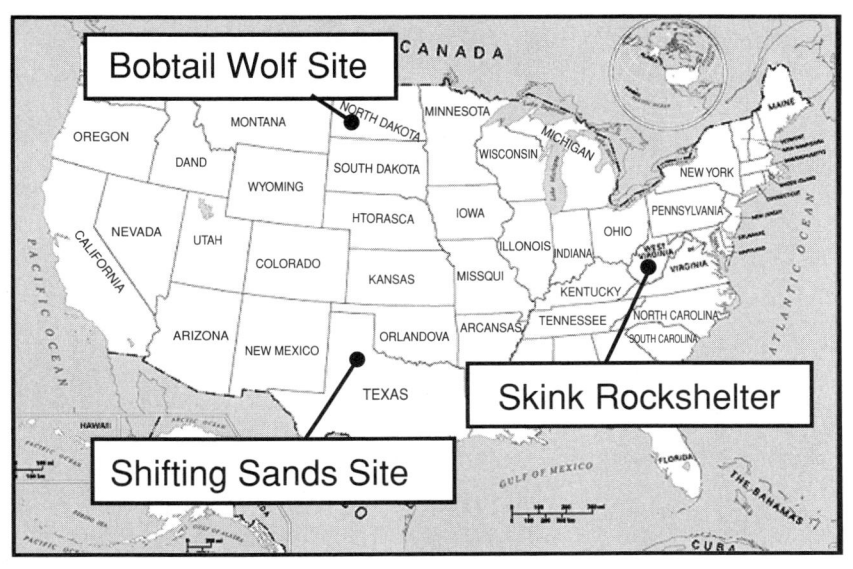

FIGURE 10.2. Location of archaeological sites discussed in text.

mobility realm – local, semilocal, and nonlocal – given various subsistence and social factors (MacDonald and Hewlett 1999; MacDonald et al. 2006) (Figure 10.1). Given the uncertain travel realms, hunter–gatherers moved freely in and out of these three mobility realms according to their needs, whether somatic or reproductive.

By identifying the sources of trace lithic raw materials at sites, archaeologists can better understand hunter–gatherer travel and trade patterns within the semilocal and nonlocal realms. For example, many Folsom-period (ca. 10,900 to 10,200 B.P.) sites in the northern plains of North America yield very small quantities of lithics from sources 100–300 miles distant (MacDonald 1999). However, at the Bobtail Wolf Site in western North Dakota (Figure 10.2), the high-quality local lithic raw material, Knife River flint, was utilized to produce all types of tools (e.g., flake tools, bifaces), whereas other local and semilocal lithics were utilized comparatively sparingly, and nonlocal lithics from distant sources are rare to nonexistent (MacDonald 1999; Root 2000). This example supports Andrefsky's (1994a, 1994b) supposition that locally available lithic materials will be used for all types of lithic tool production activities when the materials are abundant and easily accessible.

FIGURE 10.3. Skink Rockshelter, West Virginia. View south.

Another Folsom site in the plains – Shifting Sands in West Texas (Hofman et al. 1990) – also exemplifies Andrefsky's lithic technological organization predictions, but from the opposite end of the spectrum. At this site, nearly the entire lithic assemblage is composed of nonlocal lithic raw materials – Edwards chert – with lesser amounts of local and semilocal materials. Because local and semilocal lithics were scarce and of low quality, Edwards chert was imported by Folsom foragers in this toolstone-deficient environment.

SKINK ROCKSHELTER BACKGROUND

Of course, many parts of the world – such as central West Virginia – fall in between these two extremes of lithic raw material use. Within the heart of the Kanawha chert primary source area, GAI Consultants investigated Skink Rockshelter during the fall and winter of 2002–3 (MacDonald 2003) (Figures 10.3–10.4). In contrast to Knife River flint in the northern Plains and Edwards chert in the southern Plains, Kanawha chert is generally considered to be a low- to moderate-quality lithic raw material in the Middle Atlantic and Appalachian regions of eastern North America. This dark gray marine flint is a

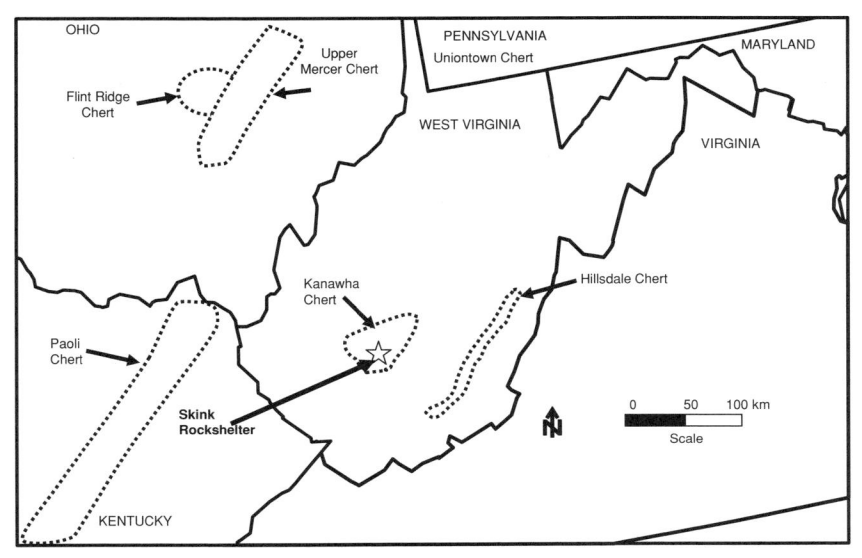

FIGURE 10.4. Location of Skink Rockshelter (46NI445), Nicholas County, West Virginia, in relation to regional lithic raw material sources.

member of the Pennsylvanian Kanawha Formation and occurs in an approximately 1,000-sq. mi. basin in parts of Boone, Kanawha, Clay, Nicholas, Webster, and Fayette Counties, West Virginia (Reger 1921: 227; Reppert 1978: 3).

Although it is of generally low quality, Kanawha chert is nevertheless abundant and occurs in knappable form across the primary source area, as well as in secondary sources throughout alluvial drainages to the west and north. Because of its widespread availability, the stone was used throughout prehistory and is found in high percentages at sites in the primary source area and vicinity in central and western West Virginia (MacDonald and Cremeens 2005).

Excavations at Skink Rockshelter recovered nearly 30,000 lithics from two horizontally stratified occupations (Figure 10.5; Table 10.1). The southern portion of the shelter contained evidence of multiple occupations during the Early Archaic (ca. 9,000–7,000 B.P.) and Late Archaic (ca. 5,500–3,800 B.P.) periods, whereas the northern portion of the shelter contained Late Woodland (ca. 1,500–1,000 B.P.) artifacts. The contrasting use of space at the shelter during the respective occupations was likely due to differential infilling from rock fall and

Table 10.1. Skink Rockshelter artifact summary by component

Component*	Debitage	Biface	Uniface	Core	Other	Pottery	Total	%
Archaic	4038	12	5	18	0	0	4073	13.7
LW	22853	69	44	91	5	41	23103	77.7
Buffer	2547	7	5	5	0	1	2562	8.6
Total	29438	88	51	114	5	42	29738	100.0
Percent	99.0	0.3	0.2	0.4	0.01	0.1	100.0	—

*Archaic includes Early and Late Archaic artifacts (9,000-3,800 BP); LW includes Late Woodland artifacts (1,500-1,000 BP); Buffer Area includes artifacts from test units separating the two horizontally stratified Archaic and Late Woodland components.

colluvium (MacDonald and Cremeens 2005). As reflected in Table 10.1, Late Woodland period (77.7% of artifacts) Native Americans used the site comparatively more intensively than their Early and Late Archaic period (13.7% of artifacts) counterparts.

Although the Late Woodland and Archaic site occupants differentially utilized space within the rockshelter, their lithic raw material

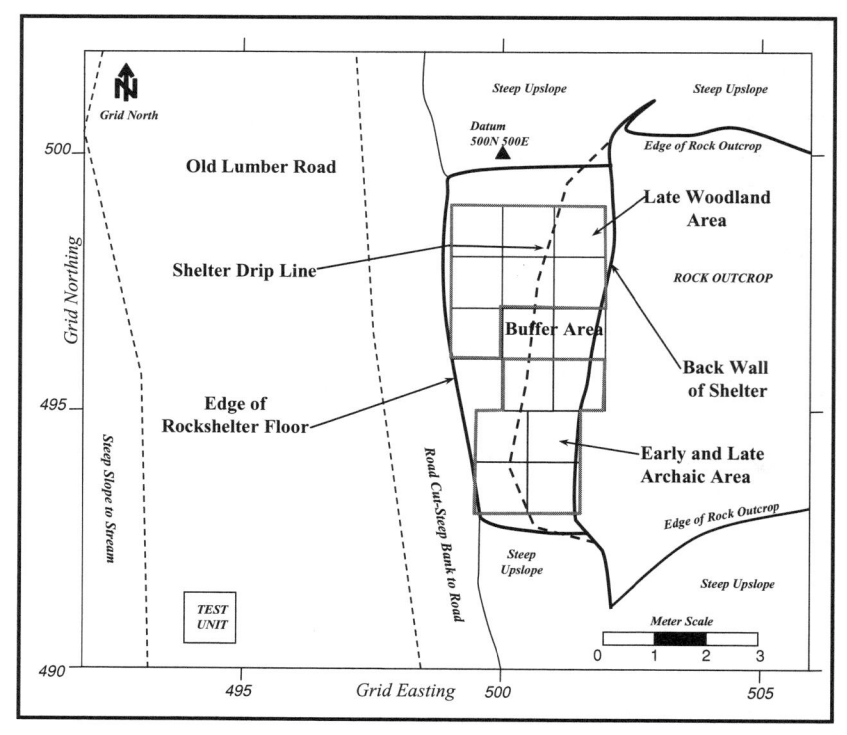

FIGURE 10.5. Differential use of space over time at Skink Rockshelter.

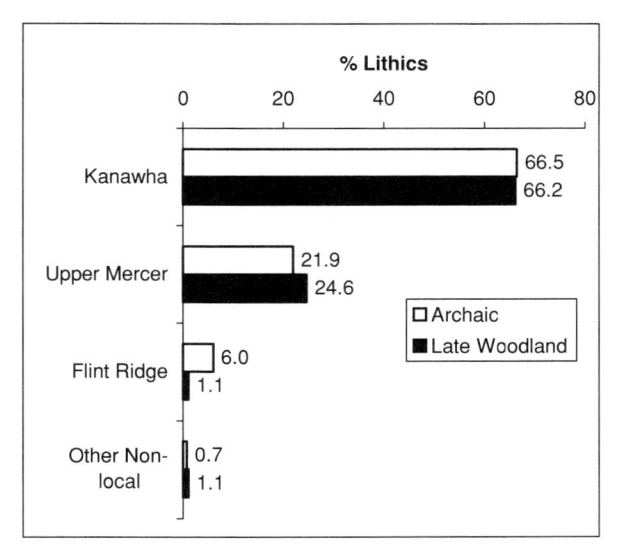

FIGURE 10.6. Lithic raw material use in Archaic and Late Woodland occupations, Skink Rockshelter.

use patterns were extremely similar (Figure 10.6). Because Kanawha chert is widely available, but of fairly poor to moderate quality, other semilocal and/or nonlocal materials were expected to occur in some quantity at Skink Rockshelter. As indicated in Figure 10.6, Upper Mercer chert from eastern Ohio occurs in fairly high percentages during both the Archaic (21.9%) and Late Woodland (24.6%) occupations, suggesting patterned movements and lithic raw material use in this region over much of the Holocene (MacDonald et al. 2006).

Upper Mercer chert derives from the Upper Mercer limestone member of the Lower Pennsylvanian system within Coshocton, Perry, and Miskingham counties of east central Ohio (Kagelmacher 2000). At their most proximate point, Upper Mercer chert primary sources are more than 85 miles northwest of Skink Rockshelter (see Figure 10.4). However, secondary sources of cobble Upper Mercer chert are likely present in far eastern Ohio and, perhaps, in western West Virginia, perhaps within 60 miles of the site. Although the current paper focuses on retouch of stone tools at Skink Rockshelter, MacDonald and Cremeens (2005) and MacDonald et al. (2006) provide additional details regarding lithic raw material sources and their differential use at Skink Rockshelter during the respective Archaic and Late Woodland occupations.

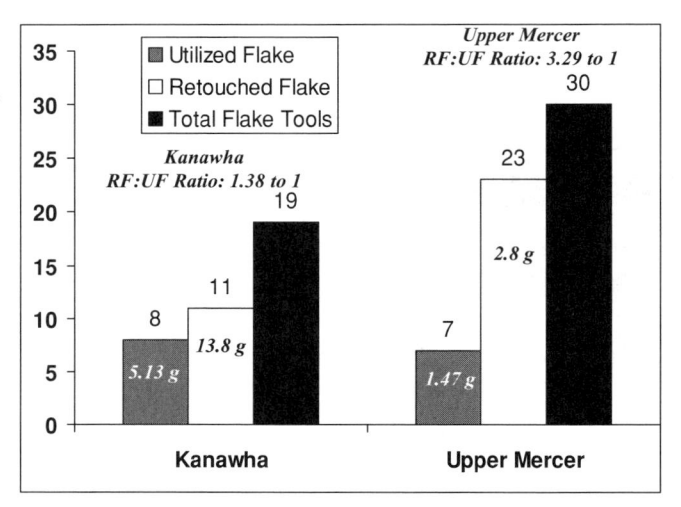

FIGURE 10.7. Ratio of retouched to unifacial tools and weights for Upper Mercer and Kanawha chert flake tools, Skink Rockshelter.

RESULTS

Analysis of stone tool data provides insight into the differential use of Kanawha and Upper Mercer cherts at Skink Rockshelter during the Archaic and Late Woodland occupations. Given its role in the curated tool kit, nonlocal Upper Mercer chert should have a higher ratio of retouched to utilized flakes, whereas the opposite would be expected for the local Kanawha chert. Upper Mercer chert tools were likely curated to the site and preserved in the tool kit via retouching, whereas Kanawha chert tools are more likely to appear as expedient utilized flake tools because of their simple replacement with widely available materials (Andrefsky 1994a; Bamforth 1986).

As shown in Figure 10.7, twenty-three Upper Mercer retouched flakes were recovered compared to only seven utilized flakes, for a ratio of 3.29:1 for the entire site assemblage. For Kanawha chert, eleven retouched flakes and eight utilized flakes were recovered, for a ratio of 1.38:1 for the entire site assemblage. Thus, as predicted, retouched flakes are more common for Upper Mercer than for Kanawha chert. However, unexpectedly, Upper Mercer chert from 60–90 miles northwest was arguably the preferred material for all flake tool use, including utilized flakes. Given its abundance in the landscape, Kanawha chert was predicted to dominate the flake tool assemblage; however, as these

THE ROLE OF LITHIC RAW MATERIAL   225

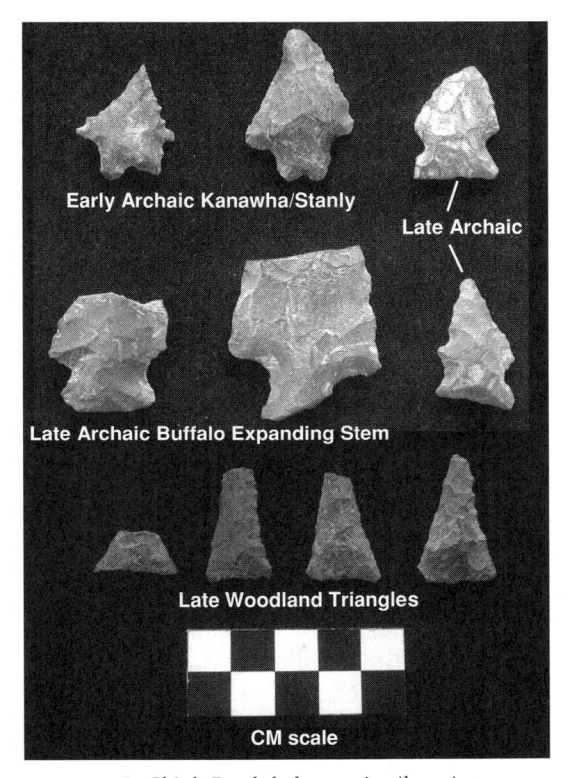

FIGURE 10.8. Skink Rockshelter projectile points.

data show, the nonlocal Upper Mercer chert ($n = 30$ flake tools) was selected for flake tool use more frequently than the local Kanawha chert ($n = 19$ flake tools). As confirmed for several other sites discussed in this volume (Bradbury et al.; Goodale et al.), tool function heavily influenced the lithic raw material selection of individuals at Skink Rockshelter.

The effects of retouch and hafted-biface reduction can also be factored into the evaluation of lithic raw material use variability and toolkit composition at Skink Rockshelter. Using methods defined in this volume and elsewhere (Andrefsky 2006), the hafted biface retouch index (HRI) was calculated for diagnostic Late Woodland and Archaic projectile points recovered from Skink Rockshelter (Figure 10.8). The formula utilized in the analysis is $\text{HRI} = \sum S_i/n$, where S is the sum of retouch indexes for the 16 projectile point segments (n). Because it is assumed that they traveled a longer distance within the foragers' tool kits, the Upper Mercer chert projectile points should have a higher HRI than points produced from locally available Kanawha chert.

The Skink Rockshelter projectile points (Figure 10.8) were largely produced from Kanawha chert ($n = 12$), with a comparatively small number of points produced from nonlocal Upper Mercer chert ($n = 3$) and semilocal Hillsdale chert ($n = 1$). At its most proximate point, Hillsdale chert is found approximately 30–40 miles east of the project area near Lewisburg in Greenbrier County and near Mill Point in Pocahontas County (see Figure 10.4) (Brashler and Lesser 1990: 199).

For the purpose of increasing sample size, the nonlocal and semilocal chert projectile points ($n = 4$) are grouped in this analysis. As predicted, the mean HRI for the Upper Mercer and Hillsdale chert projectile points is 0.578, compared to only 0.453 for Kanawha chert points (Figure 10.9). These HRI data support the hypothesis that projectile points produced from nonlocal (Upper Mercer chert) and semilocal lithic materials (Hillsdale chert) were curated and retouched more extensively than their counterparts produced from local materials (Kanawha chert).

Another measure of comparative lithic raw material use and tool retouch is size variation, including simple measures of weight and dimension. Although detailed measures of retouch, such as HRI and other indices discussed in this volume (Clarkson; Eren and Prendergast; Quinn et al.), are more precise measures of retouch, dimensional and weight measures can be used as supplemental measures of lithic tool reduction.

Given the increased distance to their sources and accompanying higher degree of retouch, we should expect that tools produced from semilocal and nonlocal lithic raw materials – such as Upper Mercer chert at Skink Rockshelter – will have generally reduced sizes compared to their counterparts produced from local materials, such as Kanawha chert in this case.

In confirmation of these predictions, Upper Mercer retouched flakes are smaller on average – 2.8 g versus 13.8 g – than Kanawha chert retouched flakes, suggesting their curation in toolkits for more extended periods (see Figure 10.7). For utilized flakes – flake tools

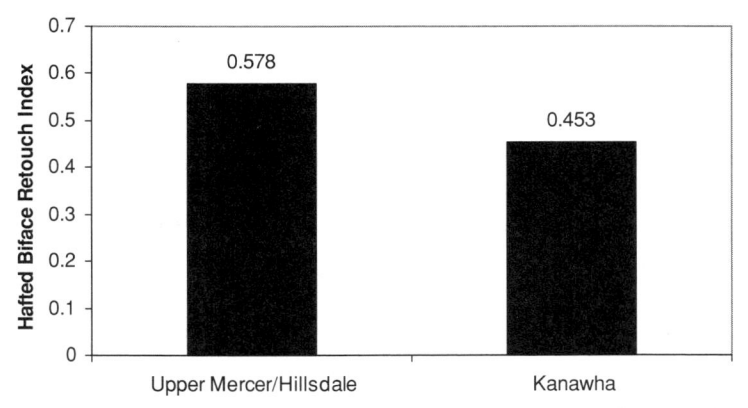

FIGURE 10.9. Comparison of hafted biface retouch index (HRI) for Kanawha chert and Upper Mercer/Hillsdale chert projectile points, Skink Rockshelter.

used for expedient tasks showing no signs of retouch – Upper Mercer chert tools weigh 1.47 g versus 5.13 g for Kanawha chert tools.

Thus, individuals at Skink Rockshelter in central West Virginia continued to use and retouch Upper Mercer chert flake tools, generally to the point where they were no longer useful and were discarded at the end of their use lives. In contrast, Kanawha chert flake tools occur in reduced quantities, despite the material's local abundance. As would be expected, Kanawha chert tools generally were discarded much earlier in their use-life history, as revealed by their larger masses and decreased use-wear and retouch indexes compared to the nonlocal Upper Mercer chert artifacts. Native Americans at Skink Rockshelter gave preferential treatment to Upper Mercer chert for daily-task activities, retouching flake tools to the point of exhaustion before using Kanawha chert. Tool function clearly influenced the differential use of Upper Mercer and Kanawha cherts at the site.

As would be expected, the increased curation distance and the accordingly higher degree of retouch and reduction resulted in significantly smaller stone tools (e.g., utilized flakes, retouched flakes, bifaces, and cores) for Upper Mercer (mean stone tool weight = 4.94) compared to Kanawha chert (mean stone tool weight = 8.49) (Figure 10.10; also see Figure 10.7). In turn, stone tools produced from the nonlocal Paoli and Flint Ridge cherts from more than 100–130 miles west (see Figure 10.4) occur in comparatively low mean weights

228   DOUGLAS H. MACDONALD

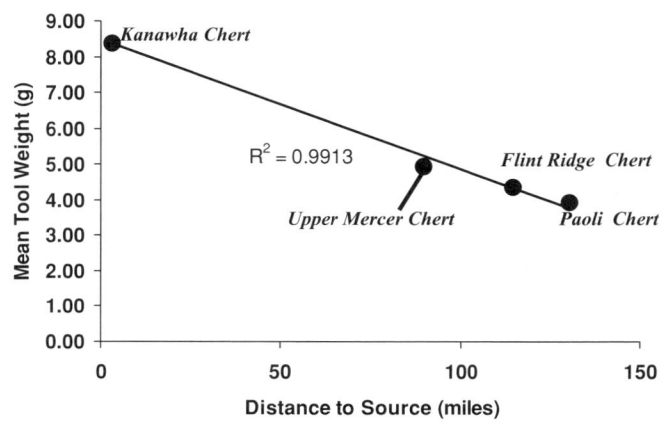

FIGURE 10.10. Differential size of stone tools based on distance to lithic material source, Skink Rockshelter.

(4.44 and 3.88 g). As reflected in Figure 10.10, regression analysis shows a strong and significant negative relationship between distance to source and mean stone tool weight for these four lithic materials at Skink Rockshelter ($F = .004$; d.f. $= 3$; $r^2 = .99$; t-stat $= 12.254$; $p = .001$). As with stone tools, the entire class of debitage should also be expected to vary by size measurements given the fall-off from distance to source, with the assumption being that the tools traveling the longer distances will be smaller due to retouch and reduction and produce accordingly smaller debitage. In this regard, mean weight for debitage is 0.58 g for Upper Mercer chert and 0.85 g for Kanawha chert. Debitage produced from the nonlocal Flint Ridge and Paoli cherts weighs less than 0.5 g each. As Figure 10.11 shows, regression analysis indicates a significant and strong relationship between distance to source and mean flake size for the five materials with known source locations in relation to Skink Rockshelter ($F = .03$; d.f. $= 4$; $r^2 = .83$; t-stat $= 6.045$; $p = .009$).

SUMMARY AND CONCLUSIONS

The main hypotheses of this paper were that, in toolstone-rich settings, the size of the curated toolkit and the degree of tool retouch will decrease in favor of expedient tool use using abundant local lithic materials. However, analysis of data collected at Skink Rockshelter in

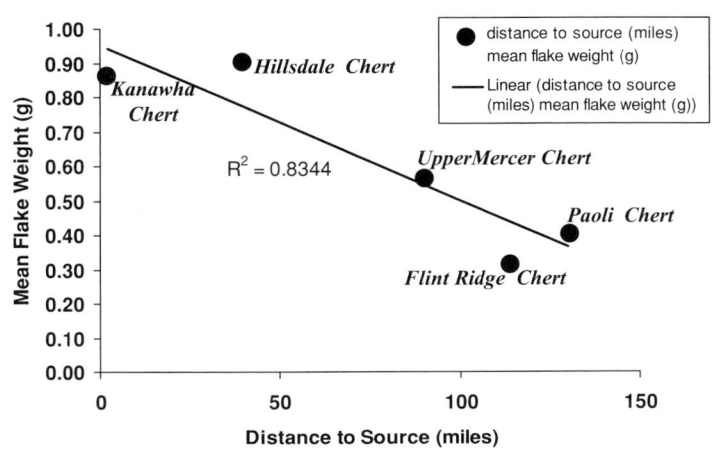

FIGURE 10.11. Differential flake size based on distance to source, Skink Rockshelter.

the heart of the Kanawha chert primary source area in central West Virginia suggests that other factors also contribute to tool kit size and degree of retouch.

Results of excavations at Skink Rockshelter indicate that Upper Mercer chert flake tools are more abundant than Kanawha chert flake tools, even though Kanawha chert is far more abundant at the site as a whole due to its local availability. Apparently, tool function influenced lithic raw material selection for Native Americans at Skink Rockshelter. As shown in Figure 10.12, although Kanawha chert represents 68% of all artifacts at the site, as well as 50% of bifaces, it represents only 37% of unifacial tools. In comparison, Upper Mercer chert represents only 24% of artifacts and 34% of bifaces, but nearly 60% of flake tools.

These stone tool data reflect a significant difference in lithic raw material use based on tool function and tool type ($\chi^2 = 8.24$, d.f. $= 1$, $p < .005$); as such, they do not support one of the original hypotheses of the paper, as reviewed above, that Kanawha chert would dominate all stone tool categories due to its ubiquitous availability near Skink Rockshelter. As discussed above and elsewhere (MacDonald and Cremeens 2005; MacDonald et al. 2006), these patterns of lithic raw material and tool use emerged during the Early Archaic period and continued until the Late Woodland period at Skink Rockshelter, suggesting patterned Native American land use and lithic technological organization for much of the Holocene.

Overall, the Skink Rockshelter data clearly indicate that raw material quality was important in determining how Native Americans differentially utilized tools at the site. In turn, tool function affected raw material selection for daily-task activities, with individuals selecting nonlocal Upper Mercer chert for flake tool use and locally available Kanawha chert for biface manufacture. As Andrefsky (1994b) suggests, simple abundance of a given lithic material does not guarantee its use for all activities. As reflected by data from Skink Rockshelter, the overall quality of the lithic material will significantly affect stone tool production activities at a given site.

As such, Skink Rockshelter lithic data also do not support another of the original hypotheses of the paper, that expedient flake tool use would increase at the expense of curated tools in the Kanawha chert primary source area. Instead, individuals curated the comparatively high-quality Upper Mercer chert tools to the site and continued to use and retouch them, rather than immediately replacing them with tools produced from the inferior, but abundant, Kanawha chert. Curation of the Upper Mercer tools resulted in their markedly reduced sizes and higher HRI measures compared to the local Kanawha chert tools. Accordingly, the size of the curated tool kit (sixty Upper Mercer chert stone tools) was very similar to that of the locally produced tool kit (sixty-three Kanawha chert stone tools). These data refute the hypothesis that the size of the curated tool kit would be reduced due to the local availability of Kanawha chert. The preference for Upper Mercer chert in daily-task activities influenced curation strategies of site occupants.

Data from Skink Rockshelter effectively reveal the impact of lithic raw material quality and tool function on tool-production and lithic raw material–selection decisions of prehistoric Native Americans in central West Virginia. Although its low quality did not dissuade users from producing bifaces and projectile points from Kanawha chert, individuals clearly believed they could not completely rely on the inferior material for use as retouched and utilized flakes in other daily-task activities. Instead, Native Americans carried stone tools produced from Upper Mercer chert to Skink Rockshelter to minimize the risks of relying upon the low-quality Kanawha chert. Tool function, thus, significantly altered lithic raw material curation patterns. As is typical of hunter–gatherer populations (Torrence 1989), such risk-minimization efforts likely cost little, but provided ample comfort to

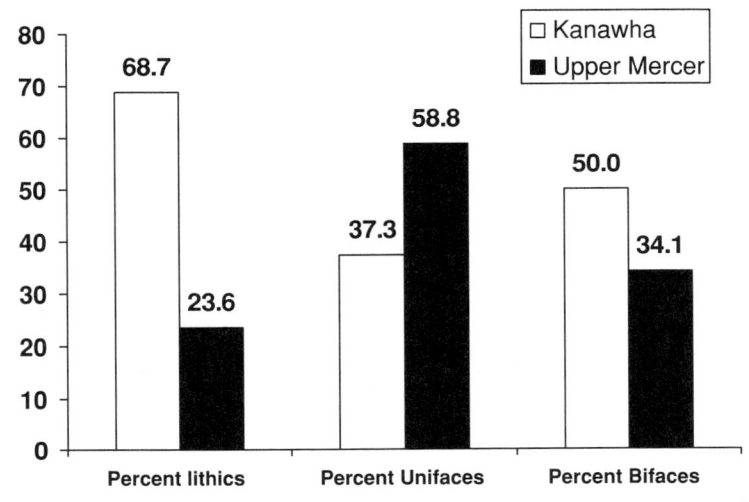

FIGURE 10.12. Differential lithic raw material use based on percentages of lithics, unifaces, and bifaces, Skink Rockshelter.

individuals with uncertain travel plans and even more uncertain access to high-quality lithic raw materials.

REFERENCES CITED

Andrefsky, William, Jr. 1994a. Raw Material Availability and the Organization of Technology. *American Antiquity* 59(1):21–34.
——— 1994b. The Geological Occurrence of Lithic Material and Stone Tool Production Strategies. *Geoarchaeology* 9(5):375–91.
——— 2006. Experimental and Archaeological Verification of an Index of Retouch for Hafted Bifaces. *American Antiquity* 71:743–58.
Bamforth, Douglas B. 1986. Technological Efficiency and Tool Curation. *American Antiquity* 51(1):38–50.
Binford, Lewis R. 1979. Organization and Formation Processes: Looking at Curated Technologies. *Journal of Anthropological Research* 35:255–73.
——— 1983. Long Term Land Use Patterns: Some Implications for Archaeology. In *Lulu Linear Punctuated: Essays in Honor of George Irving Quimby*, edited by R. C. Dunnell and D. K. Grayson, pp. 27–53. Museum of Anthropology Anthropological Papers No. 72. University of Michigan, Ann Arbor.
Brashler, J. G., and W. H. Lesser. 1990. Lithic Materials and Their Distribution in the West Virginia Highlands. In *Upland Archaeology in the East: Symposium IV*, pp. 193–207. CRM Report 92–1. USDA Forest Service, Southern Region, Atlanta.

Hofman, Jack L., Daniel S. Amick, and R. O. Rose. 1990. Shifting Sands: Folsom–Midland Assemblage from a Campsite in Western Texas. *Plains Anthropologist* 33:337–50.

Kagelmacher, Mike L. 2000. Ohio Cherts of Archaeological Interest. Unpublished manuscript on file at GAI Consultants, Pittsburgh, Pennsylvania.

MacDonald, Douglas H. 1999. Modeling Folsom Mobility, Mating Strategies, and Technological Organization in the Northern Plains. *Plains Anthropologist* 44(168):141–61.

———. 2003. Phase III Archaeological Investigations at Skink Rockshelter (46NI445), Nicholas County, West Virginia. Report submitted to Alex Energy, Summersville, West Virginia.

MacDonald, Douglas H., and David L. Cremeens. 2005. Holocene Lithic Raw Material Use at Skink Rockshelter (46NI445), Nicholas County, West Virginia. In *Upland Archaeology in the East, Symposia VIII and IX*, edited by Carole L. Nash and Michael B. Barber, pp. 133–72. Archaeological Society of Virginia Special Publication 38–7.

MacDonald, Douglas H., and Barry S. Hewlett. 1999. Reproductive Interests and Forager Mobility. *Current Anthropology* 40(4):501–23.

MacDonald, Douglas H., Jonathan C. Lothrop, David L. Cremeens, and Barbara A. Munford. 2006. Holocene Land-Use, Settlement Patterns, and Lithic Raw Material Use in Central West Virginia. *Archaeology of Eastern North America* 34:121–40.

Mandryk, Carole A. S. 1993. Hunter–Gatherer Social Costs and the Nonviability of Submarginal Environments. *Journal of Anthropological Research* 49:39–71.

Nelson, Margaret C. 1991. The Study of Technological Organization. In *Archaeological Method and Theory*, Volume 3, edited by Michael B. Schiffer, pp. 57–100. University of Arizona Press, Tucson.

Reger, D. B. 1921. *West Virginia Geologic Survey, Nicholas County*. Charleston, WV.

Reppert, R. S. 1978. Kanawha Chert: Its Occurrence and Extent in West Virginia. Open File Report 96. West Virginia Geological and Economic Survey, Charleston, WV.

Root, Matthew J., ed. 2000. *The Archaeology of the Bobtail Wolf Site*. WSU Press, Pullman.

Sampson, C. Garth. 1988. *Stylistic Boundaries among Mobile Hunter–Foragers*. Smithsonian Institution Press, Washington, DC.

Torrence, Robin A. 1989. Retooling: Towards a Behavioral Theory of Stone Tools. In *Time, Energy, and Stone Tools*, edited by Robin A. Torrence, pp. 57–66. Cambridge University Press, New York.

11 ANDREW P. BRADBURY, PHILIP J. CARR, AND
D. RANDALL COOPER

RAW MATERIAL AND RETOUCHED FLAKES

Abstract

Lithic analysts are often criticized for not engaging in theory building and for conducting particularistic studies. Such particularistic studies can be linked to theory through an organization-of-technology approach, which has great promise. However, concepts often employed in the approach, such as curation, need further refinement to become operationalized. One way to accomplish this for flake tools is to develop a method for measuring the amount of tool resharpening. One method is to determine the original flake mass and compare this to the mass of the recovered tool to determine the amount of realized use life. Here, a series of experiments in producing retouched flakes using various raw materials and two reduction modes were conducted by two flintknappers to determine how these variables influence the prediction of flake mass. Analysis indicates that raw material type is important for estimating original flake mass, but a tripartite division of quality may be sufficient to account for the variation. No significant differences are evident between the two knappers. Equations used to calculate original flake mass from retouched flakes must be derived with a consideration of raw material.

We would like to thank Bill Andrefsky for inviting us to participate in the SAA session at which this paper was originally presented and for his efforts in seeing the project through to publication. Comments by the discussants and one anonymous reviewer aided the revisions, but we are responsible for any errors.

INTRODUCTION

A decade ago, George Odell discussed the "particularism" that was apparent in lithic studies and the need to link these studies to theory and to prehistoric behavior (Odell 1996a: 2–3). This is something of an echo of the characterization of lithic analysis made by David Hurst Thomas a decade prior as "in danger of chasing rainbows rather than providing archaeology with the theory so obviously lacking" (Thomas 1986: 247). Instead of following the ten-year cycle of berating lithic analysts, we are pleased to say that the development and use of an organization-of-technology approach is one way to link lithic studies to theory and prehistoric behavior, as well as being in touch with some mainstream archaeologies and avoiding the chasing of rainbows. Further, the particularistic studies engendered by this approach fit the conception of "normal science" (sensu Kuhn 1962) and are moving lithic studies forward methodologically and in terms of understanding the complexity of lithic assemblage formation. In this vein, detractors recognize the utility of case studies that employ an organization-of-technology approach (e.g., Clark 1999; Torrence 1994). For Robin Torrence, "analyses of technological organization are here to stay because they provide data relevant to the goals of North American archaeology" (Torrence 1994: 123). John Clark states, "I find much in *Stone Tools* (an edited volume with clear ties to an organization of technology approach, if not always explicitly recognized) to be enthusiastic about" and describes individual chapters as "interesting, innovative, and useful" (Clark 1999: 127–30).

This does not mean that advances in the theoretical underpinnings, conceptualization, and application of an organization-of-technology approach are unnecessary. Rather, we would argue that they are vital for the continued utility of the approach and that something of a "scientific crisis" (sensu Kuhn 1962) exists in terms of employing an organization-of-technology approach in answering the myriad questions asked by lithic analysts, particularly those interested in social questions. There are a number of detractors of an organization-of-technology approach (Simek 1994; Torrence 1994) and the state of theory-building in lithic analysis more generally (Clark 1999). We disagree with some specifics of their critique, although we find others on target. We have confidence that progress can be made as analysts push and pull at an organization-of-technology approach in its application to understanding the complex suite of behaviors that contributed to the formation of the archaeological record.

Jan Simek, in an overview of a volume dedicated to an organization-of-technology approach, calls for "dialogue between Americans and Europeans" concerned with "integrating lithic studies into their wider social and economic contexts... experimental replication in the service of model building... and refitting" (Simek 1994: 120). The European chaîne opératoire method is contrasted by Simek with the American concept of "reduction sequence," and the latter is found wanting. This is not the proper comparison, as the chaîne opératoire should be measured against the framework of the organization-of-technology approach, as thoughtfully diagrammed by Nelson (1991). Nelson's (1991: 59) diagram of an organization-of-technology approach mirrors the chaîne opératoire as described by Simek and is the proper equivalent to chaîne opératoire, not the more limited reduction sequence. In the diagram, artifact form and artifact distribution are at the base and one moves up a level to design and activity distribution, respectively. In combination, these allow investigation of technological strategies and subsequently of social/economic strategies. Environmental conditions top the diagram and this demonstrates the close relationship between social/economic strategies and the environment. With other means of reconstructing past environments, lithic analysts are in the position of working from both ends of the diagram to understand social/economic strategies.

Simek (1994: 119) is right in pointing out that "where the notion of chaîne opératoire differs significantly from *reduction sequence* [emphasis added] is that tool use is also part of the concept." However, an organization-of-technology approach, which is much broader than reduction sequence, as defined by Nelson (1991) and repeated by Carr (1994a), does encompass stone tool use, it is part of "activity distribution" in Nelson's diagram, and stone tool use data are employed by Odell (1994) in his case study. Reduction sequence appears as something of a straw man in this case and is easily shown to be inadequate. Finally, the chaîne opératoire is lauded by Simek (1994: 119), because "lithic technology is seen as embedded in other aspects of economic and social behavior," which is obviously Nelson's goal and that of many others who employ an organization-of-technology approach. In this regard, Shott (2003) suggests that chaîne opératoire and reduction sequence may differ in semantics but not in substance and are essentially the same concept. Strangely enough, one area of significant difference between the two approaches involves the extension of the chaîne opératoire by some French prehistorians into the "cognitive realm," but Simek (1994: 120) is cautious not to expand on this, as he considers it to "go rather far beyond the data." This cognitive realm is likely what Clark (1999) would want to see lithic analysts further explore as he calls for employment of praxis theory as an alternative to the evolutionary ecological theory upon which some studies of technological organization are based.

Robin Torrence's critique of the contributions to the same volume mainly revolves around a perceived lack of theory-building. The normal science air of the set of case studies made "some concepts appear to have become embedded within the normal procedure of lithic analysis and are no longer questioned" (Torrence 1994: 123). We agree that concepts in organization-of-technology studies need to be questioned and, perhaps more importantly, clarified and refined. For example, key concepts such as curation, expediency, reliable, and maintainable are too often used in disparate manners (see discussions in Hayden et al. 1996; Nash 1996; Odell 1996a). It should be noted that Clark (1999: 130) sees a major strength of evolutionary ecology, and by extension of an organization-of-technology approach, as its being "fairly easy to operationalize." Unfortunately, lithic analysts have not realized complete agreement and clarity in a basic concept such as curation, as recognized by Clark (1999: 127). If these represent the easy concepts to operationalize, much work remains for lithic analysts employing an organization-of-technology approach and especially for those employing praxis theory.

Torrence (1994) is correct that archaeologists employing an organization-of-technology approach have generally been unengaged in theory-building at the highest level, but like Clark recognizes utility in the studies conducted. We accept this criticism and point to the relatively short use of the approach and only hope the baby is not thrown out with the bath water, but rather allowed to further develop. If an organization-of-technology approach does not reach its full potential upon maturity, certainly lithic analysts will turn to new paradigms.

Clark (1999) is the most recent critic of the organization-of-technology approach discussed here, and again, the lack of theory-building is the key problem, but more specifically the reliance on evolutionary ecology as opposed to praxis theory. For Clark, praxis theory makes humans into "agents" as opposed to "automatons" and as agents they act "meaningfully and purposefully in a meaningful world" (Clark 1999: 131). In our opinion, the inclusion of social strategies as a key aspect in Nelson's diagram demonstrates the potential for considering social aspects in organization-of-technology studies. However, we must agree that the potential has not been generally realized. Perhaps Charles Cobb has come closest in his "rethinking the organization of technology" (Cobb 2000: 70–82). Cobb does not think that organization-of-technology studies must do away with their methodology, but rather that these studies should "be placed in a context that systematically links lithic analyses with other realms of material culture and more holistic research questions – particularly questions that address the organization of labor" (Cobb 2000: 83). We are not in the position to judge whether Cobb's specific questions regarding the organization of labor fit with Nelson's diagram or with Clark's call for the employment of praxis theory. Certainly studies of technological organization must not all fit with Nelson's heuristic model, but we believe there is room for such accommodation. Carr and Bradbury (2006) have gone as far as modifying Nelson's original diagram to more explicitly detail aspects of the life cycle of a tool, and other such explications would likely enhance its utility. Two final points made by Clark are his criticism of the term "middle range theory" and his correct assertion that that building theory can occur from "careful assembly of minute observations of artifact traits" is "fallacious" (Clark 1999: 132–3). We agree with both points and have attempted to situate our own work in Schiffer's (1988) theoretical framework, which more explicitly outlines the place of middle range theory and its relationship to high-level theory.

After this discussion of organization-of-technology studies, one might wonder how we could employ such an approach that is in the apparent "state of becoming" that we described, and at our paradoxical characterization of the approach as both in normal science and scientific crisis. An organization-of-technology approach has had successes and continues to be applied (Andrefsky 1994; Bradbury 1998; Carr 2005; Kelly 1988; Shott 1989b), expanded (Cobb 2000), and refined (Carr and Bradbury 2006). We consider this a healthy sign, but agree that it must continue to be applied in particularistic ways so that concepts are operationalized, its utility is measured against the archaeological record, and it is linked to theory. Further, an organization-of-technology approach needs to continue to be expanded, especially in consideration of the importance of social aspects and strategies, and its advocates must engage more explicitly in theory-building at all levels. Here, we do not address all of these issues, but rather attempt a particularistic study to better our understanding of curation and the reconstruction of flake tool use life.

CURATION

The concept of curation, introduced by Binford (1973), has received considerable attention and is often used by analysts employing an organization-of-technology approach (e.g., Carr 1994b; Kelly 1988; Nelson 1991; Shott 1989a). However, it is clear that the term curation is used in multiple ways and encompasses a variety of dimensions (e.g. Andrefsky; Eren and Prendergast; Hiscock and Clarkson; Quinn et al., all this volume). Odell (1996a) examines five aspects of curation and the manner in which each aspect can be operationalized with archaeological materials. The aspect of interest for this study is tool maintenance. Odell (1996a: 60–62) states that "to maintain a tool is to resharpen it" and that examining retouch on a "specific subset of the assemblage" is one means of measuring tool maintenance. If we can devise a method for determining the amount of resharpening, then we can better understand the curated nature of specific tools, as reflected by tool maintenance. Continued refinement of the concept and specification of its various aspects are needed, but operationalization of tool maintenance allows one measure to be successfully applied to understanding prehistoric behavior.

For retouched flake tools, especially scrapers, there is a considerable literature regarding the determination of the amount of tool resharpening as a mean of indicating the amount of use-life that has been realized. Prehistoric stone tool users made decisions about when to discard a tool, and knowing its potential remaining use life given a particular environment can shed light on the social and economic strategies employed by that user. A variety of means have been employed to measure the degree of resharpening of flake tools, from characterizing the amount of retouch as "no retouch, light retouch, medium retouch," etc. (Nash 1996: 90–91) to Kuhn's (1990) geometric index and from employing equations to predict original flake mass (Dibble and Pelcin 1995; Pelcin 1998) to using experiments in determine how much edge is lost when a scraper is resharpened (Morrow 1997). Harold Dibble has long investigated the continued reduction of retouched flakes – examining implications for traditional typologies (Dibble 1984), developing a "scraper reduction index" (Dibble 1997), and deriving formulae to estimate original flake size (Dibble and Peclin 1995). Michael Shott (1995) asked the provocative question "How Much is a Scraper?" as a means to explore the amount of tool-using behavior represented by individual scrapers in an assemblage. He employed ethnographic and ethnoarchaeological data to estimate the number of resharpening flakes struck from scrapers in a Paleoindian assemblage and in combination with flake debris data demonstrated that much of the use and resharpening of end scrapers occurred at other sites. In conclusion, he calls for a "program of experimentation" involving the production of resharpening flakes and the stone tools from which these were struck.

This call for experimental work was answered both by those conducting highly controlled experiments and those engaged in freehand knapping experiments. The experiments of concern here are those addressing whether platform variables can be used to accurately predict flake mass. Platform variables are the focus because many flake tools recovered from the archaeological record retain those variables, and thus allow estimation of flake size despite being reduced. Importantly, the use of this approach is not restricted to end scrapers, but is applicable to a wide range of flake tool forms. This is a worthy endeavor because if the original size of the flake can be determined from platform variables, then we can gain a sense of the amount of tool use life expended and a measure of curation.

HOW MUCH FLAKE TOOL RESHARPENING WOULD A LITHIC ANALYST MEASURE, IF A LITHIC ANALYST COULD MEASURE FLAKE TOOL RESHARPENING?

We would answer this query with a question of our own, "Is the stone tool made of low-, medium-, or high-quality raw material?" That is, simple relationships may not exist between attributes measurable on flake tools and the amount of resharpening. Knapping modes and lithic raw material type may need to be controlled in attempts to develop equations for predicting original flake size in the service of measuring flake tool resharpening. Here, we review previous studies and then use data from freehand flintknapping experiments to examine the role played by the knapper, knapping mode, and raw material type.

In an examination of fracture mechanics, Dibble and Pelcin (1995) examined the effects of the mass and velocity of the percussor on the resulting flake mass. In their highly controlled experiments, glass pane cores were held at specific angles and ball bearings were dropped onto the cores to produce flakes. They developed a formula for predicting flake mass based on platform thickness and exterior platform angle. Other authors have more recently examined the original equation (e.g., Davis and Shea 1998; Shott et al. 2000) and suggested other measures.

Davis and Shea (1998) tested the hypothesis that flake mass could be predicted using platform thickness and exterior platform angle by conducting experiments designed to be analogous to the use and curation of Paleolithic stone tools. Obsidian flakes were modified for use as handheld tools in woodworking and butchery tasks. These tasks were not actually conducted, but rather the edges were dulled using an abrader and then the flakes were resharpened. Davis and Shea used the Dibble and Pelcin formula and found that, although predicted original sizes of many flakes approached their empirical values, many estimates were moderately low and several were extremely high. They attribute these results to differences in materials knapped and in experimental design – that is, highly controlled experiments versus actualistic knapping. In addition, they note the importance of platform width in affecting flake mass. Platform width was held constant in Dibble and Pelcin's (1995) experiments, but is practically impossible to control in freehand reduction.

In his response, Pelcin (1998) suggested that the equation derived for plate glass flakes may not be appropriate for use with other materials. However, he disputed the conclusion that platform width influenced flake size, instead being a threshold variable. To predict original mass, Pelcin advocated experiments to calibrate relationships with various raw materials and derivation of separate equations for flakes produced by bending initiation or soft hammer reduction and those produced by conchoidal initiation or hard hammer reduction. Subsequently, he added theoretical platform thickness as a variable useful in some equations.

More recently, Shott et al. (2000) attempted to extend this approach to other experimental assemblages that included not only flakes suitable for tools, but also those that span the reduction process. It was determined that none of the various equations were acceptable models for these assemblages. This is no shortcoming of any assemblage or equation, merely recognition that the conditions present in Pelcin's highly controlled experiments do not occur elsewhere. They suggest the need to refine the equations and to calibrate predictions to various tool stones and knapping modes.

Hiscock and Clarkson (2005) conducted the freehand reduction of thirty flakes as a means to assess Kuhn's geometric index of reduction and other quantitative methods for determining how much retouch has been applied to flakes. Their experiment involved "highly variable blank forms reduced in a standard way by unifacial retouching one lateral margin" (Hiscock and Clarkson 2005: 18). They found that Kuhn's geometric index of reduction performed well, especially as compared to other indices that "performed very badly, such as Dibble's (1995) surface area to platform area index which explains as little as 6.7% of variation." Although a quick perusal of this study may leave the reader with the notion that flake size cannot be accurately predicted from platform variables, formulae for predicting flake mass developed in the studies discussed above were not part of the comparison. Before Kuhn's geometric index of reduction is employed exclusively by analysts, further work is needed to examine the utility of equations for predicting flake mass under various experimental conditions.

EXPERIMENTS

In a preliminary attempt to address Dibble and Pelcin's suggestion that flake mass can be accurately predicted based on platform measurements (Carr and Bradbury 2005), one of us produced sixty-eight flakes from seven raw materials. Our results indicated that there were differences that could be attributed to raw material type; therefore, no single equation could be derived that could be applied to all of the materials examined.

FIGURE 11.1. Examples of flakes produced during the experiments, along with the two percussors used.

To further examine this issue, we added to our original study by conducting a second series of flintknapping experiments using the same seven raw materials, but included two knapping modes and two knappers. From lowest to highest quality, the materials used in these experiments were Kanawha, Kaolin, Fort Payne, Ste. Genevieve, Flint Ridge, and Cobden cherts and obsidian (Figure 11.1). These materials were selected because they exhibit considerable variability in what might generally be referred to as knapping quality. Kanawha is a coarse, grainy chert, whereas Cobden and Flint Ridge are of much higher quality. Obsidian has properties similar to those of the plate glass used by Dibble and Pelcin (1995) in the original work referred to here and was chosen by Davis and Shea (1998) for their experiments.

In our experiments, flakes were removed by one of two knapping modes, hard hammer or soft hammer percussion, and one of two knappers (APB and DRC). Both knappers used the same quartzite hammer stone for the hard hammer reduction and the same antler billet for the soft hammer reduction. To lessen the possible effects of internodule variability, both the hard hammer and the billet were

Table 11.1. Flake summary data

Material		Knapper	Hard Hammer	Billet
Kanawha	1	9	10	19
	2	9	11	20
Kaolin	1	3	4	7
	2	5	7	12
Fort Payne	1	7	9	16
	2	0	4	4
Ste. Gen.	1	5	5	10
Flint Ridge	1	4	3	7
Cobden	1	5	8	13
	2	0	9	9
Obsidian	1	13	6	19
	2	9	10	19
TOTAL		69	86	155

used to remove flakes from the same nodule. For example, flakes were removed from a nodule of obsidian with the hammerstone and flakes from the same obsidian nodule were removed with the antler billet. In addition, both knappers removed flakes from the same nodules. For raw materials for which more than one nodule was used, each knapper removed flakes from each of the nodules with both the hard hammer and the billet. In all cases, an attempt was made to produce a flake that could be used as an expedient tool or a blank that could be used for further reduction. Flakes that were part of the initial experiment were included here. To increase the sample size, the second knapper produced flakes from several of the raw material types representing the range of material quality (Kanawha, Kaolin, Cobden, and obsidian). Additional flakes were also removed by the first knapper.

A total of 155 flakes was included in the analysis (Table 11.1). For material type, the number of flakes ranges from 7 to 39 and they are essentially split for knapping mode, with 69 hard-hammer flakes and 86 billet flakes. The first knapper (APB) removed a total of 91 flakes and the second knapper (DRC) removed the remaining 64. The variation in numbers for knapping mode, knapper, and material is largely the result of the restriction in using the same nodule for each material throughout. For each flake, the following variables were recorded: weight, exterior platform angle, platform depth, platform

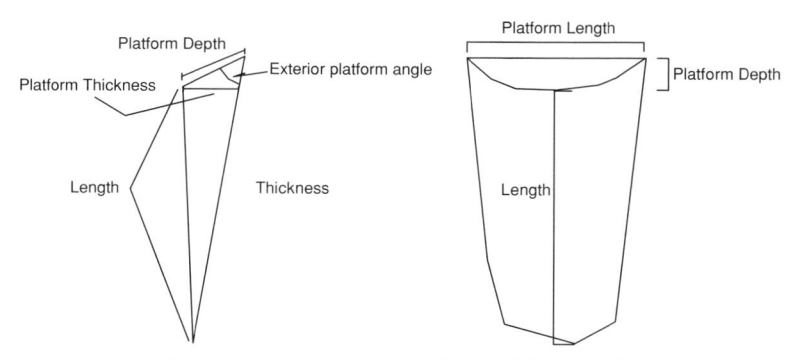

FIGURE 11.2. Flake measurements recorded, after Dibble and Pelcin (1995).

thickness, platform length, flake length, maximum thickness, knapping mode, and raw material (Figure 11.2). Exterior platform angle and platform thickness were used in Dibble and Pelcin's (1995) original formula to predict flake mass. Subsequent to that formulation, Dibble (1998: 612) argues that "If the goal is to determine the amount of mass loss during reduction, then it is preferable to base the reconstruction *on the basis of whatever measures are available*, and not *only* on the basis of platform variables" (emphasis in original). He suggests using flake thickness, as it is not usually affected by retouch, but can be an indicator of mass. The other variables recorded are those that have been used in various studies or are thought to have some relation to flake mass. To lessen bias from interobserver error, all measurements were recorded by just one of us. Finally, it should be noted that flakes with feather, hinge, and step terminations were included in the analysis.

RESULTS

As an initial exploration of the data, we graphed each variable against flake weight and saw similar plots in that there is a relationship between each variable and flake weight. For example, platform thickness (Figure 11.3) has a general linear relationship with flake weight. However, graphs of individual raw materials show a variety of slopes for a given variable (Figure 11.4). Our previous analysis suggested that raw material differences can have an effect on the accurate prediction of flake mass. Also, it should be noted that the graphs of the present data set

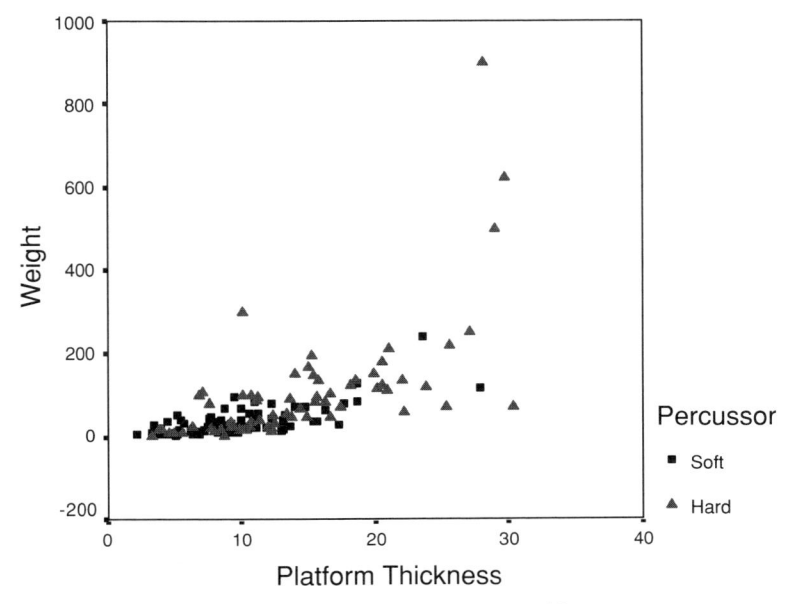

FIGURE 11.3. Platform thickness by weight, differentiated by percussor.

suggest the possibility of three outliers in the upper right. All three are hard hammer flakes of obsidian produced by the same knapper, which are very large (between 500 and 900 g).

One of the goals of the current analysis was to further examine material differences. In addition, we added a second knapper to explore potential differences that relate to variation in knapper. Flake counts were also increased to allow an examination of percussor (hard versus soft hammer) differences. If percussor and knapper differences are negligible, then equations can be derived for various raw materials recovered from archaeological sites. If such differences cannot be controlled for, then alternative methods may be needed.

For this examination of the data, we used the general linear model in SPSS to examine potential influences of knapper, raw material, percussor, and various interactions between the variables. We examined exterior platform angle and platform thickness, as both variables were used in Dibble and Pelcin's (1995) original model. For exterior platform angle, it was determined that knapper had no effect ($F = 0.966$, $p = .328$). However, percussor ($F = 5.574$, $p = .02$) and material type ($F = 3.99$, $p = .01$) did have an effect. There was also an interaction

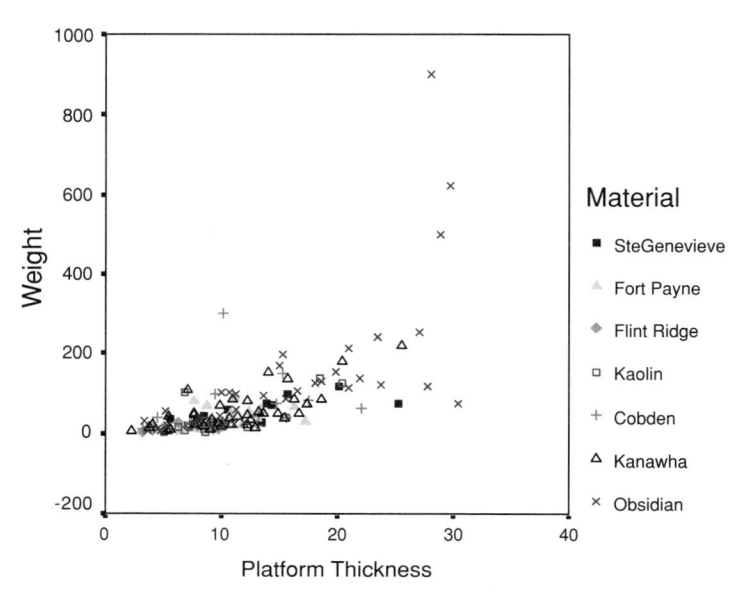

FIGURE 11.4. Platform thickness by weight, differentiated by material type.

between percussor and material ($F = 2.247$, $p = .087$). Similar results were seen when the three obsidian outliers were removed. The main difference was that there was no interaction between percussor and material.

For platform thickness, we again found no differences between knappers ($F = 0.01$, $p = .92$), but did find differences between materials ($F = 5.274$, $p = .002$) and percussors ($F = 19.634$, $p < .001$). An interaction between knapper and percussor was also found ($F = 3.833$, $p = .053$). Similar results were seen when the three obsidian outliers were removed. The main difference was that there was no interaction between percussor and knapper. A possible interaction between knapper, material, and percussor was seen ($F = 2.989$, $p = .055$).

The above results suggest that differences due to knapper are negligible. However, there are observable differences in raw material and possibly percussor. Such results are not too surprising given the design of the experiment and other flintknapping experimental results.

Ignoring these differences for now, we utilized the entire assemblage of seven raw materials, two knapping modes, and both knappers to examine the usefulness of platform thickness and exterior platform angle in predicting original flake mass. These two variables were

Table 11.2. Results of regression analysis using platform thickness and exterior platform angle as independent variables

Material	R^2	Std error
All cases	.439	78.5804
Kanawha	.563	33.1065
Kaolin	.635	26.0835
Fort Payne	.185	24.3136
Ste. Gen.	.451	26.485
Flint Ridge	.9	3.1507
Cobden	.128	64.6893
Obsidian	.462	135.741

regressed on flake weight. R^2 for the model was .439 and the ANOVA results indicated that the relationship was significant ($F = 59.543$, $p < .001$). Slightly better results were obtained when the three possible obsidian outliers noted above were removed ($r^2 = .46$). These results indicate that platform thickness and exterior platform angle provide some information on flake mass. However, the amount of variation explained is not particularly large. That is, other factors need to be considered to explain the remaining variation.

To further explore the relationship between raw material and the two variables, a separate regression analysis was performed for each individual material (Table 11.2). The r^2 values vary widely; ranging from a low of .128 for Cobden to a high of .9 for Flint Ridge. The ANOVA results do not indicate a significant relationship for Cobden, Ste. Genevieve, or Fort Payne. We note that one possible outlier (a hard hammer flake) existed in the Cobden sample. With this flake removed, the r^2 value was .333 and the ANOVA statistics ($F = 4.492$, $p = .026$) indicate a significant relationship. Although we do not want to make too much from these small sample sizes, clearly raw material type must be considered in making predictions of flake mass. Interestingly, our assessment of quality does not correlate with the R^2 value when these two variables are used.

Following in this same line of investigating specific raw material types, a stepwise method was used to determine the best predictor variables for each raw material type (Table 11.3). That is, we wanted to see if certain variables were consistently chosen by statistical means

Table 11.3. Results of the regression analysis using a stepwise method to find variables

Material	R^2	Std error	Variables
Kanawha	.808	21.9593	max thk, plat depth
Kaolin	.914	12.6348	max thk, plat len
Fort Payne	.807	11.8326	max thck, flk len
Ste. Gen.	.711	17.9831	max thick
Flint Ridge	.845	3.4957	Ex angle
Cobden	.823	29.172	Max thk, plat depth
Obsidian	.708	99.9854	max thk, flk length

as the best predictors of flake mass for different raw materials. They were not, though maximum thickness was chosen for five out of the seven material types. Of the six predictor variables, only platform thickness was not chosen for any of the raw materials, and exterior platform angle was only chosen for one raw material. Importantly, the lowest R^2 value is .708 for obsidian, which means in each case that well over half of the variation is being explained by the chosen variable or variables, with five explaining more than 80% of the variation in the data.

Given that the ANOVA analysis indicated possible percussor effects in addition to raw material effects, we reran the regression analyses and separated the four main materials by percussor. Separate regression analyses were conducted for hard hammer and billet flakes (Tables 11.4 and 11.5). The results are mixed when compared to the analysis that included both percussors. That is, with obsidian, the r^2 for the billet model is higher than that of the mixed model, whereas that for the hard hammer model is lower. The opposite is seen with Kanawha chert. For Kaolin, the hard hammer model is essentially the same as the mixed model, but the soft hammer model is lower. For Cobden, the hard hammer model is slightly higher than the mixed model, but the billet model is lower. Given these results, and the difficulty in distinguishing between hard hammer and billet flakes, it would be best to use the mixed model in archaeological cases.

To investigate the possibility of separating the raw materials into groups of similar quality, we reran the regression analyses using all materials. Kanawha was considered low-quality. Kaolin, Fort Payne, and Ste. Genevieve were considered medium-quality. Cobden and

Table 11.4. Results of the regression analysis on hard hammer flakes using a stepwise method to find variables

Material	R^2	Std error	Variables
Kanawha	.861	24.556	max thick, plat thick
Kaolin	.909	17.9215	Max thick
Cobden	.857	52.2298	flake len
Obsidian	.698	38.2199	Max thick, flk len, plat len

Flint Ridge were considered high-quality. Obsidian was considered as its own group. The results of these analyses indicate that combining raw materials creates relationships similar to those when the raw materials are considered individually (Table 11.6). That is, it might be possible to combine raw materials of similar quality in an analysis of archaeological materials.

SUMMARY AND CONCLUSIONS

For the experiments presented here with the specific goal of producing flakes for tool use, knapper influence on flake mass is negligible, but the type of material knapped is a significant factor for predicting flake mass from the variables used. The use of percussor appears negligible in our experiments. However, other studies with a greater number of knappers using a variety of percussors would do much to refute or strengthen these observations. This preliminary study is good news for the investigation of resharpening via predicting original flake mass, because controlling for knappers with an archaeological assemblage would be tenuous at best. Additionally, the determination of whether a flake was produced by a hard or soft hammer remains a difficulty (see similar results in Redman 1998). Our results suggest that hard

Table 11.5. Results of the regression analysis on billet flakes using a stepwise method to find variables

Material	R^2	Std error	Variables
Kanawha	.697	14.3698	max thick
Kaolin	.709	5.6688	plat thick
Cobden	.704	15.1069	max thick
Obsidian	.795	27.6	max thick

Table 11.6. Regression analysis with materials broken into four groups

Material	R^2	Std error	p	Variables
Low	.808	21.9593	<.001	max thk, plat depth
Medium	.856	13.4195	<.001	max thick, flk len, plat len
High	.812	26.7951	<.001	max thick, plat depth
Obsidian	.708	99.9854	<.001	max thk, flk length

and soft hammer flakes can be considered together in the equations. The data presented here indicate that application of equations to predict original flake mass for archaeological materials will have to be developed based on experiments conducted with either the same raw material used by the prehistoric knappers, or a very similar quality raw material. If several materials are combined as one, then some means of testing whether this is justified or not will be needed. Alternatively, one could select complete, unretouched flakes from the archaeological assemblage and develop equations based on these materials to be applied to the retouched archaeological materials. Separate equations would be needed for each raw material in the assemblage, or at least very similar raw materials combined as one.

The concept of "curation" as related to stone tool design, as opposed to what we do with our collections of excavated artifacts, was introduced to archaeologists over thirty years ago by Binford (1973). This concept has received considerable attention since this time and has played a role in how archaeologists make inferences about human behavior at specific prehistoric sites and across regions (i.e., Bamforth 1986; Carr 1994b). However, there is disagreement over the usage of this concept and the use of design principles in general (Hayden et al. 1996). Our work with tool maintenance here is not meant to solve these problems, but does demonstrate that with continued methodological refinement, tool maintenance can be accurately measured and used as a means to monitor one aspect of curation. Knowing that one assemblage has tools with an average of 5 g lost whereas another has an average of 10 g lost of that same material allows comparisons with considerations given to the overall organization-of-technology framework. For example, if each of these sites is located in the same environment, then the difference in the amount of resharpening is likely due to either social or economic reasons. Perhaps the site showing a greater amount of resharpening (1) was occupied longer; (2) was part of a different settlement pattern; (3) evidenced greater amount of processing so more resharpening of tools; or (4) was affected by a combination of these and other factors. In addition, different raw materials would need specific consideration with regard to how quickly each dulls in different tasks and how much mass is lost in resharpening.

Although such specific research appears to be far removed from prehistoric human behavior, such details are critical if we are ever to reach the promise of a scientific archaeology. There is too much we still do not know about the complexities of lithic assemblage formation. If all of our time is spent at higher levels of theory-building or criticizing lithic analysts for not building theory, we will not advance our understanding of prehistoric technology or of lithic assemblage formation.

REFERENCES CITED

Andrefsky, William, Jr. 1994. Raw Material Availability and the Organization of Technology. *American Antiquity* 59:21–35.

Bamforth, Douglas B. 1986. Technological Efficiency and Tool Curation. *American Antiquity* 51(1):38–50.

Binford, Lewis R. 1973. Interassemblage Variability: The Mousterian and the "Functional" Argument. In *The Explanation of Culture Change: Models in Prehistory*, edited by C. Renfrew, 227–54. Duckworth, London.

Bradbury, Andrew P. 1998. The Examination of Lithic Artifacts from an Early Archaic Assemblage: Strengthening Inferences through Multiple Lines of Evidence. *Midcontinental Journal of Archaeology* 23(2):263–88.

Carr, Dillion H. 2005. The Organization of Late Paleoindian Lithic Procurement Strategies in Western Wisconsin. *Midcontinental Journal of Archaeology* 30(1):3–36.

Carr, Philip J. 1994a. The Organization of Technology: Impact and Potential. In *The Organization of North American Chipped Stone Tool Technologies*, edited by P. J. Carr, pp. 1–8. International Monographs in Prehistory, Ann Arbor.

——— 1994b. Technological Organization and Prehistoric Hunter–Gatherer Mobility: Examination of the Hayes Site. In *The Organization of North American Chipped Stone Tool Technologies*, edited by P. J. Carr, pp. 35–44. International Monographs in Prehistory, Ann Arbor.

Carr, Philip J., and Andrew P. Bradbury. 2005. Raw Material and Retouched Flakes: More Complicated Than We Thought. Paper presented at the

70th Annual Meeting of the Society for American Archaeology, Salt Lake City.

——— 2006. Learning from Lithics. Paper presented at the 71st Annual Meeting of the Society for American Archaeology, San Juan, Puerto Rico.

Clark, John E. 1999. On *Stone Tools: Theoretical Insights into Human Prehistory* by G. H. Odell. *Lithic Technology* 24(2):126–35.

Cobb, Charles R. 2000. *From Quarry to Cornfield: The Political Economy of Mississippian Hoe Production*. University of Alabama Press, Tuscaloosa.

Davis, Zachary J., and John J. Shea. 1998. Quantifying Lithic Curation: An Experimental Test of Dibble and Pelcin's Original Flake-Tool Mass Predictor. *Journal of Archaeological Science* 25: 603–10.

Dibble, Harold L. 1984. Interpreting Typological Variation of Middle Paleolithic Scrapers: Function, Style, or Sequence of Reduction? *Journal of Field Archaeology* 11(4):431–6.

——— 1995. Middle Paleolithic Scraper Reduction: Background, Clarification, and Review of the Evidence to Date. *Journal of Archaeological Method and Theory* 2(4):299–368.

——— 1998. Comment on Quantifying Lithic Curation: An Experimental Test of Dibble and Pelcin's Original Flake-Tool Mass Predictor. *Journal of Archaeological Science* 25:611–13.

Dibble, Harold L., and Andrew W. Pelcin. 1995. The Effect of Hammer Mass and Velocity on Flake Mass. *Journal of Archaeological Science* 22: 429–39.

Hayden, Brian, Nora Franco, and Jim Spafford. 1996. Evaluating Lithic Strategies and Design Criteria. In *Stone Tools: Theoretical Insights into Human Prehistory*, edited by G. H. Odell, pp. 9–45. Plenum Press, New York.

Hiscock, Peter, and Chris Clarkson. 2005. Measuring Artifact Reduction – An Examination of Kuhn's Geometric Index of Reduction. In *Lithics "Down Under": Australian Perspectives on Lithic Reduction, Use and Classification*, edited by C. Clarkson and L. Lamb, pp. 7-20. British Archaeological Reports, Oxford, England.

Kelly, Robert L. 1988. The Three Sides of a Biface. *American Antiquity* 53:717–34.

Kuhn, Steven L. 1990. A Geometric Index of Reduction for Unifacial Stone Tools. *Journal of Archaeological Science* 17:583–93.

Kuhn, Thomas. 1962. *The Structure of Scientific Revolutions*. University of Chicago Press, Chicago.

Morrow, Juliet. 1997. End Scraper Morphology and Use-Life: An Approach for Studying Paleoindian Lithic Technology and Mobility. *Lithic Technology* 22:70–85.

Nash, Stephen E. 1996. Is Curation a Useful Heuristic?. In *Stone Tools: Theoretical Insights into Human Prehistory*, edited by G. H. Odell, pp. 81–99. Plenum Press, New York.

Nelson, Margaret C. 1991. The Study of Technological Organization. In *Archaeological Method and Theory*, Vol. 3, edited by M. B. Schiffer, pp. 57–100. University of Arizona Press, Tucson.

Odell, George H. 1994. Assessing Hunter–Gatherer Mobility in the Illinois Valley: Exploring Ambiguous Results. In *The Organization of North American Chipped Stone Tool Technologies*, edited by P. J. Carr, pp. 70–86. International Monographs in Prehistory, Ann Arbor.

1996a. Economizing Behavior and the Concept of "Curation." In *Stone Tools: Theoretical Insights into Human Prehistory*, edited by G. H. Odell, pp. 51–80. Plenum Press, New York.

1996b. Introduction. In *Stone Tools: Theoretical Insights into Human Prehistory*, edited by G. H. Odell, pp. 1–5. Plenum Press, New York.

Pelcin, Andrew W. 1998. The Threshold Effect of Platform Width: A Reply to Davis and Shea. *Journal of Archaeological Science* 25:615–20.

Redman, Kimberly L. 1998. *An Experiment-Based Evaluation of the Debitage Attributes Associated with "Hard" and "Soft" Hammer Percussion*. M.A. thesis, Washington State University, Pullman, WA.

Schiffer, Michael B. 1988. The Structure of Archaeological Theory. *American Antiquity* 53:461–85.

Shott, Michael J. 1989a. On Tool Class Use-Lives and the Formation of Archaeological Assemblages. *American Antiquity* 54:9–30.

1989b. Technological Organization in Great Lakes Paleoindian Assemblages. In *Eastern Paleoindian Lithic Resource Use*, edited by C. J. Ellis and J. C. Lothrop, pp. 221–38. Westview Press, Boulder.

1995. How Much Is a Scraper? Uniface Reduction, Assemblage Formation, and the Concept of "Curation." *Lithic Technology* 20:53–72.

2003. Chaine Operatoire and Reduction Sequence. *Lithic Technology* 28:95–105.

Shott, Michael J., Andrew P. Bradbury, Philip J. Carr, and George H. Odell. 2000. Flake Size from Platform Attributes: Predictive and Empirical Approaches. *Journal of Archaeological Science* 27(10):877–94.

Simek, Jan. 1994. Some Thoughts on Future Directions in the Study of Stone Tool Technological Organization. In *The Organization of North American Chipped Stone Tool Technologies*, edited by P. J. Carr, pp. 118–22. International Monographs in Prehistory, Ann Arbor.

Thomas, David H. 1986. Contemporary Hunter–Gatherer Archaeology in America. In *American Archaeology Past and Present*, edited by D. J. Meltzer,

D. D. Fowler, and J. A. Sabloff, pp. 237–56. Smithsonian Institution Press, Washington, DC.

Torrence, Robin. 1994. Strategies for Moving On in Lithic Studies. In *The Organization of North American Chipped Stone Tool Technologies*, edited by P. J. Carr, pp. 123–31. International Monographs in Prehistory, Ann Arbor.

PART FOUR

EVOLUTIONARY APPROACHES TO LITHIC TECHNOLOGIES

12 ANNA MARIE PRENTISS AND DAVID S. CLARKE

LITHIC TECHNOLOGICAL ORGANIZATION IN AN EVOLUTIONARY FRAMEWORK: EXAMPLES FROM NORTH AMERICA'S PACIFIC NORTHWEST REGION

Abstract

The organization of lithic technology as a field of study prescribes ecological explanations for variation in the ways that people made, used, and transported stone tools. Although these models go far in providing insight into the economic rationale behind lithic production and use systems, they do not emphasize the historical and evolutionary nature of change in stone tools. The paper offers an approach to integrating ecological and evolutionary views of lithic technology. Two case studies from North America's Pacific Northwest region seek to illustrate how change in aspects of chipped stone tool retouch patterns corresponds with shifts in local subsistence tactics, despite socioeconomic stability on the higher organizational scale.

INTRODUCTION

In 1994, Jan Simek gently chided archaeologists seeking to reconstruct and explain the organization of lithic technology for failing to integrate Darwinian evolutionary thinking into their models. This

---

We thank Bill Andrefsky for inviting us to contribute this paper. Field work at the Keatley Creek site was funded by grants from the National Science Foundation (BCS-0108795), the Wenner–Gren Foundation for Anthropological Research Inc. (GR. 6754), and the University of Montana. Field work was conducted under provincial permit and under the invitation of the Pavilion (Ts'kway'laxw) Indian Band. We are grateful to Bill Andrefsky, Doug MacDonald, and two anonymous reviewers for their comments. Figures 12.1 and 12.8 were created by Nathan Goodale.

came at a time when evolutionary thinking in archaeology was beginning to be visible as a significant new paradigm in American archaeology (e.g., O'Brien and Lyman 2003b). Archaeologists interested in evolution have since offered methodological and theoretical advances spanning elements of material culture such as pots and projectile points (O'Brien et al. 1999) to change in entire systems or packages of cultural behavior (Spencer 1995). Yet research into the organization of lithic technology has typically remained comfortably in the realm of ecology and its search for general principles. Although this work has at times been very sophisticated (e.g., Kuhn 1994) its reliance on ecology as the explanatory framework has not yet led to a comprehensive evolutionary theory of stone tools.

Archaeologists interested in the organization of lithic technology typically seek to explain variation in production, use, and transport of lithic tools, often in the light of mobility regimes, subsistence resource conditions, and access to lithic sources (Andrefsky 1994, this volume; Bamforth 1986; Binford 1979; Shott 1987; Torrence 1989). They generally argue that economic logic will strongly dictate the tactics chosen and that these can be predicted using general theoretical models often based (implicitly or explicitly) on the microeconomic logic of optimal foraging theory (e.g., Binford 2001). Although these models go a long way toward helping us to understand the economics of human decision-making, they do not always fully explain the evolutionary/historical processes that give rise to variation in chipped stone tools (Simek 1994). The fundamental issue is that humans act within a cultural framework defined by past history. Consequently, technological decisions are made within parameters defined by culturally designated modes of operation (operationalized by some lithic technologists [e.g., Close 2006] as *chaine opératoire*). Yet change does happen, varying in scale from tactical shifts in tool retouch patterns to reorganization of production accompanying grander changes on higher socioeconomic scales (e.g., a hunter–gatherer group's annual mobility and subsistence strategy). In this paper we provide a framework in which we can examine variation in stone tool manufacture, use, and maintenance in its ecological and historical/evolutionary context. Our intent is to outline an approach within which to consider relationships between human decision-making as tied to membership within larger evolutionary structures and also as a consequence of local ecological contingencies.

To help accomplish this, we provide two case studies from the Pacific Northwest region of North America in which changes in lithic tool manufacture and retouch occurred within stable socioeconomic strategies. First, we examine the effect of changing environment on a technological system, focusing on the Early Holocene expansion of microblade-using peoples from interior Alaska to the central Northwest Coast. Here we argue that although the basic economic system moved relatively intact, variation in seasonal resource conditions had significant effects on prevailing approaches to some stone tool production and use systems. Although this example provides insight into broad changes in technological repertoires, current data do not permit precise measurement of historical rates of change. This is primarily due to having a relatively small sample of sites with adequate dating from such an extensive landscape. We can get a closer look at rates of change from our second case study, which examines change in patterns of tool retouch at a late prehistoric village located in interior British Columbia. In this case, we can recognize rapid shifts in preferred tool types and patterns of retouch tangential with changes in predation and associated mobility patterns. By combining evolutionary and ecological ideas, we are able to offer new ideas about change and stability in these lithic tool manufacturing and use regimes.

THEORETICAL ISSUES: LITHIC TECHNOLOGICAL ORGANIZATION IN AN EVOLUTIONARY FRAMEWORK

Archaeologists have long studied lithic tools as markers of variability in human socioeconomic systems (e.g., Binford and Binford 1966). Measurement of variability in tool manufacture, form, use, and resharpening provides insight into such cultural practices as mobility strategy, subsistence behavior, and elements of social organization associated with the organization of labor. In recent years some archaeologists (e.g., O'Brien and Lyman 2003b) have argued that tool designs can be better understood as targets of a Darwinian selection–like process that gives rise to histories of particular forms. It has even been shown that this can be modeled using the method and theory of cladistics (O'Brien and Lyman 2003a). Although this is an important development, we cannot abandon our knowledge that lithic tools also represent parts of more complex cultural systems. If they are parts, they may be affected by changes on higher scales of integration, rather than simply existing in isolation as competitive entities in and of themselves.

As argued by Eldredge (1985), evolution occurs on multiple genealogical scales. In biology we recognize the evolution of genes, organisms, and even species. In cultural contexts we must also recognize change in a range of entities of varying degrees of complexity (Eldredge 2000). Chatters and Prentiss (2005; Prentiss and Chatters 2003) have described a simple hierarchical model linking informational (or genealogical) and behavioral (ecological or phenotypic) elements of culture. In the genealogical framework, culture is best understood as a complex information system ranging from simple traits to integrated packages of traits. Resource management strategies (RMS) are complex packages of integrated information that define organized behavior associated with acquiring, distributing, and consuming food and nonfood resources (Chatters and Prentiss 2005). Although all human groups operate within and, in essence, possess an RMS, it is probably impossible for any single person to possess the total information associated with an RMS (Bettinger 2003). Consequently, the RMS is an emergent character (e.g., Vrba and Eldredge 1984) visible only on the scale of organized activities associated with human groups. Individuals inherit and otherwise come to possess enough information to contribute to their respective groups within this matrix.

The RMS is expressed through behavioral manifestations termed tactics by Chatters and Prentiss (2005). Integrated behavior or tactics among hunter–gatherers takes the form of such things as food acquisition and processing behaviors, group movements, and technological activities. Fortunately for archaeologists, tactics, being phenotypic, leave fossil remnants in the archaeological record, including food remains, portions of houses, and stone tools and flaking debris. Analysis of the archaeological record of these items permits archaeologists, using middle range theory (e.g., Binford 1977, 1981), to recognize key dimensions (or substrategies) of a given community's RMS (e.g., mobility and foraging strategy, scheduling, technological organization, and social organization as it pertains to resource distribution). As tactics involve energy exchange, greater and lesser degrees of success are possible, and consequently it is here that natural selection can act to effect change, stasis, or extinction in resource management strategies (Prentiss and Chatters 2003).

The formation of variability in artifacts in the archaeological record therefore results from transmission of information, practice of organized behavior coded by that information, and in the long term, feedback (selection) on the human individual and group engaging in that behavior. Variation may occur due to other factors as well, including such things as errors and innovations (e.g., Eerkens and Lipo 2005). Hunter–gatherers may employ a complex repertoire of inherited (learned) technologies in their standard resource-gathering activities, yet, as argued by Binford (1978), they also must respond to contingencies, sometimes making alterations to specific tools or creating situational tools to serve in particular circumstances. We can imagine that in rare situations, the solving of dramatic contingencies could trigger rapid and major changes in scheduling, resource procurement strategies, and ultimately entire RMS, in a process Prentiss and Chatters (2003) term niche reorganization. However, far more typically, technology will evolve in such a manner that it helps to preserve a standard of living for its human users, but also has the effect of preserving or, put another way, maintaining cultural stasis at the higher RMS level. We provide two examples of the latter process in the following case studies.

THE NORTHWEST MICROBLADE TRADITION

The expansion of microblade-using groups from Alaska down the Northwest Coast (Figure 12.1) during the terminal Pleistocene and early Holocene provides an excellent example of variability in lithic retouch tactics under changing ecological conditions. Although archaeologists have associated the North Coast Microblade tradition (e.g., Matson and Coupland 1995) with the Denali Complex of interior Alaska and Diuktai culture of northeast Siberia (Ackerman 1992, 1996; Carlson 1998; Matson and Coupland 1995) due to strong typological similarities in microblade cores, burins, and bipointed or leaf-shaped bifaces, there is not total agreement regarding the cultural origins of the similarly dated southern coastal manifestation known as Old Cordilleran. Matson and Coupland (1995) argue for a link to Clovis-derived interior cultures, presumably on the basis of geographic proximity, coupled with frequent bifaces and little evidence for microblades. However, Chatters et al. (2006; see also Carlson 1998), demonstrate that coastal and interior (e.g., Cascade phase) Old Cordilleran manifestations are strongly similar to early Alaskan complexes on the basis of lithic tool manufacture (bifaces, microblades, heavy-duty woodworking tools), settlement, and subsistence tactics, as well as human craniofacial morphology. For the purposes of this paper, we agree with Chatters et al. that the Old Cordilleran pattern is so different from the earlier Paleoindian Stemmed Point or Protowestern tradition (e.g., Cressman 1977) that it can only reflect a replacement of these populations on the southern coast and interior by populations originating to the north in Alaska, as marked by the Denali Complex and North Coast Microblade tradition.

Although data from much of the greater Pacific Northwest and interior Alaska remain relatively sparse (Fedje et al. 2004; Mason et al. 2001), it is possible to recognize at least a rudimentary pattern of mobility, subsistence, and technological organization employed by members of this greater Northwest Microblade tradition (we combine Clark's [2001] Northwest Microblade tradition and Dixon's [2001] Northwest Coast Microblade tradition). Denali complex populations were relatively sparse and apparently highly mobile, as indicated by transport of nonlocal lithic raw materials (Mason et al. 2001). Living in a terminal Pleistocene and early Holocene landscape with generally dry conditions with intense summer storms, fluctuating cold and warm periods, and highly patchy game distributions (Mason et al. 2001), the Denali Complex favored settlement tactics that emphasized relatively short-term occupation of small to larger residential camps, often on river terraces, but rarely in rockshelters (Mason et al. 2001; Yesner 2001). As reflected at sites such as Dry Creek (Powers and Hoffecker 1989) and Onion Portage (Anderson 1988), Denali peoples may have periodically aggregated in larger groups. It is also possible that some of these places were frequently reoccupied. Although subsistence tactics are not very well understood (Mason et al. 2001), sites such as Broken Mammoth hint that Denali peoples may have taken a flexible approach to a seasonally variable, immediate-return diet incorporating resources spanning fish, waterfowl, and small rodents, but also including large numbers of caribou and likely sheep, elk, and bison (Yesner 2001). Denali peoples continued to employ a lithic technological

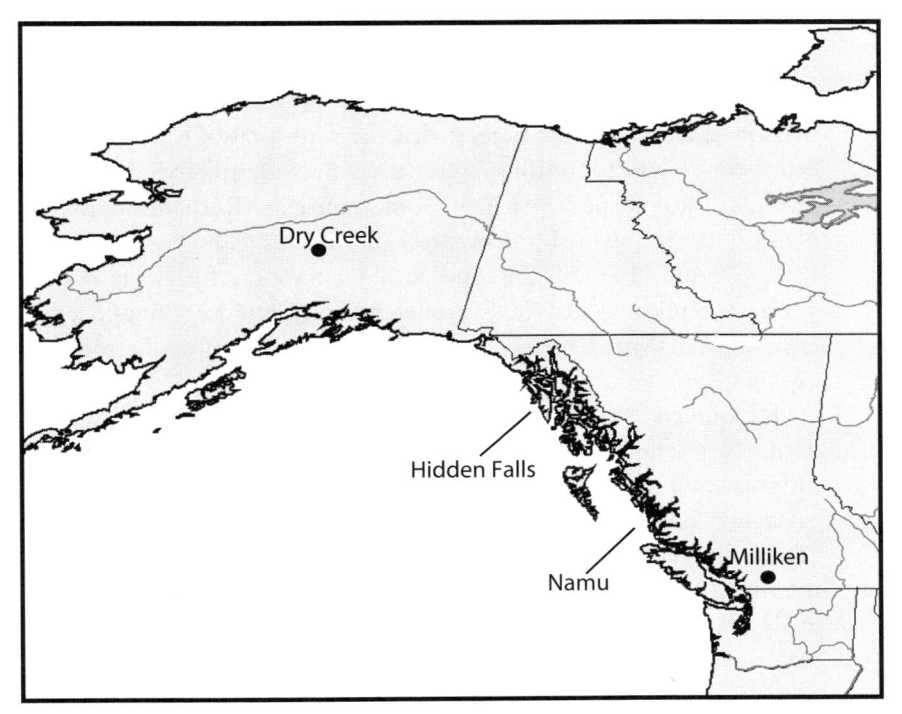

FIGURE 12.1. Map of the Northwest Coast.

organization that probably evolved in North China, Mongolia, and eventually Siberia (Diuktai), emphasizing production of composite weapons and other tools using modular parts such microblades and variously shaped bifaces and unifaces (Hoffecker 2005). Burins were frequently manufactured to aid in production of hafts for microblades and other uses. This technological strategy was clearly designed to help its users overcome contingencies associated with life in a cold but highly seasonal interior environment and also to increase the likelihood of success in hunting using "overdesigned" (e.g., Bleed 1986) and modular tool/weapon systems.

Although it is not clear how Denali peoples moved to the northern Northwest Coast, it is clear that by about 10,000 B.P., some of these groups had made the move south. Some, such as Fedje et al. (2004), suggest that the first arrivals may not have been microblade users and perhaps were derived from some other nonmicroblade culture (e.g., the Nenana complex). However, the sample of sites from the early

Northwest Coast is very low and sampling error could be a possibility. Archaeologists should also recognize that there is similar variation in interior Alaskan Denali sites and that not all have microblades (Mason et al. 2001). We suggest that the sum total of early northern Northwest Coast tools still substantially reflect Denali-style technologies (e.g., leaf-shaped bifaces in some contexts [Richardson Island, Namu]; wedge-shaped microcores and burins elsewhere [Hidden Falls, Ground Hog Bay 2]). This basic technological package is substantially replicated in Old Cordilleran sites, with a continued presence of leaf-shaped bifaces and microblades (the latter in reduced numbers).

Resource conditions on the Northwest Coast were obviously substantially different from those of interior Alaska. Limited research into marine productivity during the early Holocene suggests extreme variability but overall high productivity between 7000 and 9000 B.P. (Tunnicliffe et al. 2001). Terrestrial environments on larger islands and the mainland likely also offered a wide array of resources under generally dry conditions (Fedje et al. 2004). Salmon numbers, however, may have been substantially reduced during this era due to dry conditions, particularly on the interior (Chatters et al. 1995). Seasonality was not extreme (possibly less so than today) and winters were relatively mild, placing little pressure on mobile groups to anticipate shortages between seasons.

Upon arrival in this generally rich environment, early groups appear to have maintained their settlement tactics, focused on frequent, and sometimes long-distance (e.g., Andrefsky 1995; Prentiss and Chatters 2003), moves and relatively short stays in small camps located on ridges or beach fronts overlooking water. With the occasional exception (e.g., On Your Knees Cave [Dixon et al. 1997]), rockshelters and caves were very rarely used. Predictably, subsistence was dramatically diverse, with evidence for use of a variety of terrestrial fauna, anadromous resources such as salmon, various inshore shellfish and fish, and deep-water resources such halibut (Carlson 1998; Fedje et al. 2004; Matson and Coupland 1995). However, there remains no firm evidence for food storage at any site throughout the region pre-dating around 5000 B.P. (but see Cannon and Yang 2006). In essence, the residentially mobile, non-storage-oriented Denali RMS had moved south, remaining structurally intact, but varying at some

Table 12.1. Northwest Coast and interior Alaska site locations

Site	Specific location	General location
Dry Creek	Nenana Valley	Interior Alaska
Namu	Fitzhugh Sound	Coastal British Columbia
Hidden Falls	Baranof Island	Southeast Alaska
Milliken	Lower Fraser Canyon	Southwestern British Columbia

tactical levels associated with altered resource types and changes in seasonal access.

CHANGE IN RETOUCHED LITHIC TOOLS

In order to better understand technological aspects of the Denali RMS expansion, we examine change in lithic tool frequencies across a series of sites spanning from the Denali through the Old Cordilleran cultural complexes. We accomplished this by selecting a sample of sites with unmixed and reasonably dated early components that also have large and adequately described lithic assemblages in order to obtain geographic representation spanning from interior Alaska to southern British Columbia and from ca. 10,500 to 6500 B.P. These strict criteria provided us with only a limited number of site components, including Dry Creek Component II (eastern Beringia at ca. 9,300–10,600 B.P.) (Powers and Hoffecker 1989), Namu Components Ia and Ib (northern Northwest Coast at ca. 8000–10,000 B.P.) (Carlson 1996), Hidden Falls, Component 1 (northern Northwest Coast at ca. 8600–9500 B.P.) (Davis 1989), and the Milliken's (central Northwest Coast) Milliken (ca. 8000–9000 B.P.), Mazama (ca. 7000–7500 B.P.), and Gravel (ca. 6700–7300 B.P.) components (Mitchell and Pokotylo 1996). The Milliken components are identified elsewhere in the paper as A (Milliken), B (Mazama), and C (Gravel). Plotting tool frequencies by tool forms and functional categories permits us to gain some insight into changes in lithic retouch tactics as reflected in the data from these sites (Tables 12.1 and 12.2).

First we divided lithic tool assemblages into three functional classes on the basis of functional analysis, ethnographic descriptions of similar

Table 12.2. Northwest Coast and interior Alaska site data

	Hunt/ butcher	Light duty	Heavy duty	Microblade index	Shaped tool index	Uniface index	Biface index
Dry Creek	0.23	0.32	0.45	0.9	0.65	0.11	0.23
Namu	0.48	0.18	0.34	0.1	0.5	0.16	0.44
Hidden Falls	0.01	0.89	0.1	0.02	0.12	0.07	0.01
Milliken A	0.11	0.79	0.1	0.03	0.17	0.06	0.11
Milliken B	0.19	0.56	0.25	0.02	0.18	0	0.19
Milliken C	0.03	0.62	0.34	0	0.07	0	0.04

tools (e.g., Teit 1900), and studies in design theory (e.g., Hayden et al. 1996b). Shaped bifacial projectile points and knives (as opposed to lightly retouched flake edges) were placed into the group we term hunting tools most effectively applied to killing and butchering animals. Larger, steeply retouched unifacial scrapers and bifacial drills and boring tools were classified as heavy-duty and assumed to be more commonly associated with applications to hard materials such as wood or antler. All other flake tools and lightly retouched scrapers (e.g., end scrapers) were classified as light-duty and generally associated with tasks linked to processing of softer plant and animal materials (hides, bark, etc.). None of these classes include microblades, as these are not generally retouched tools, though they can be associated with hunting and light-duty activities.

Plotting the data for our site sample suggests a potentially high degree of tool production/use variability in the earliest groups, but ultimately major shifts away from the Denali pattern (Figure 12.2). Early Namu is little different from Dry Creek, whereas Hidden Falls is more typical of the pattern common to later occupations further down the coast. The primary change at Hidden Falls comes in the sudden shift toward high frequencies of light-duty tools and the dramatic reduction in hunting and heavy-duty tools. Although this could be partially due to changes in resources now requiring nets and baskets rather than spears and heavy knives, we suggest that it probably also reflects differences in annual access to lithic raw materials. Denali groups had to carefully plan across the long term to maintain access to toolstone, whereas coastal groups generally had more ubiquitous access to lithics. This does not mean that some finer raw materials were not transported (e.g., Mt. Edziza obsidian) or that some places were

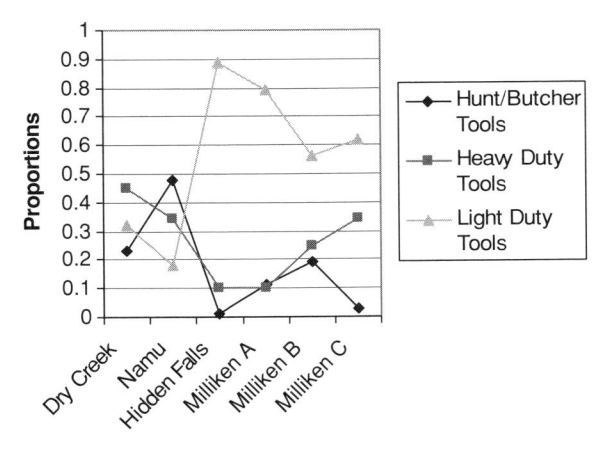

FIGURE 12.2. Plot of functional artifact classes associated with early sites in Beringia (Dry Creek) and on the Northwest Coast.

not bereft of lithic sources (Namu); just that lower grade materials are widespread on the coast, and in the Early Holocene, there were not the same seasonal constraints on access.

These conclusions are further supported by data regarding the persistence of particular tool forms. The use of microblades declined precipitously as groups moved south on the Northwest Coast, so that by the time of the latest Milliken occupations, no microblades at all were in use (Figure 12.3). This does not mean that no more microblades were made or used in the region. Chatters et al. (2006) document Old Cordilleran microblade use on the Columbia Plateau between 8000 and 5000 B.P. Recent research at the Maccallum site in the Lower Fraser Valley of southern British Columbia documents microblades associated with a late (ca. 5000–6000 B.P.) Old Cordilleran context (Lepofsky and Lenert 2005). However, microblade use was certainly infrequent in the south at the later dates. Formal shaped tools (tools other than lightly retouched and/or used flakes) declined in tandem with microblades (Figure 12.4). This was primarily due (with the exception of Namu) to a decline in specific unifacial scraper forms (Figure 12.5). Bifaces fluctuate less predictably (Figure 12.6). We return to the bifaces in our discussion section.

We consider it likely that the Denali RMS, with its strong focus on shaped modular tools, arrived intact on the Northwest Coast

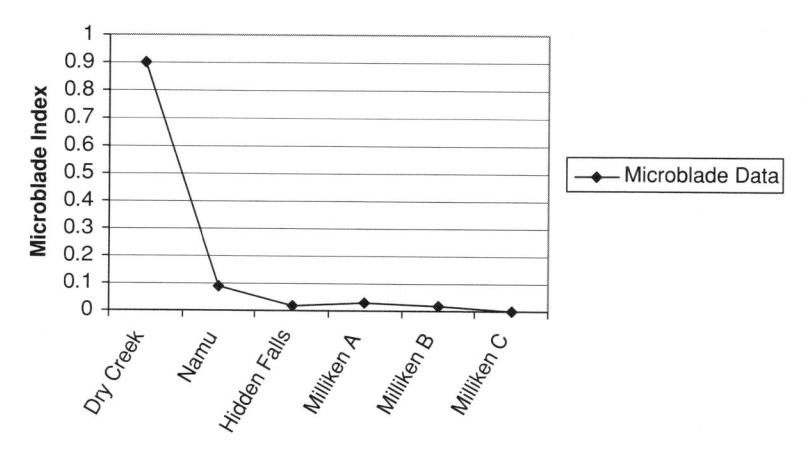

FIGURE 12.3. Plot of microblade index (total microblades divided by total lithic tools and microblades) for early sites in Beringia (Dry Creek) and on the Northwest Coast.

sometime shortly after 10,000 B.P. Lithic technologies changed as some local groups experimented with new preferred technologies to meet the new conditions, producing the more variable early pattern. We can imagine that selective forces could have subsequently narrowed the range of variation, generating the well-recognized North Coast Microblade tradition and Old Cordilleran cultural complex. In essence, the stable later pattern was preceded by a short-lived period

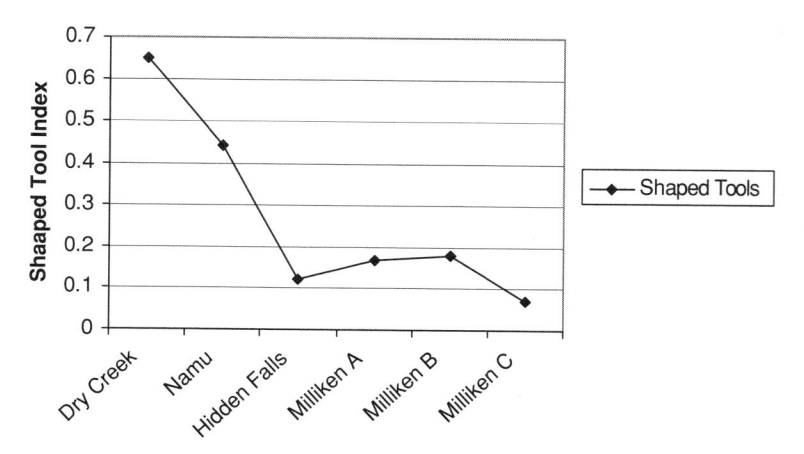

FIGURE 12.4. Plot of shaped tools index (total shaped tools divided by shaped and nonshaped tools) for early sites in Beringia (Dry Creek) and on the Northwest Coast.

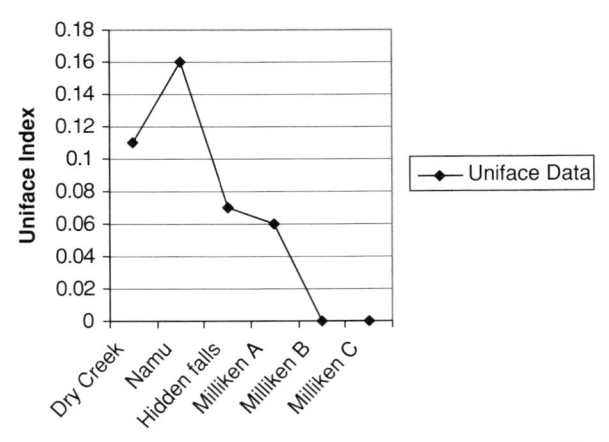

FIGURE 12.5. Plot of uniface index (total unifaces divided by total unifaces and other tool forms) for early sites in Beringia (Dry Creek) and on the Northwest Coast.

of diversification, in this case, as groups found themselves in a new and more benign environment without the previous harsh penalties for mistakes in experimentation. However, this experimentation did not drastically affect mobility and subsistence strategies (despite many specific resource changes). Indeed, it probably permitted them to maintain their familiar socioeconomic package. Thus, although the

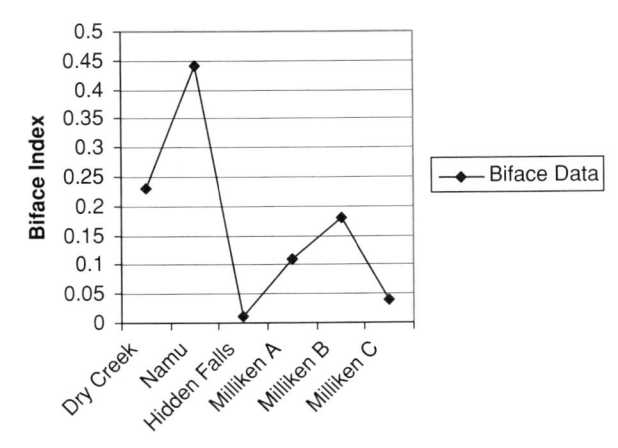

FIGURE 12.6. Plot of biface index (total bifaces divided by total bifaces and other tool forms) for early sites in Beringia (Dry Creek) and on the Northwest Coast.

evolutionary process created some change at the scale of technological organization, it failed to precipitate new RMS. Indeed, it more likely worked to maintain stability in a way similar to that described by some evolutionary biologists (e.g., Leiberman and Dudgeon 1996). Major change was not to come until between 3500 and 4500 B.P., when the forager RMS of the Northwest Microblade users was replaced on the Northwest Coast by populations with collector strategies (Chatters and Prentiss 2005). It had declined at least two millennia earlier in Alaska with climate change (Mason et al. 2001) and replacement by groups with new strategies moving in from the west (Sumnagin [Ackerman 1992; Mason et al. 2001]) and southeast (Paleoindian [Dixon 2001]).

Although these data do have implications for understanding the evolutionary process, they are vague on specific rates of change. They hint at rapid adjustments, as indicated at Hidden Falls, but specifics are currently difficult to reconstruct without more excavated sites. To gain some appreciation for how rapidly major technological changes of this nature (retouch tactics) can occur within in situ cultural systems, we now turn to the winter village collectors of the Canadian Plateau in the Late Prehistoric period.

COMPLEX HUNTER-GATHERERS OF THE MIDDLE FRASER CANYON

Housepit villages such as the Keatley Creek site (Figure 12.7) of the Middle Fraser Canyon in southern British Columbia were occupied by large groups who were very likely ancestral to today's Stl'atl'imx (Lillooet) peoples. Throughout the history of each village (ca. 1800–200 cal. yr. B.P.), occupants relied upon a winter village collector RMS (Hayden 1997; Prentiss et al. 2003, 2006). Lithic resources were relatively abundant within a 20-mile range of both villages. Patterns of access to these sources probably did not change much over time (Hayden et al. 1996a). Keatley Creek knappers maintained a stable technological repertoire of lithic tools centered on unifacial and bifacial tool production from small transported block cores (Hayden et al. 1996b).

Despite stability in many areas, subsistence tactics did change during the lifespans of the Mid-Fraser villages. The detailed chronology of Housepit 7 at Keatley Creek provides us with the opportunity

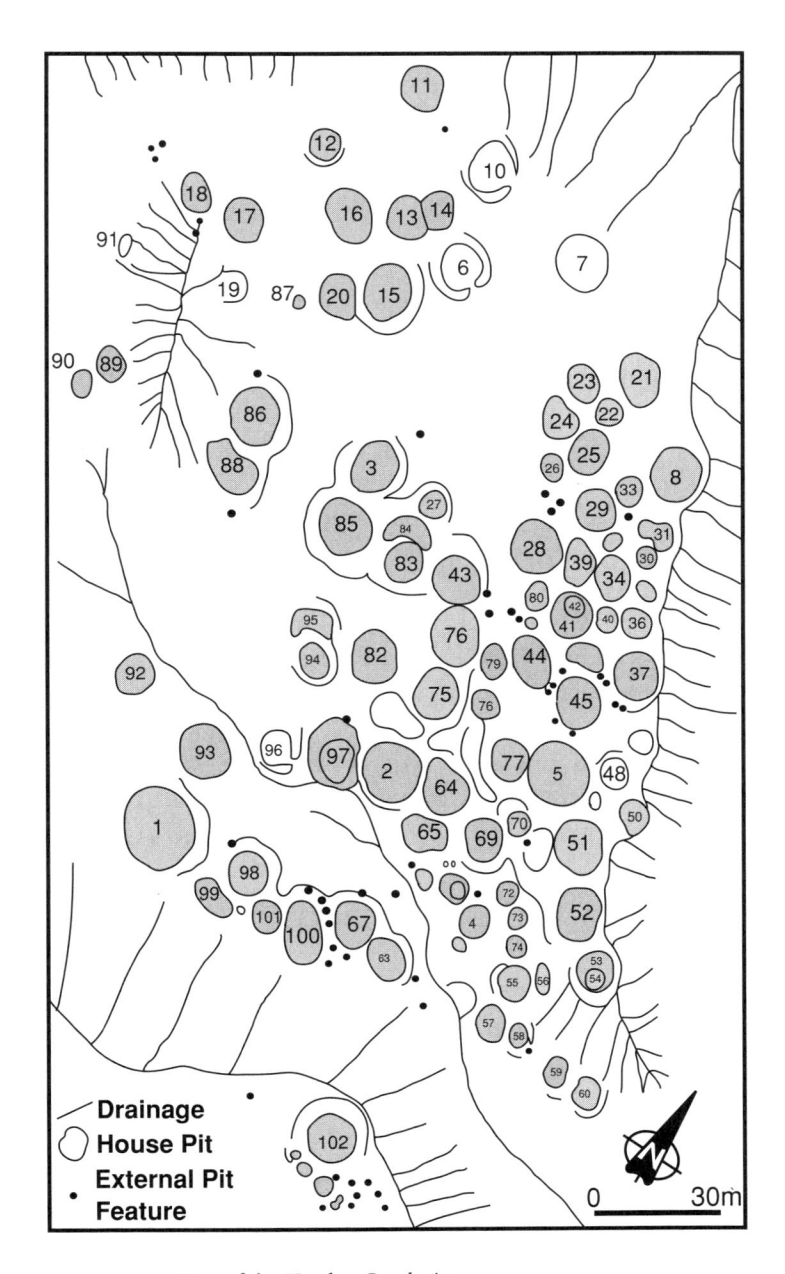

FIGURE 12.7. Map of the Keatley Creek site.

to hold RMS constant again and yet to still assess changes in tool retouch tactics and to address with greater precision rates of change. Recent research at Keatley Creek indicates that after its emergence by ca. 1600 cal. yr. B.P., the village went through a significant subsistence transition at ca. 1200 cal. yr. B.P. (Prentiss et al. 2006). Faunal data suggest that prior to this time village subsistence centered on salmon. However, after 1200 cal. yr. B.P., mammals were increasingly added into the diet. Deer element frequencies change from a nearly complete carcass transport pattern in the earlier period to one of selective transport featuring primarily limbs (especially lower limbs) in the later period. Plant remains also suggest frequent use of more distant higher elevation resources after 1200 cal. yr. B.P. Prentiss et al. (2007) suggest that this pattern reflects local resource depression requiring subsistence diversification and that more frequent and longer-distance logistical mobility excursions occurred during the post–1200 cal. yr. B.P. period.

There are two probable proximate causes for this subsistence transition. First, it is well known that Eastern Pacific productivity declined rapidly at this point (Tunnicliffe et el. 2001), causing declines in salmon in the Columbia (Chatters et al. 1995) and very likely in the Fraser. Second, human populations in the Mid-Fraser had probably peaked at ca. 1200–1300 cal. yr. B.P., with concomitant pressures on local resources. It is likely that it was this pattern of economic stress that led to the famous Keatley Creek abandonment by ca. 800 cal. yr. B.P. (e.g., Hayden and Ryder 1991; Kuijt 2001). The site was not clearly occupied again as an aggregated village.

CHANGE IN RETOUCHED LITHIC TOOLS

The lithic tool assemblage from Keatley Creek was derived from tools collected during excavations of the Housepit 7 rim (e.g., Prentiss et al. 2003) (Figure 12.8) in the 1999, 2001, and 2002 field seasons. We chose all tools from the Early Housepit 7 strata reflecting the early period (ca. 1600–1250 cal. yr. B.P.) associated with intensive salmon use and the Rim 4 strata (ca. 1150–980 cal. yr. B.P.) associated with the later subsistence diversification process. Assemblages from both contexts consisted entirely of chipped stone tools (no ground stone).

Table 12.3. Keatley Creek data (R = richness; D = Shannon diversity index; J = Pielou's evenness index; N = sample size for scrapers only)

	Hunt/ butcher	Light duty	Heavy duty	Scraper index	Retouch index	R	D	J	N
Early Housepit 7	0.33	0.22	0.44	0.7	0.039	3	.8	1.09	10
Late Housepit 7	0.43	0.24	0.33	0.44	0.046	7	1.6	1.89	55

We relied on several measures of variability in stone tools from Keatley Creek (Table 12.3).

In order to explore general trends in chipped stone tool production and use, we examined variation in the three functional classes defined above. During the earlier (lower-mobility and more salmon-dependent) period, lithic tools are most commonly classified as being designed for heavy-duty/woodworking activities, and most of these

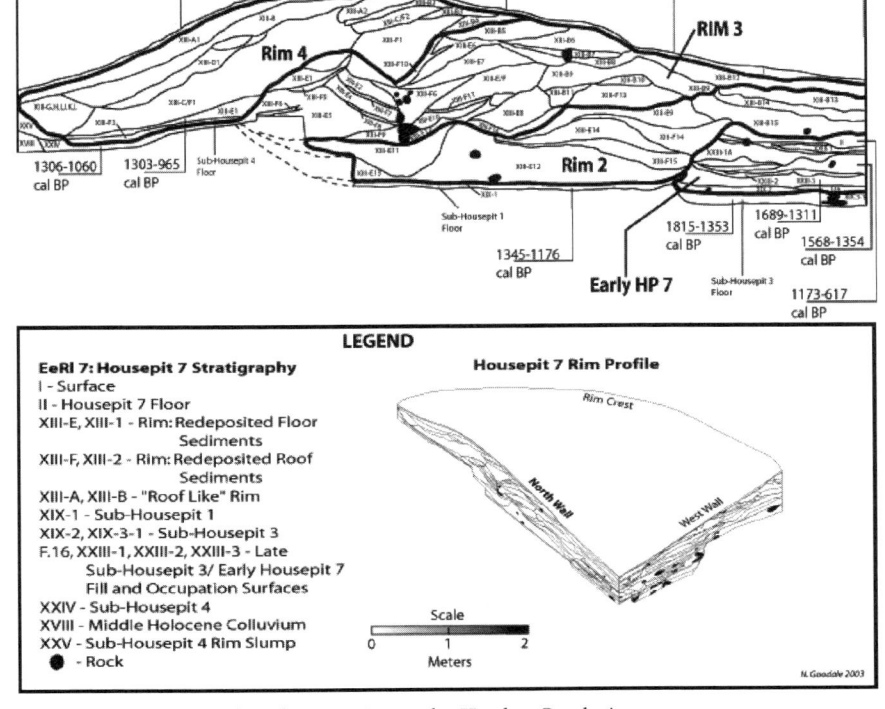

FIGURE 12.8. Stratigraphy of Housepit 7 at the Keatley Creek site.

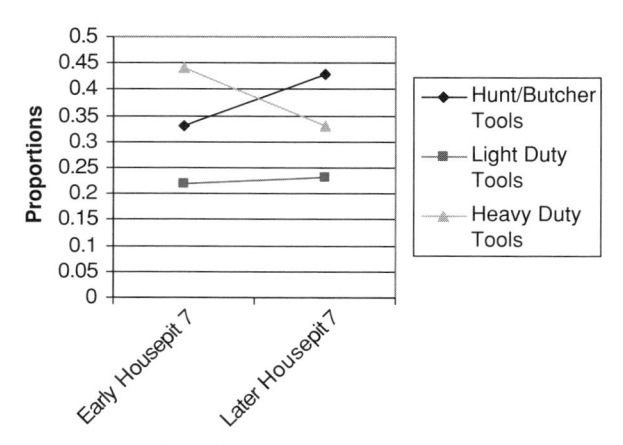

FIGURE 12.9. Plot of functional artifact classes from Housepit 7 strata.

are heavy-duty retouched scrapers (Figure 12.9). Hunting and butchery and light-duty tools occur in slightly lower numbers. After 1200 cal. yr. B.P., the assemblage is dominated by more frequent hunting/butchery tools and somewhat reduced numbers of heavy- and light-duty items (Figure 12.9). Although not dramatic, these patterns make sense if we assume that, all things equal, tool frequencies reflect major subsistence-related tasks. Mid-Fraser salmon fishing required large wooden dip nets, wooden platforms for fisherman, wooden fish traps, and long wood/bone/antler fish leisters (pronged fish spears) (e.g., Romanoff 1992). Preparation of these tools occurred within winter villages and undoubtedly required a wide array of heavy-duty lithic tools. Consequently, the greater the role of salmon fishing, the more frequent heavy-duty tools should be.

To provide a closer look at retouch behavior, we examined variability in retouch patterns, using three additional data sets. There is no significant difference in frequencies of shaped unifaces and bifaces versus unshaped flake tools ($\chi^2 = 2.8$, $p > .1$, d.f. $= 2$). This is probably a by-product of extreme stability in the underlying structure of the winter-village collector RMS and the lack of change in annual access to lithic raw materials. However, there are interesting changes in retouch patterns on scrapers. We calculated a ratio of scrapers with single retouched edges to all scrapers (single, double, convergent, alternating, end, etc.) as an initial indicator of general degree of retouch among scrapers, under the assumption that single scrapers would be

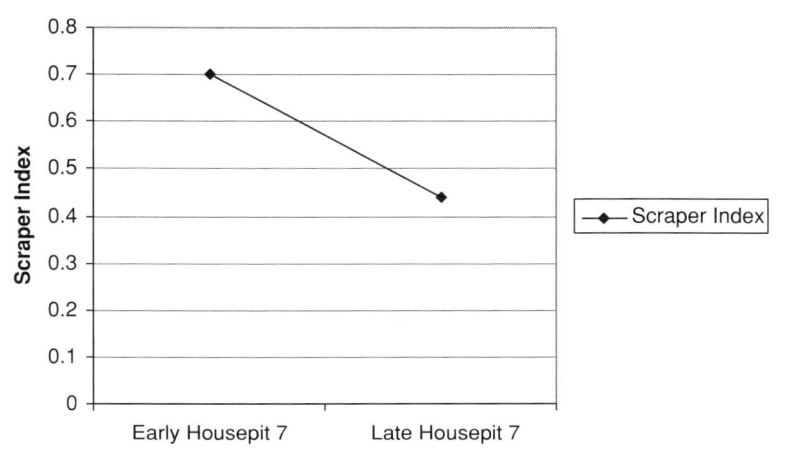

LITHIC TECHNOLOGICAL ORGANIZATION AND EVOLUTION 275

FIGURE 12.10. Plot of single scraper index (single scrapers divided by single scrapers plus all other scrapers) from Housepit 7 strata.

the least intensely modified forms. The single scraper index declines between the early and later period, suggesting a shift away from a focus on single scrapers to a wider range of retouch patterns in the later period (Figure 12.10). This conclusion is supported by diversity indices suggesting that no matter how it is measured (richness, Shannon index, evenness using Pielou's J), the late scraper group is more than twice as diverse (Table 12.2). Then, using methods outlined by Carper (2005), we calculated a ratio of retouch area to total area for all chipped stone scrapers. This index measures variation in the degree of invasive facial flaking on each stone tool. Generally speaking, the more invasively flaked the higher the index score. The data suggest that as scraper form diversified, the degree of facial flaking increased slightly and became much more variable (Figure 12.11). Although these results are anticipated by differences in sample size between the early and later deposits, we suggest that they also make sense in light of theoretical expectations regarding relationships between mobility and tool manufacture. Groups reliant on more frequent long-distance moves associated with specific resource targets in patchy environments are expected to produce a wider range of more formally shaped or extensively modified lithic tools (MacDonald, this volume; Parry and Kelly 1986).

Subsistence and logistical mobility regimes did have important impacts on patterns of Mid-Fraser stone tool production and use. Keatley Creek peoples used a wider range of retouched scrapers, along with a variety of shaped hunting and butchering tools, under conditions of subsistence diversification and, likely, more frequent and geographically extensive logistical mobility. In contrast, under conditions of reduced logistical mobility and specialization in subsistence, these people had been rewarded by a reduction system that tended to produce more examples of simple single-edge scrapers and fewer examples of hunting/butchering tools. We suggest that the Mid-Fraser complex hunter–gatherers shifted technological priorities in order to solve economic contingencies while, as in the case of the Northwest Microblade tradition, effectively preserving their basic way of life (e.g., RMS).

The rate of change in retouch tactics appears to have been high. A period of around 100 years separates the Early Housepit 7 strata from that of the late period. This suggests that the transition required a maximum of perhaps three generations. However, if the decline in salmon documented elsewhere (Chatters et al. 1995) occurred rapidly, as hinted in some paleoecological studies (e.g., Finney et al. 2002), the shift to a tool strategy less tied to sedentary fishing and more to logistically mobile broad-spectrum collecting may have been substantially quicker, perhaps much less than a generation. This implies that even within a stable technoeconomic regime, ecologically linked contingent technological decisions may have rapidly and permanently affected tool retouch tactics in this village. Extending these conclusions to our Northwest Coast case study, we can imagine that Denali complex and early Northwest Coast Microblade tradition peoples could have operated in much the same way, rapidly solving contingencies as needed, and thus producing an archaeological record conforming in many ways to predictions of the organizational theorists.

DISCUSSION

As a field of research, the organization of lithic technology has offered substantial contributions toward development of explanatory (usually ecological) understanding of variability in lithic technology. Many of our results (this study) are anticipated by the organizational models.

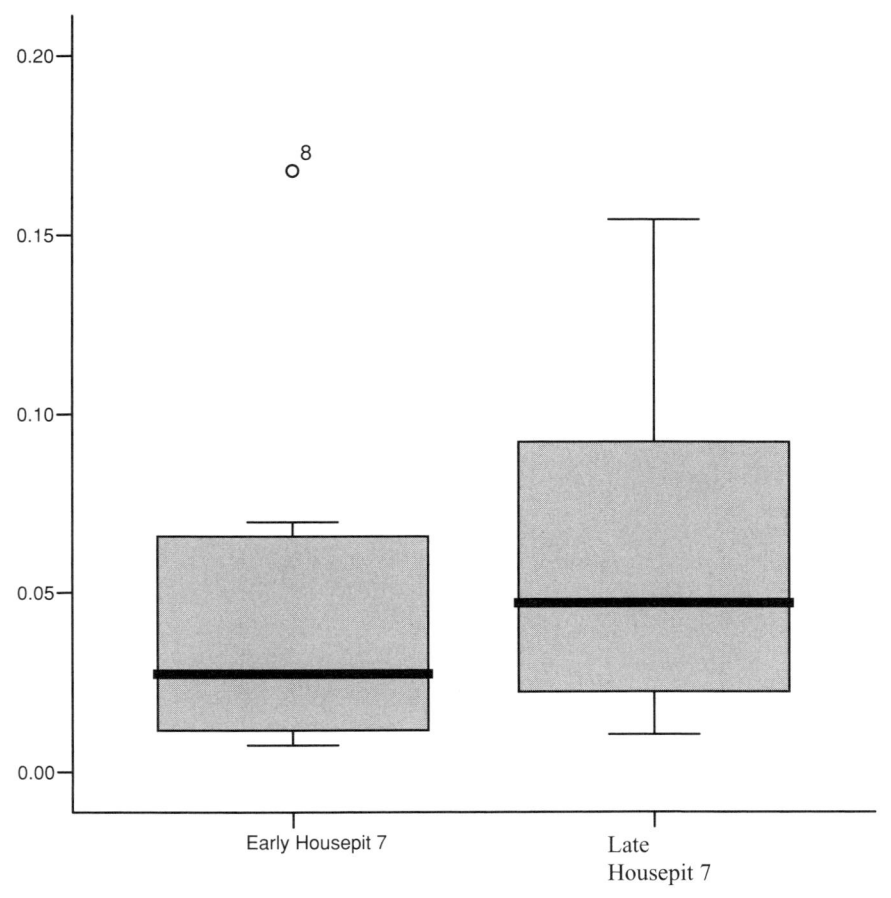

FIGURE 12.11. Box plot of scraper retouch index for Housepit 7 strata.

Residentially mobile hunter–gatherers in seasonal environments with annually variable access to lithic resources tended to use more carefully shaped tools that were often highly transported (e.g., Binford 1979). Holding mobility constant, the prevalence of situational tools rose as annual risk of access to lithic sources declined (e.g., Andrefsky 1994). Sedentary winter-village groups relied on small flake cores, probably obtained and stockpiled during the warm season (e.g., Parry and Kelly 1986). However, as anticipated by Simek (1994), there is still variation in our data that seems to confound the predictions of the organizational models. For example, why did a package of technologies (e.g., large leaf-shaped bifaces and microblades), originally designed for optimal performance on the cold steppes, taigas, and tundras of East Asia, persist on the more temperate Northwest Coast for over 4000 years? Surely there were more nearly optimal technological alternatives!

In this chapter, we have sought to explore this problem by integrating evolutionary thinking into the organization of lithic technology. We argued that although variability in lithic assemblages is a consequence of the kinds of optimal situational and anticipatory decision-making favored by most ecologically minded modelers, it is also the product of long histories of technological learning (information transmission), decision-making (phenotypic action), and feedback (e.g., selection). Thus, lithic technological regimes are by-products of evolutionary history and cannot be fully understood without consideration of this dimension.

We argued that the knowledge associated with lithic technology, when viewed from an organizational and evolutionary perspective, could be viewed as a substrategy within a community's greater resource management strategy. Successful application of that knowledge would serve ecologically to maintain a stable energy management system for the human population, although it might also, from a genealogical standpoint, prevent breakdown and extinction or change at the RMS level (similar to the impact of stabilizing selection in biological contexts). Consequently, we recognize variation in application of lithic reduction tactics often directly tied to historically contingent decisions and contexts, but also constrained by past history.

When examined from this standpoint, the Northwest Microblade tradition can be viewed as reflecting a complex cultural lineage with its evolutionary origin in the Upper Paleolithic of interior East Asia (Yesner 2001). As the originally Asian RMS moved on to the coast and further south, its basic structure or *Bauplan* did not change; yet, as also recognized in the Mid-Fraser case study, users shifted food harvest and lithic reduction tactics to fit the new resource configurations. Consequently, there were some major changes in the frequencies of some forms of tool production and retouch. Yet ancient technological hallmarks such as microblades and biconvex bifaces were retained. Were they vestigial? Could they have persisted merely as markers or symbols of group membership but without a significant functional contribution linked to the specific design? This seems unlikely, given the great span of time and the incredibly wide geographic range associated with these groups. If these were significant group/ethnic markers, within-group transmission processes should have generated greater between-group stylistic variation (e.g., Boyd and Richardson 1985, 1987) during that time span. Another possibility could be that these technologies evolved as critical parts of the strategy, integrated in such a way that loss could be critically damaging, much like loss of a key piece of anatomy in an organism. In other words, they became locked (*sensu* Gould 2002) in place through a selective process that designated them as the tools for acquiring specific resources (specifically some larger mammals) during an annual hunting and gathering cycle. If they served primarily as parts of hunting and butchering tool kits, then their frequency should have declined as group diets became substantially more marine in orientation. Yet their persistence could still be linked to their role in providing terrestrial resources. This hypothesis seems to have some support in the microblade data in particular (Figure 12.3). However, bifaces fluctuate widely (Figure 12.5) in numbers, suggesting that some other factor may have played a role in their persistence. If biconvex or leaf-shaped bifaces evolved within a hunting society, probably assuming roles as highly reliable tools for killing and butchering large mammals, then their continued persistence at fishing sites, sometimes in large numbers (Namu and Milliken, for example) is indeed perplexing. We offer two alternative possibilities. First, it is possible that, despite obvious formal shaping (e.g., targets of a reduction process per Kelly [1986]), these tools had always been designed for flexibility in actual application, perhaps associated with contexts where a high degree of resource processing was required. If this is the case, then there would be no surprise in finding them at kill/processing sites in Beringia and fishing sites on the Northwest Coast (e.g., Namu and Milliken). However, and in partial contrast, we also suggest that, following, these tools, originally designed for game processing, may have been coopted as fish processing knives, particularly associated with warm season salmon harvests. Thus, their persistence is not entirely explained through simple adaptation; exaptation (*sensu* Gould and Vrba 1982) may have played an important role. This could have set the stage for the emergence of the first (similarly shaped) ground slate knives/points (Matson and

Coupland 1995) after ca. 5000 B.P. during the subsequent period of major cultural changes in the Pacific Northwest (Chatters and Prentiss 2005).

Lithic technology can play a critical role in the rarer incidents of radical cultural change – the kind that produce new RMS, as in the emergence of collectors on the Northwest Coast (Prentiss and Chatters 2003), the development of agricultural societies in the Near East (Bar-Yosef and Meadow 1993), or some Paleoindian to Archaic transitions (e.g., Andrefsky 2004). Major RMS changes typically involve radical shifts in the nature and scheduling of food (and other resources) production and distribution tactics, which are typically associated with reorganized labor patterns (Chatters and Prentiss 2005). Lithic tools can, of course, play a critical role in this process. Stylistic alterations can also occur during periods of radical change as groups develop new social and ideological formations marked by new learning tactics (e.g., Bettinger and Eerkens 1999) and applications of new symbolic markers of identity and ideology (prestige technologies [Hayden 1998] and symbolically loaded hafting and flaking patterns such as fluting [MacDonald 1998]). However, as this paper has shown, functional and stylistic change probably occurs even more often during the long periods of RMS stability characterizing many cultural histories. If simple frequencies of lithic tools measured using functional and stylistic typologies are an ambiguous marker of major change, then how can lithics contribute? We suggest that the answer still lies, at least in part, in some of the original thinking of the organizational theorists (e.g., Torrence 1989) who emphasized gaining an understanding of tool design, production systems, and scheduling as integrated into broader socioeconomic systems. However, if we are ever to build a truly comprehensive model, these concepts must be revisited now in an evolutionary framework. Chapters in this volume (e.g., Goodale et al.) suggest that this process is already under way.

REFERENCES CITED

Ackerman, Robert E. 1992. Earliest Stone Industries on the North Pacific Coast of North America. *Arctic Anthropology* 29:18–27.
    1996. Early Maritime Culture Complexes of the Northern Northwest Coast. In *Early Human Occupation in British Columbia*, edited by Roy

L. Carlson and Luke Dalla Bona, pp. 123–32. University of British Columbia Press, Vancouver, B.C.

Anderson, Douglas D. 1988. Onion Portage: The Archaeology of a Stratified Site from the Kobuk River, Northwest Alaska. *Anthropological Papers, University of Alaska* 22:1–163.

Andrefsky, William, Jr. 1994. Raw Material Availability and the Organization of Technology. *American Antiquity* 59:21–35.

———. 1995. Cascade Phase Lithic Technology: An Example from the Lower Snake River. *North American Archaeologist* 16:95–115.

———. 2004. Materials and Contexts for a Culture History of the Columbia Plateau. In *Complex Hunter–Gatherers: Evolution and Organization of Prehistoric Communities on the Plateau of Northwestern North America*, edited by William C. Prentiss and Ian Kuijt, pp. 23–35. University of Utah Press, Salt Lake City.

Bamforth, Douglas B. 1986. Technological Efficiency and Tool Curation. *American Antiquity* 51:38–50.

Bar-Yosef, Ofer, and Richard Meadow. 1993. The Origins of Agriculture in the Near East. In *Last Hunters–First Farmers*, edited by T. Douglas Price and Anne B. Gebauer, pp. 39–94. School of American Research Press, Santa Fe.

Bettinger, Robert L. 2003. Comment on "Cultural Diversification and Decimation in the Prehistoric Record" by William C. Prentiss and James C. Chatters. *Current Anthropology* 44:48–49.

Bettinger, Robert L., and Jelmer W. Eerkens. 1999. Point Typologies, Cultural Transmission, and the Spread of Bow-and-Arrow Technology in the Prehistoric Great Basin. *American Antiquity* 64:231–42.

Binford, Lewis R. 1977. General Introduction. In *For Theory Building in Archaeology*, edited by Lewis R. Binford, pp. 1–10. Academic Press, New York.

———. 1978. *Nunamiut Ethnoarchaeology*. Academic Press, New York.

———. 1979. Organization and Formation Processes: Looking at Curated Technologies. *Journal of Anthropological Research* 35:255–73.

———. 1981. *Bones: Ancient Men and Modern Myths*. Academic Press, Orlando.

———. 2001. *Constructing Frames of Reference*. University of California Press, Berkeley.

Binford, Lewis R., and Sally R. Binford. 1966. A Preliminary Analysis of Functional Variability in the Mousterian of Levallois Facies. *American Anthropologist* 68:238–95.

Bleed, Peter. 1986. The Optimal Design of Hunting Weapons: Maintainability or Reliability. *American Antiquity* 51:737–47.

Boyd, Robert, and Peter J. Richerson. 1985. *Culture and the Evolutionary Process*. Chicago: University of Chicago Press.

1987. The Evolution of Ethnic Markers. *Cultural Anthropology* 2:65–79.

Cannon, Aubrey, and Dongya Y. Yang. 2006. Early Storage and Sedentism on the Pacific Northwest Coast: Ancient DNA Analysis of Salmon Remains from Namu, British Columbia. *American Antiquity* 71:123–40.

Carlson, Roy L. 1996. Early Namu. In *Early Human Occupation in British Columbia*, edited by Roy L. Carlson and Luke Dalla Bona, pp. 83–102. University of British Columbia Press, Vancouver, B.C.

1998. Coastal British Columbia in the Light of North Pacific Maritime Adaptations. *Arctic Anthropology* 35:23–35.

Carper, Raven G. 2005. On the Use of Biface Symmetry to Assess Biface Production Goals. *Lithic Technology* 30:127–44.

Chatters, James C., Virginia L. Butler, Michael J. Scott, David M. Anderson, and Duane A. Neitzel. 1995. A Paleoscience Approach to Estimating the Effects of Climatic Warming on Salmonid Fisheries of the Columbia Basin. *Canadian Special Publication in Fisheries and Aquatic Sciences* 21, 489–96.

Chatters, James C., Steve Hackenberger, and Bruce Lenz. 2006. From Paleo-Indian to Archaic in the Pacific Northwest: Transition or Replacement? Paper presented at the 2006 Northwest Anthropological Conference, Seattle, WA.

Chatters, James C., and William C. Prentiss. 2005. A Darwinian Macroevolutionary Perspective on the Development of Hunter–Gatherer Systems in Northwestern North America. *World Archaeology* 37:46–65.

Clark, Donald W. 2001. Microblade-Culture Systematics in the Far Interior Northwest. *Arctic Anthropology* 38:64–80.

Close, Angela E. 2006. *Finding the People Who Flaked the Stone at English Camp (San Juan Island)*. University of Utah Press, Salt Lake City.

Cressman, Luther S. 1977. *Prehistory of the Far West*. University of Utah Press, Salt Lake City.

Davis, Stanley D. 1989. *The Hidden Falls Site, Baranof Island, Alaska*. Brockport: Alaska Anthropological Association Monograph Series.

Dixon, E. James. 2001. Human Colonization of the Americas: Timing Technology and Process. *Quaternary Science Reviews* 20:277–99.

Dixon, E. J., T. H. Heaton, T. E. Fifield, T. D. Hamilton, D. E. Putnam, and F. Grady. 1997. Late Quaternary Regional Geoarchaeology of Southeast Alaska Karst: A Progress Report. *Geoarchaeology* 12:689–712.

Eerkens, Jelmer W., and Carl P. Lipo. 2005. Cultural Transmission, Copying Errors, and the Generation of Variation in Material Culture and the Archaeological Record. *Journal of Anthropological Archaeology* 24:316–34.

Eldredge, Niles. 1985. *Unfinished Synthesis*. Oxford University Press, New York.

2000. Biological and Material Cultural Evolution: Are There Any True Parallels? *Perspectives in Ethology* 13:113–53.

Fedje, Darrel W., Quenton Mackie, E. James Dixon, and Timothy H. Heaton. 2004. Late Wisconsin Environments and Archaeological Visibility on the Northern Northwest Coast. In *Entering America: Northeast Asia and Beringia before the Last Glacial Maximum*, edited by David B. Madsen, pp. 97–138. University of Utah Press, Salt Lake City.

Finney, Bruce P., Irene Gregory-Eaves, Marianne S. V. Douglas, and John P. Smol. 2002. Fisheries Productivity in the Northeastern Pacific Ocean over the Past 2,200 Years. *Nature* 416:729–33.

Gould, Stephen J. 2002. *The Structure of Evolutionary Theory*. Belknap Press, Cambridge, MA.

Gould, Stephen J., and Elizabeth S. Vrba. 1982. Exaptation – A Missing Term in the Science of Form. *Paleobiology* 8:4–15.

Hayden, Brian. 1997. *The Pithouses of Keatley Creek*. Harcourt Brace College Publishers, Fort Worth, TX.

1998. Practical and Prestige Technologies: The Evolution of Material Systems. *Journal of Archaeological Method and Theory* 5:1–55.

Hayden, Brian, Ed Bakewell, and Robert Gargett. 1996a. The World's Longest-Lived Corporate Group: Lithic Analysis Reveals Prehistoric Social Organization near Lillooet, British Columbia. *American Antiquity* 61:341–56.

Hayden, Brian, Nora Franco, and James Spafford. 1996b. Evaluating Lithic Strategies and Design Criteria. In *Stone Tools: Theoretical Insights into Human Prehistory*, edited by G. Odell, pp. 9–45. Plenum Press, New York.

Hayden, Brian, and June Ryder. 1991. Prehistoric Cultural Collapse in the Lillooet Area. *American Antiquity* 56:50–65.

Hoffecker, John F. 2005. *A Prehistory of the North: Human Settlement of the Higher Latitudes*. Rutgers University Press, New Brunswick, NJ.

Kelly, Robert L. 1986. The Three Sides of a Biface. *American Antiquity* 53:717–34.

Kuhn, Steven L. 1994. A Formal Approach to the Design and Assembly of Mobile Toolkits. *American Antiquity* 59:426–42.

Kuijt, Ian. 2001. Reconsidering the Cause of Cultural Collapse in the Lillooet Area of British Columbia: A Geoarchaeological Perspective. *American Antiquity* 66:692–703.

Lepofsky, Dana, and Michael Lenert. 2005. Report on the 2004 Excavations of the Maccallum Site (DhRk 2), Aggassiz, B.C. Report on file, Archaeology Branch, Ministry of Small Business, Tourism and Culture, Victoria, British Columbia.

Lieberman, B. S., and S. Dudgeon. 1996. An Evaluation of Stabilizing Selection as a Mechanism for Stasis. *Palaeogeography, Palaeoclimatology, Palaeoecology* 127:229–38.

MacDonald, Douglas H. 1998. Subsistence, Sex, and Cultural Transmission in Folsom Culture. *Journal of Anthropological Archaeology* 17:217–39.

Mason, Owen K., Peter M. Bowers, and David M. Hopkins. 2001. The Early Holocene Milankovitch Thermal Maximum and Humans: Adverse Conditions for the Denali Complex of Eastern Beringia. *Quaternary Science Reviews* 20:525–48.

Matson, R. G., and Gary Coupland. 1995. *The Prehistory of the Northwest Coast*. Academic Press, San Diego.

Mitchell, Donald, and David L. Pokotylo. 1996. Early Period Components at the Milliken Site. In *Early Human Occupation in British Columbia*, edited by R. L. Carlson and L. Dalla Bona, pp. 65–82. University of British Columbia Press.

O'Brien, Michael J., John Darwent, and R. Lee Lyman. 1999. Cladistics is Useful for Reconstructing Archaeological Phylogenies: Palaeoindian Points from the Southeastern United States. *Journal of Archaeological Science* 28:1115–36.

O'Brien, Michael J., and R. Lee Lyman. 2003a. *Cladistics and Archaeology*. University of Utah Press, Salt Lake City.

O'Brien, Michael J., and R. Lee Lyman, eds. 2003b. *Style, Function, Transmission: Evolutionary Archaeological Perspectives*. University of Utah Press, Salt Lake City.

Parry, William J., and Robert L. Kelly 1986. Expedient Core Technology and Sedentism. In *The Organization of Core Technology*, edited by J. K. Johnson and C. Morrow, pp. 285–304. Westview Press, Boulder.

Powers, W. Roger, and John F. Hoffecker. 1989. Late Pleistocene Settlement of the Nenana Valley, Central Alaska. *American Antiquity* 54:263–87.

Prentiss, Anna Marie, Natasha Lyons, Lucille E. Harris, Melisse R. P. Burns, and Terry Godin. 2007. The Emergence of Status Inequality in Intermediate Scale Societies: A Demographic and Socio-economic History of the Keatley Creek Site, British Columbia. *Journal of Anthropological Archaeology* 26:299–327.

Prentiss, William C., and James C. Chatters. 2003. Cultural Diversification and Decimation in the Prehistoric Record. *Current Anthropology* 44:33–58.

Prentiss, William C., Michael Lenert, Thomas A. Foor, Nathan B. Goodale, and Trinity Schlegel. 2003. Radiocarbon Dating at Keatley Creek: The Chronology of Occupation at a Complex Hunter–Gatherer Village. *American Antiquity* 68:719–36.

Romanoff, Stephen. 1992. Fraser Lillooet Salmon Fishing. In *A Complex Culture of the British Columbia Plateau*, edited by B. Hayden, pp. 222–65. University of British Columbia Press, Vancouver.

Shott, Michael. 1987. Technological Organization and Settlement Mobility: An Ethnographic Examination. *Journal of Anthropological Research* 42:15–51.

Simek, Jan F. 1994. Organization of Lithic Technology and Evolution: Notes from the Continent. In *The Organization of North American Prehistoric Chipped Stone Tool Technologies*, edited by Philip J. Carr, pp. 118–22. International Monographs in Prehistory, Archaeological Series 7, University of Michigan Press, Ann Arbor.

Spencer, Charles S. 1995. Evolutionary Approaches in Archaeology. *Journal of Archaeological Research* 5:209–64.

Teit, James. 1900. The Thompson Indians of British Columbia. *Memoirs of the American Museum of Natural History* 2(4).

Torrence, Robin. 1989. Re-tooling: Towards a Behavioral Theory of Stone Tools. In *Time, Energy and Stone Tools*, edited by Robin Torrence, pp. 57–66. Cambridge University Press, Cambridge.

Tunnicliffe, V., J. M. O'Connell, and M. R. McQuoid. 2001. A Holocene Record of Marine Fish Remains from the Northeastern Pacific. *Marine Geology* 174:197–210.

Vrba, Elizabeth S., and Niles Eldredge. 1984. Individuals, Hierarchies and Processes: Towards a More Complete Evolutionary Theory. *Paleobiology* 10:146–71.

Yesner, David R. 2001. Human Dispersal Into Interior Alaska: Antecedent Conditions, Mode of Colonization, and Adaptations. *Quaternary Science Reviews* 20:315–27.

13 CHRIS CLARKSON

CHANGING REDUCTION INTENSITY, SETTLEMENT, AND SUBSISTENCE IN WARDAMAN COUNTRY, NORTHERN AUSTRALIA

Abstract

The reduction of stone materials to produce functional tools has formed a vital part of hunter–gatherer technology and land use in Australia for at least the past 45,000 years. Measuring reduction is therefore a vital component of understanding past technology and behavior in Australia, and requires that we develop effective procedures for quantifying reduction for all classes of materials. In this paper, a range of reduction measures are presented for cores, flakes, and different kinds of retouched flakes. These are used to determine the extent to which varying levels of reduction intensity have shaped assemblage composition through time. Changing artefact reduction is tied to systems of land use and provisioning over the past 15,000 years by examining fluctuating occupational intensities, raw material movement, scavenging and recycling of artefacts, and technological diversity. Changes in all of these dimensions of past behavior are linked to Holocene climatic fluctuations and the onset of intensified El Niño/Southern Oscillation (ENSO) events in the past 3,500 years.

My thanks to Bill Andrefsky for inviting me to present a paper at the 2006 SAA session in Puerto Rico, which I unfortunately could not attend, and for then kindly inviting me to contribute a chapter to this volume. The Australian National University, AIATSIS, and the Wenner–Gren Foundation for Anthropological Research funded excavations and fieldwork in Wardaman Country. The Centre for Archaeological Research (Australian National University) and the Australian National Science and Technology Organisation funded many radiocarbon dates. Alex Mackay, Kelvin Hawke, Garry Estcourt, Darren Rousel and Catriona Murray provided valuable field assistance. Bruno David and Josephine Flood helped obtain access to many relevant collections. The author would especially like to thank members of the Wardaman

INTRODUCTION

It is now widely held that differences in extent of reduction condition can explain a great deal of the variation observed within and between lithic assemblages. We need no longer debate this proposition given the numerous published quantitative, substantive and rigorous demonstrations that this is unequivocally the case in many times and places (Andrefsky 2007; Blades 1999; Clarkson 2002, 2005; Dibble 1987, 1988, 1995; Gordon 1993; Hiscock and Attenbrow 2002, 2003, 2005; Hiscock and Clarkson 2005a, 2005b, this volume; Holdaway 1991; Wilson and Andrefsky, this volume). It is now the job of archaeology to explore in detail what changing reduction intensity might mean in terms of changing mobility, subsistence, ways of dealing with risk, and cognitive capacity, different strategies for conserving and extending the use life of tools, and the relationships between reduction and functionality of tools. These are all avenues of research begun on a theoretical level long ago, but only recently have the methodological tools caught up with our intellectual forays into these areas.

This paper presents methods for examining changing levels of core reduction, stages of flake removal, and levels of retouch intensity that will be used to explore major changes in mobility and land use, the organization of technology, technological investment, and the extension of artefact use life in Wardaman Country over time. These changes are argued to stem from varying levels of economic risk and climatic instability in this part of northern Australia since the terminal Pleistocene, and also to show broad parity with proxy measures of population size and occupational intensity over time. Fluctuations in reduction intensity are further linked to alterations in the provisioning of raw materials from local and exotic sources, changes in raw material selection, and changes in implement design that suggest that people sought greater functional specificity, as well as longer periods of functionality, from their tools around the mid-Holocene. These changes in mobility, tool design, and provisioning can be linked to changing world views and systems of symbolic engagement with places, as seen through periods of intensive artistic activity and the caching of valued objects

Aboriginal community – Bill Harney, July Blootcher, Lilly Gin.ginna, and Oliver Raymond – for generously supporting my research in their country.

in rockshelters. The emerging view is one of cultural dynamism and constant change and innovation in this part of northern Australia.

THE MANY FORMS OF ARTEFACT REDUCTION

Core Reduction

The reduction of stone must begin with cores (Hiscock 2007), yet few sophisticated schemes for the analysis of reduction intensity among cores exist. Here a set of observations are made on an assemblage of cores from Wardaman Country using the number of platform rotations as a guide to reduction intensity (Clarkson and O'Connor 2005). Number of core rotations is plotted against various morphological features of cores in Figure 13.1. A consistent series of transformations takes place in the morphology of cores, confirming the utility of the number of core rotations as a measure of core reduction intensity in this region.

First of all, the number of scars found on cores increases with each rotation, as does the percentage of platforms that have more than one conchoidal scar, resulting from the use of a previous core face as the new platform (Figure 13.1A). The percentage of scars found on the core showing step and hinge terminations also increases as core rotation proceeds, as does the minimum external angle of the last platform used on cores. The use of overhang removal also increases steadily throughout the remainder of the reduction sequence. Overhang removal was presumably used to strengthen the platform to better receive the forceful blows required to remove flakes from smaller cores with increasingly high-angled platform edges.

In contrast to these increasing trends, cortex diminishes rapidly in the early stages of reduction, indicating that more surface material was removed prior to the first rotation than at any stage subsequent to it (Figure 13.1B). This idea is supported by the rapid reduction in the weight of cores over the first few rotations. Platform size and the length of flake scars also decrease as rotations increase in number. As length decreases, so too does the elongation of flake scars. Finally, the used portion of the platform edge first increases and then decreases as the viable platform perimeter decreases. This is no doubt largely due to irregularities, left on the core face and platform by previous rotations,

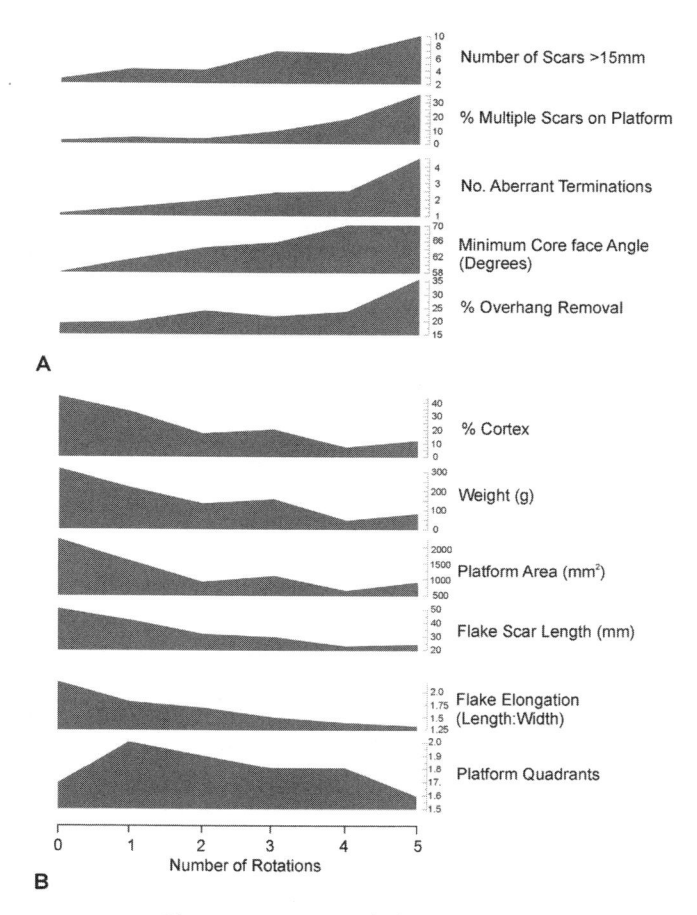

FIGURE 13.1. Changes in core morphology over the reduction sequence.

that restrict flaking to certain areas, but may also reflect decreasing control over force variables that allow successful flake detachments.

Flake Production Stages

Flakes can also be ordered into reduction stages according to the nature of the platform surface, and changes in flake morphology examined as the reduction process progresses confirm directional trends in these features. The four platform types used to order flakes are cortical platforms, representing the first stages of core reduction, platforms formed from a single conchoidal flake scar, representing early to middle stages,

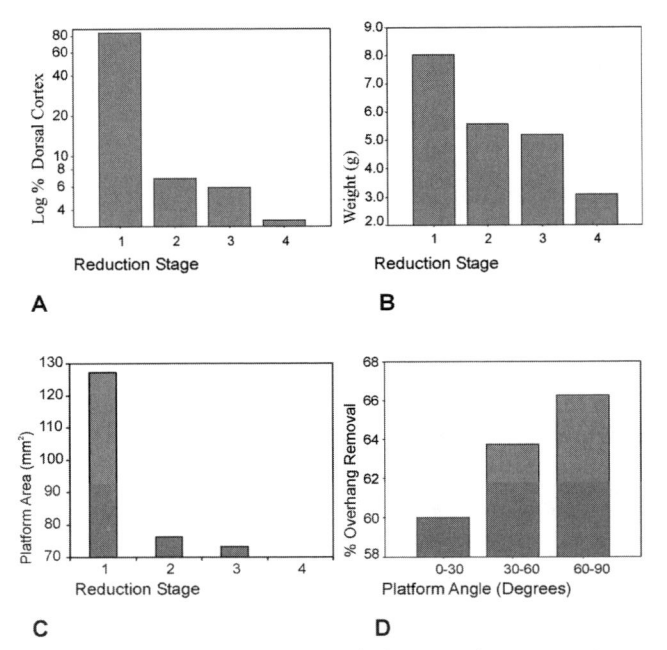

FIGURE 13.2. Changes in flake morphology as reduction continues. Reduction stage is measured using four platform types: cortical, single conchoidal, multiple conchoidal, and bipolar. Changes in morphology include (A) % dorsal cortex, (B) mean weight, (C) platform area, and (D) frequency of overhang removal as platform angle increases.

platforms with multiple conchoidal scarring, representing later stages of freehand percussion, and crushed bipolar platforms, representing the last stage of the reduction continuum.

Figure 13.2 tracks the sorts of changes in flake characteristics that accompany each stage of reduction as inferred from platform surface type, including reductions in cortex, mean weight, and platform area that are consistent with the changes seen in core reduction above.

Retouch Intensity

Retouched flakes are most commonly the subject of detailed reduction analyses. A wide range of techniques have been proposed in recent years (Andrefsky 2006; Eren and Prendergast, this volume; Hiscock and Clarkson 2005a, 2005b; Quinn et al., this volume), but many

Segment Score = 1

Segment Score = 0.5

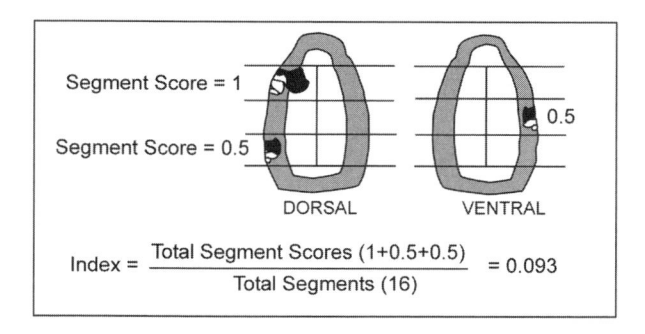

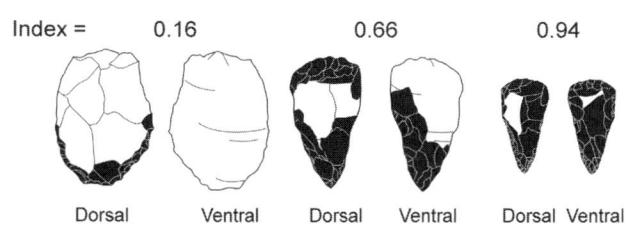

0.5

DORSAL   VENTRAL

$$\text{Index} = \frac{\text{Total Segment Scores } (1+0.5+0.5)}{\text{Total Segments } (16)} = 0.093$$

Index =   0.16   0.66   0.94

Dorsal   Ventral   Dorsal   Ventral   Dorsal Ventral

FIGURE 13.3 The index of invasiveness (from Clarkson 2002).

remain untested in terms of their performance as absolute measures of reduction (i.e., in terms of the percentage of original mass lost from specimens). Two procedures that have been tested in recent years offer a means of measuring flake reduction for two very different forms of retouching (see Clarkson 2002; Wilson and Andrefsky, this volume).

The first of these is a procedure for measuring flake reduction using estimation of retouch scar coverage (Clarkson 2002). This "index of invasiveness" (II) calculates intensity of retouch by estimating the extent of retouching around the perimeter of a flake as well as the degree to which retouch scars encroach onto the dorsal and ventral surfaces. This index has since been modified by Andrefsky (2006) to measure further reduction of artefacts that are already fully bifacial. The index is calculated by conceptually dividing an artefact into eight segments on each face. Each segment is then further divided into an inner "invasive" zone, ascribed a score of 1, and an outer "marginal" zone, ascribed a score of 0.5. Scores of 0 (no retouch), 0.5 (marginal), or 1 (invasive) are allocated to each segment according to the maximum encroachment of scars into one or other of these zones (Figure 13.3). The segment scores are then totaled and divided by 16 to give an index between 0 and 1. Experimental evidence indicates a strong

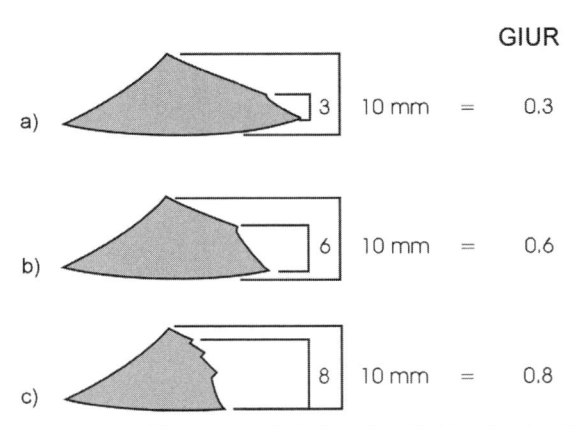

FIGURE 13.4. The geometric index of unifacial reduction (GIUR).

and significant positive relationship exists between the index and the number of retouch blows and the percentage of original weight lost from flakes, which is linear when the percentage of original mass lost is square root transformed.

In assemblages with only unifacial retouch, alternative measures of reduction may be more appropriate, such as Kuhn's (1990) index of reduction. Kuhn's (1990; 1995) "geometric index of unifacial reduction" (GIUR) is a fast and sophisticated quantitative measure of flake margin attrition. The GIUR calculates the ratio between retouch height and flake thickness, expressed as a figure between 0 and 1 (Figure 13.4). Although it is theoretically sensitive to variation in the cross-sectional shape of flakes (and particularly "flat flakes" – see Dibble 1995), recent independent experimental testing has revealed that the GIUR is a robust and reliable measure of dorsal unifacial reduction that is linear when the percentage of original weight lost from specimens is log transformed. The GIUR has been shown to outperform most other measures of unifacial reduction currently available (Hiscock and Clarkson 2005a, 2005b).

The II can therefore be measured on all retouched flakes (but is best suited to assemblages with invasive bifacial retouch), whereas the GIUR can be used to measure unifacial dorsal retouch. When the GIUR is used to measure marginal unifacial "scraper retouch" and the II is used to measure invasive bifacial retouching, the two indices give almost identical results in terms of the rate at which both indices increase relative to the percent of original mass lost from

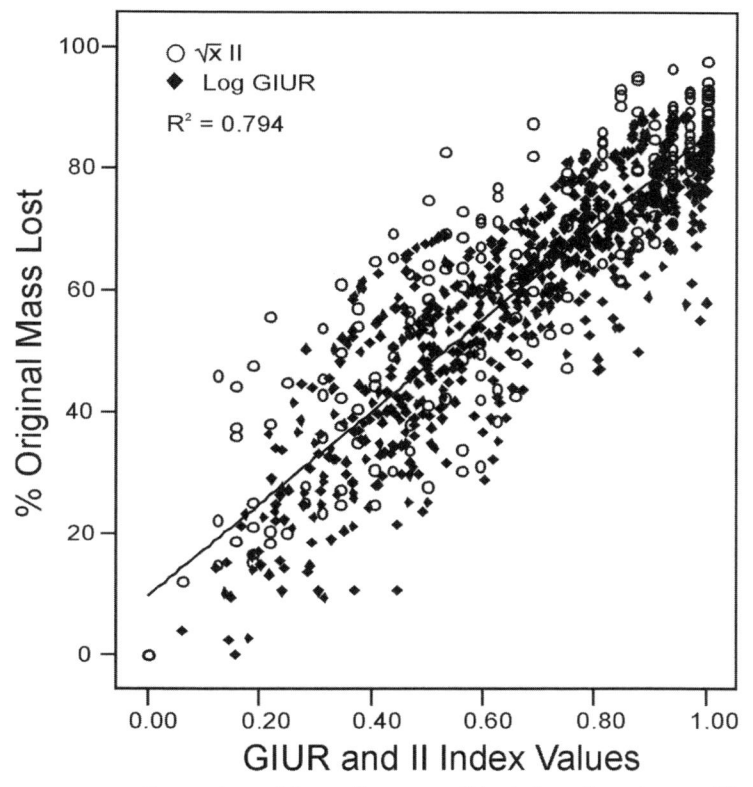

FIGURE 13.5. Comparison of the performance of the index of invasiveness (II) and the geometric index of unifacial reduction (GIUR) from experimental specimens. When transformed using different mathematical procedures (log and square root), the two data sets provide much the same results in terms of rate of index increase for percentage of original mass lost.

flakes. The results of this comparison are shown in Figure 13.5. There is total overlap in the results of both indices, and a linear regression analysis returns a very high Spearman's product–moment correlation coefficient (d.f. = 897, r^2 = .794, F = 3445, p < .005), indicating that 79% of mass lost is explained by one or other of these indices. In combination, they therefore offer the potential to measure reduction in much the same way but on flakes reduced using quite different retouching methods. When used appropriately (i.e., when the GIUR is used to measure unifacial marginal retouch and the II is used to measure bifacial invasive retouch), these techniques provide a powerful means of quantifying retouch intensity on assemblages that contain retouched implements produced using varied reduction techniques. These two measures of retouch intensity will be employed in this manner below.

THE WARDAMAN COUNTRY SITES

To examine changing reduction intensity over time in Wardaman Country, assemblages of cores, flakes, and retouched flakes from four rockshelter sites in Wardaman Country were analyzed to determine if parity exists in changing levels of reduction intensity, and what effects this change might have on assemblage form. The four sites chosen for analysis are Ingaladdi, Garnawala 2, Jagoliya, and Gordolya, all sandstone rockshelters located within ca. 20 km of each other around 250 km south of Darwin (Figure 13.6). These rockshelters have been excavated by various archaeologists over the past 30 years and the results published in a series of reports (Attenbrow et al. 1995; Clarkson 2004, 2007; Clarkson and David 1995; Clarkson and Wallis 2003; David 1991; David et al. 1990, 1992, 1994, 1995; McNiven 1992; Mulvaney 1969). Regression analysis of dated charcoal from sites shows that linear methods provide the best fit between depth and age data. Linear regression allows the basal ages of sites to be determined at ca. 10,000 cal. yr. B.P. for Ingaladdi, ca. 15,000 cal. yr. B.P. for Gordolya, ca. 6,500 cal. yr. B.P. for Jagoliya, and ca. 13,000 cal. yr. B.P. for Garnawala 2 (see Clarkson 2004, 2007 for details). Only the last 10,000 years of technological change are examined in this paper, as sample size can be too small before this time.

TECHNOLOGICAL CHANGE AND REDUCTION SEQUENCES IN WARDAMAN COUNTRY

Large stone artefact assemblages were recovered from excavations at each rockshelter (more than 10,000 artifacts) and these showed enormous technological diversity and the use of a wide range of raw materials. Analysis of the sequences at these sites shows continuous technological change from first occupation to the present, involving the gradual introduction of new retouched forms such as points, tulas, burins, burrens, and lancet flakes, along with other implements such as grindstones and axes, as well as declining emphasis on on-site core reduction and scraper production (Clarkson 2004, 2007). Broad

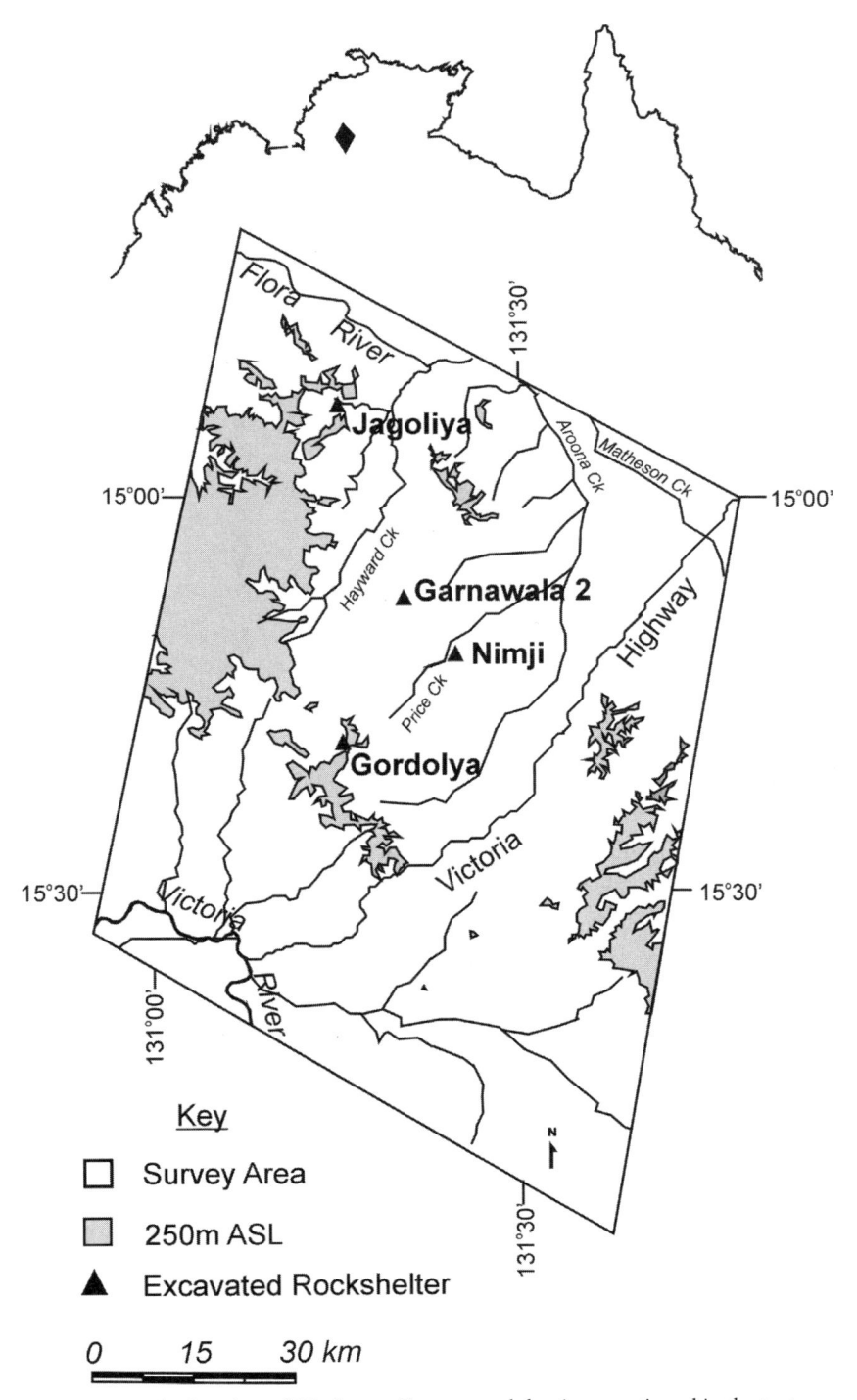

FIGURE 13.6. The location of Wardaman Country and the sites mentioned in the text.

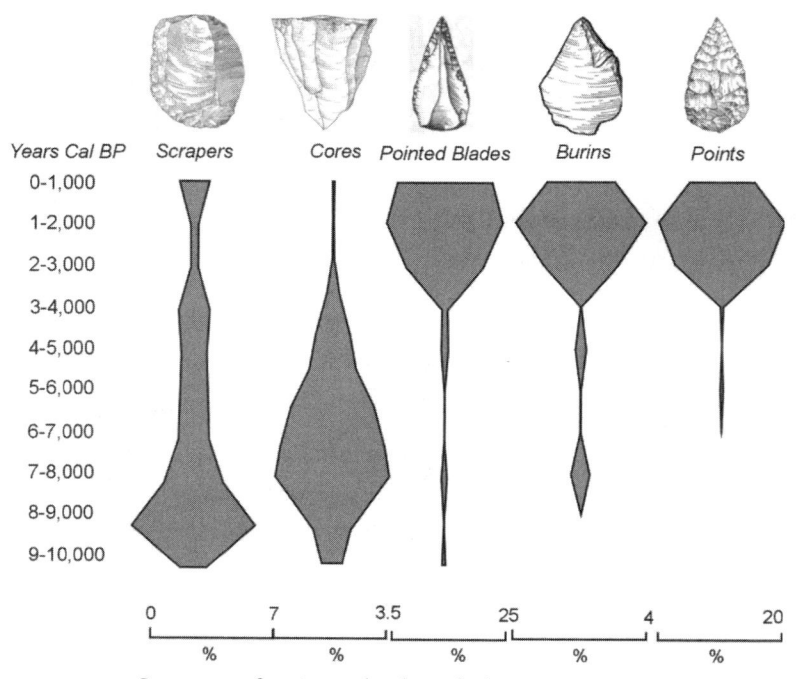

FIGURE 13.7. Summary of major technological changes in Wardaman Country during the Holocene.

technological changes observed at these sites are summarized in Figure 13.7. The various retouched implements graphed in Figure 13.7 represent distinctive production and retouching strategies, and do not represent stages of a single reduction sequence, though some exchange between sequences is likely. Reduction sequences have been determined for these implements using quantitative analysis of changing artefact morphology as retouch increases, and analyses of this material have been presented elsewhere (Clarkson 2002, 2004, 2005, 2006). A summary of typical reduction sequences for Wardaman retouched implements is provided in Figure 13.8, illustrating the distinctive stages and end products that result from these separate reduction pathways.

OCCUPATIONAL INTENSITY

Before exploring changing reduction intensity, it is useful to first examine changing discard rates at rockshelters in Wardaman Country that might reflect changes in frequency of visitation, the size of visiting

| Burrens | Tulas | Grindstones | Axes | Leliras | Kimberley Points |

| 20 | 3.5 | 0.1 | 0.7 | 0.2 | 0.14 | 0.14 |
| % | % | % | % | % | % |

FIGURE 13.7 (*continued*)

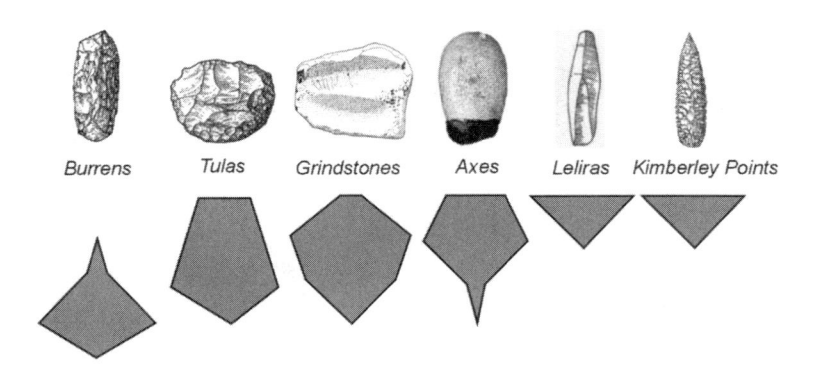

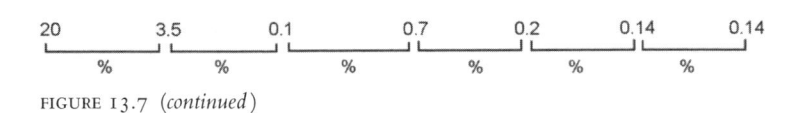

groups, length of stay, or regional population size more generally. This helps place changing reduction intensity within the context of changing land use practices. Figure 13.9 plots the changes in pooled stone artefact discard rates for all four rockshelters over the past 10,000 years. Numbers of complete flakes greater than 2 cm in length are also plotted to determine to what extent fragmentation or small debris size might be partly driving the trends. Weight of bone from one shelter with a large and well-preserved faunal assemblage (Gordolya) is also plotted as a proxy measure of the amount of animal food consumed in shelters through time. Weight of charcoal and burnt earth is also plotted to give an indication of the frequency and intensity of firing conducted in the shelters through time. The quantities of these materials all correspond closely, indicating peaks in discard at 1,500 and 7,000 cal. yr. B.P., perhaps pointing to periods of increased occupational intensity at these times. Fragmentation and other taphonomic processes do not appear to have had a serious effect on artefact numbers.

The first of the distinct peaks in artefact discard coincides with the "early Holocene optimum" or a period of warm, wet, and very stable

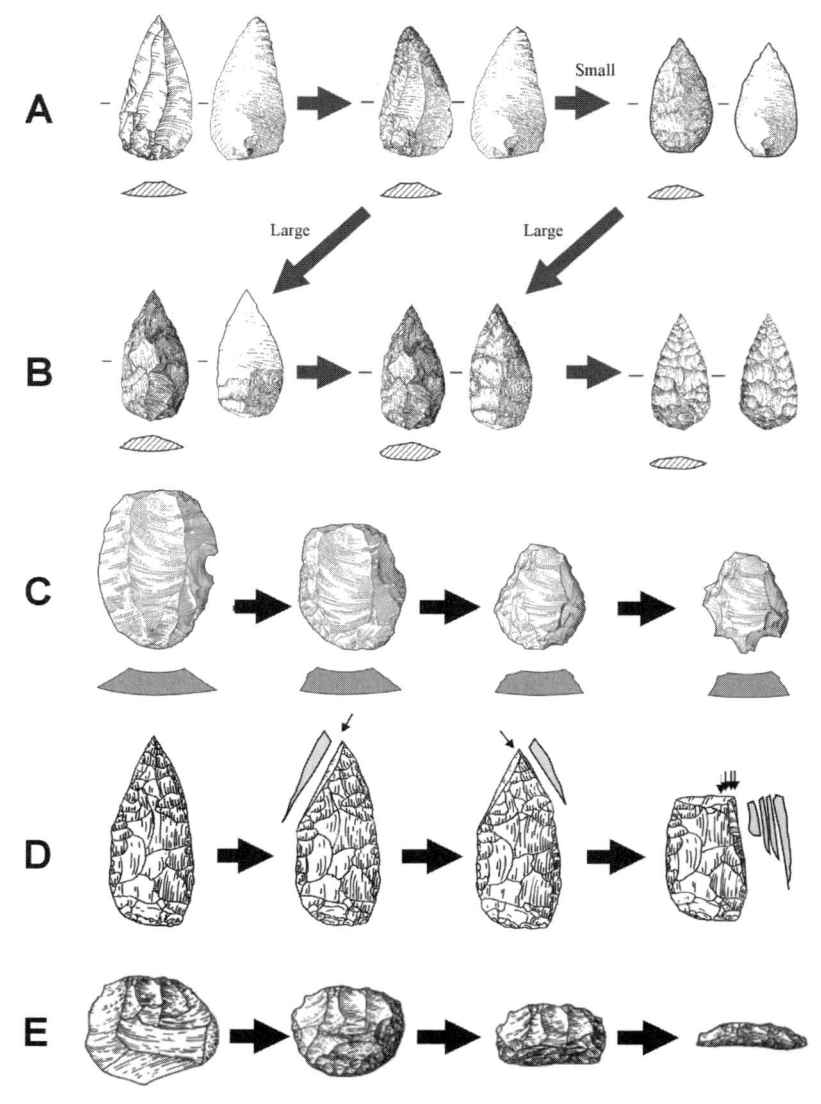

FIGURE 13.8. Typical reduction sequences for common retouched artefacts in Wardaman Country, determined from quantitative analysis of changing flake morphology as indices of reduction intensity increase (see Clarkson 2004, 2006). (A) Unifacial points, (B) bifacial points, (C) unifacial scrapers, (D) burins, and (E) tulas.

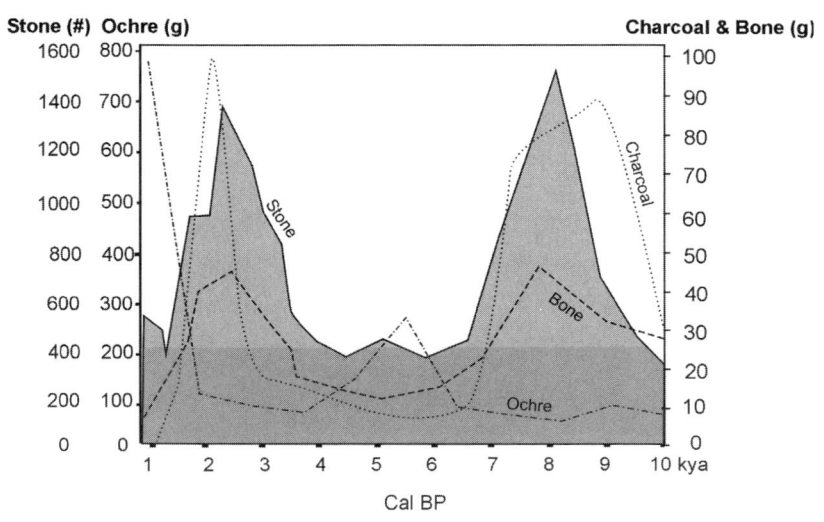

FIGURE 13.9. Changing levels of artistic activity in sites as frequency of site visitation and the nature of land-use changes over time.

climatic conditions. At this time, population or visitation frequency to rockshelters appears to have increased as foragers enjoyed favorable conditions. The trough between the earlier and later peaks coincides with the onset of an intensified El Niño Southern Oscillation (ENSO) climatic regime, resulting in heightened interannual variability in rainfall, with recurrent periods of extreme drought and flood and overall much reduced effective precipitation (Gagan et al. 1994, 2004; Haberle and David 2004; Kershaw 1995; Koutavas et al. 2002; McGlone et al. 1992; Nott and Price 1999; Schulmeister and Lees 1995). This was likely a period of increased economic risk in Wardaman Country of a kind not experienced since people first occupied the region after the Last Glacial Maximum, giving rise to a likely demographic response in the form of reduced population size and site visitations. The second peak in discard rates takes place in the late Holocene. El Niño events decreased in severity markedly around 1,500 years ago (Schulmeister and Lees 1995), and populations appear to have visited sites more frequently or in larger numbers. The past 1,000 years saw diminishing use of the landscapes again as El Niño settled into its modern pattern, with very high interannual variation still characterizing the region today (Dewar 2003).

REDUCTION INTENSITY

We can now examine changes in technology against this backdrop of oscillating occupational intensity. Figure 13.10 plots pooled mean reduction intensity for the four rockshelter sites through time, showing changes in core reduction (Figure 13.10a), flake production stages (Figure 13.10b), and levels of flake retouch intensity (Figure 13.10c). Changes to these indices of reduction are superimposed over changing artefact discard rates for the region. Changes in the reduction intensity at rockshelters in Wardaman Country show the inverse of fluctuations in stone artefact discard through time, with reduction peaking when artefact discard is lowest. The nature of changes in reduction intensity is almost identical for each artefact class and each index of reduction, with the exception of retouched flakes, which retain high levels of reduction after the second peak in stone artefact discard at 1,500 cal. yr. B.P.

Changes in reduction intensity are reflected in typological changes in the region that coincide with shifts up and down the reduction intensity scale for each reduction sequence. The earliest peak in reduction intensity, prior to the 7000 cal. yr. B.P. peak in artefact discard, coincides with high frequency of what are commonly called "steep edge" scrapers, "discoids," and "double side and double end scrapers," which generally represent heavily reduced scrapers. Between the peaks in reduction intensity a new suite of reduction sequences appear, and reduced forms of scrapers, points, tulas, and burins appear at this time. Thus, we see at this time a peak in burrens – or heavily reduced scrapers that typically appear after 5,000 cal. yr. B.P. Highly reduced bifacial point forms are also most common after 4,000 cal. yr. B.P. Tulas are also at their most reduced after 4,000 cal. yr. B.P., with "slug" forms most common between 2,500 and 2,000 cal. yr. B.P. Burins also reach their most reduced stages between 3,000 and 2,000 cal. yr. B.P., with cases of between nine and twelve spalls removed from burins at this time. In some cases, these more reduced forms represent vast extensions of the viable use life of artefacts, with late stage bifacial points, for instance, extending point reduction by up to 60% more mass loss than that which can be taken from unifacial points. Tula slugs also represent incredible reduction of up to 80% of the original mass of implements, indicating that periods of high retouch intensity,

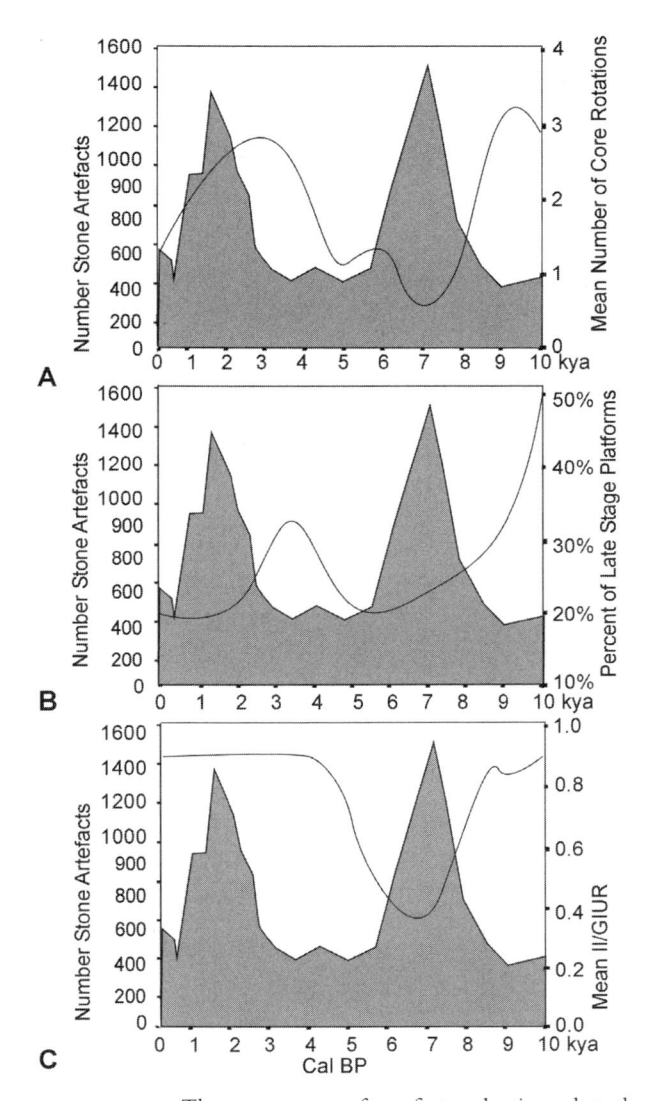

FIGURE 13.10. Three measures of artefact reduction plotted against changes in artefact discard. (A) Numbers of core rotations, (B) percentage of late stage flake platforms, and (C) mean retouch intensity for either the GIUR or the index of invasiveness.

particularly between 2,000 and 4,000 years ago, witnessed incredible levels of implement exhaustion, presumably reflecting strong pressures to curate retouched implements. Bipolar cores and bipolar flakes are also most common during periods of high reduction intensity, reflecting the extension of core reduction into the last stages of core reduction sequences.

Fluctuations in reduction intensity probably reflect several aspects of the organization of technology. The first of these is the intention to extend the reduction of raw materials to get more use from artefacts before they are discarded. This could reflect shortages of raw materials at sites, limited access to replacement stone due to unpredictability of past and future movements (Kuhn 1995; Nelson 1991), or attempts to recoup the manufacturing costs of implements that require significant investments in time and materials to make (Clarkson 2004; Ugan et al. 2003). In all cases, periods of increased reduction intensity point to increased demands on the technological system to keep tools and cores functional longer. It is argued here that the peaks in reduction intensity seen between peaks in artefact discard (interpreted as increased occupational intensity) reflect periods of high mobility when the opportunities to reprovision with replacement raw materials were reduced, and where long use-life artefacts were desirable due to longer periods between provisioning. Interestingly, the new retouched implements that appeared over the past 5,000 years are highly retouched, standardized forms that were no doubt hafted in most cases. Investment in these implements would no doubt have been far greater than in the simpler retouched flakes that were dominant before 5,000 years ago, and hence sustained high levels of retouch intensity after 3000 cal. yr. B.P. may partly reflect the need to recoup the cost of manufacturing these specialized and standardized items.

ASSEMBLAGE DIVERSITY

Toolkit diversity probably reflects a number of important features of past subsistence and technological systems. For instance, different levels of toolkit diversity are often argued to reflect limits on transportation capacity as well as different forms of mobility employed by human groups. The technological literature commonly associates low diversity, multifunctional toolkits with high residential mobility, whereas high diversity toolkits are typically associated with high levels of task specificity and time-limited foraging typical of logistical mobility (Binford 1979; Bleed 1986; Shott 1986; Torrence 1986). Toolkit diversity cannot be measured directly, as "tools" cannot easily be differentiated from "non-tools" in archaeological contexts without conducting use-wear and residue studies. However, it may be possible to measure diversity in this study by counting the number of reduction sequences in existence in Wardaman Country at any one time. Figure 13.7 showed the temporal occurrence of each of the major reduction sequences found in Wardaman Country over the past 10,000 years. Diversity can be seen to increase gradually between 10,000 and 4,000 years ago, and then more dramatically after 4,000 cal. yr. B.P.

One way that this trend may be interpreted is as a gradual shift from greater residential mobility (where fewer, perhaps more multifunctional tools were employed) to logistical mobility (where many specialized tools were employed) through time. This shift can probably be linked to increasing patchiness and a rise in mobile/clumped resources as rainfall became more variable in the past 5,000 years. Increasing the number of specialized tools in the tool kit would presumably have reduced time-stress and subsistence risk by increasing the efficiency of tools as well as the chances of successful resource capture in more time-limited encounters with resources in patchier and less productive environments.

ARTIFACT RECYCLING AND TRANSFORMATION

Another dimension of reduction intensity is the transformation and recycling of artifacts. This practice might give an indication of changes in technological versatility and the use of situational gear (Binford 1979, 1980; Nelson 1991). Situational gear can be important in ensuring successful resource capture when replacement tools fail or are unavailable, and a rise in the frequency of its use may point to an increase in time-limited foraging, and increased constraints on the transported supply. In this case, scavenged flakes with fresh retouch over old weathered surfaces and broken implements that have been recycled or repaired through retouching of the break are considered likely to be one archaeologically visible form of situational gear.

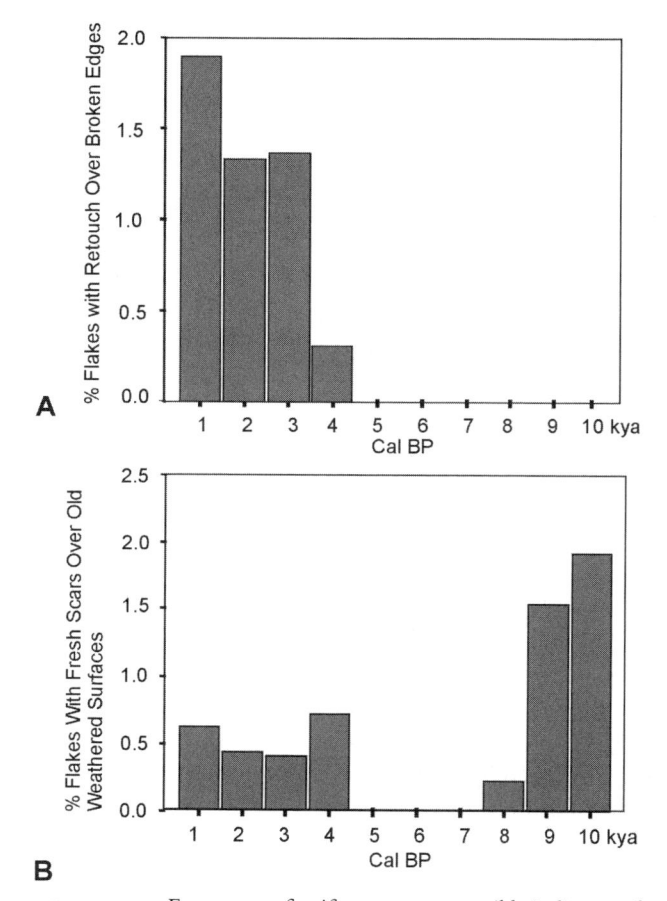

FIGURE 13.11. Frequency of artifact reuse as a possible indicator of the use of situational gear. (A) Frequency of retouch over broken edges and (B) flakes with fresh retouch scars over old weathered surfaces.

Figures 13.11A and 13.11B show the percentages of these kinds of implement recycling in sites through time. There is a clear increase in retouched implement scavenging before 8,000 cal. yr. B.P. and after ca. 4,000 cal. yr. B.P. (Figure 13.11B). This suggests that situational gear only became important at times of greater subsistence stress associated with fluctuating climate and aridity, presumably as mobility, risk, and time-limited foraging increased.

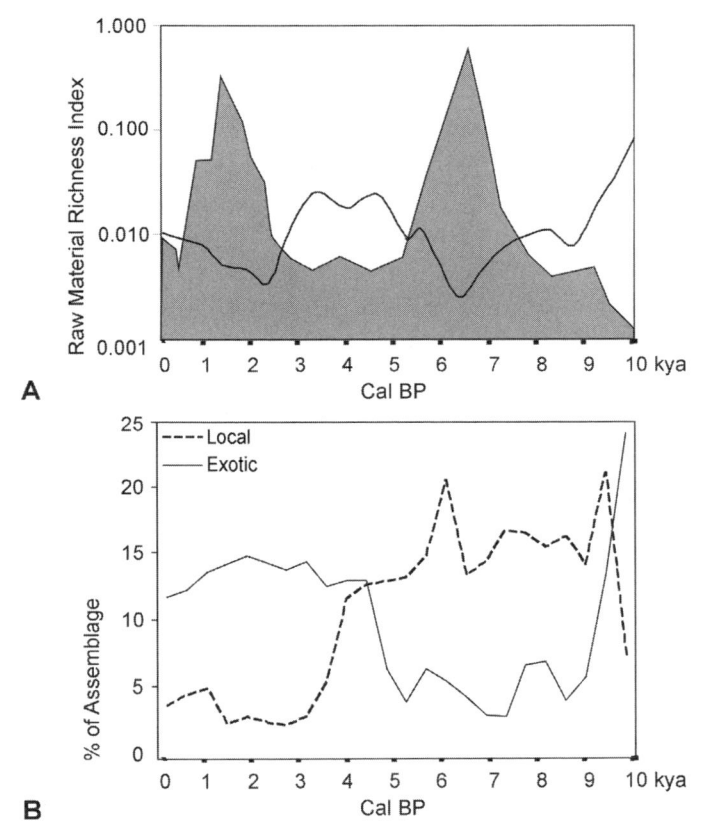

FIGURE 13.12. Changes in mean raw material richness and transport for all four sites. (A) Raw material richness (number of raw material types/sample size) and (B) proportions of local and exotic raw materials.

RAW MATERIAL SELECTION AND TRANSPORT

If raw material diversity reflects the number of locations (or patches) visited by foragers, as it is often argued to do (Binford 1980; Brantingham 2003; Gould and Saggers 1985; MacDonald, this volume), then changes in raw material richness (i.e., number of types of raw materials/sample size) may provide a measure of the diversity of places and stone sources visited, and hence the overall level of mobility and of foraging range. As seen in Figure 13.12A, raw material richness is highest during periods of lower stone artifact discard, suggesting that mobility and patch visitation are also highest at these times.

Changing procurement patterns can also be explored by examining the changing proportions of local (< 10 km) versus exotic stone (> 10 km) over time. Following initially high proportions of exotic stone, local stone dominates the assemblage from ca. 9,000 cal. yr. B.P. until around 4,000 years ago, after which a huge increase in the use of exotic stone takes place. This likely indicates that people were travelling over much greater distances in the past 4,000 years than any time previously. Furthermore, these raw materials are typically of much higher quality than the locally available quartzites at most sites, and are mostly represented by cherts and chalcedonies. This suggests that people tried wherever possible to procure and retain high-quality raw materials after 4,000 cal. yr. B.P., perhaps to meet the higher demands on artifact performance and maintenance after this time (Goodyear 1989). If raw material procurement was embedded, then foragers must have been more mobile to have encountered these more distant, higher-quality raw materials more regularly. If procurement was organized into specialized visits to quarries, the pattern likely indicates greater investment in long-distance journeys to procure higher-quality materials.

PROVISIONING AND TECHNOLOGICAL ORGANIZATION IN WARDAMAN COUNTRY

The organization of technology prior to the first peak in occupational intensity at 7,000 cal. yr. B.P. in Wardaman Country appears to indicate high reduction of cores, and the use of high frequencies of exotic stone for the manufacture of highly reduced scrapers. Raw material richness, as a proxy measure of patch visitation, and the frequency of exotic raw materials, as a measure of foraging range, both indicate high mobility and long-range foraging prior to 7,000 cal. yr. B.P. Technological diversity, however, remains low, suggesting that few if any specialized implements were being manufactured at this time. The signature for this early period appears most consistent with low-diversity provisioning of individuals with small cores and retouched flakes (Kuhn 1995). This combination of portable artifacts and indications of high-frequency, relatively long-range mobility is what would be expected of a highly residentially mobile system of land use, where resources tend to be stable and evenly spaced rather than mobile and clumped (Clarkson 2004, 2007; Horn 1968; Smith 1983).

An apparent reversal in land use and provisioning practices takes place around 7,000 cal. yr. B.P. This coincides with a major peak in stone artefact deposition, which is argued to reflect an increase in occupational intensity. Reduction intensity decreases for cores, flakes, and retouched flakes at this time, as does the proportion of exotic raw materials and raw material richness. The proportion of larger, lightly reduced cores in the assemblage increases, and the proportion of retouched flakes declines (Figure 13.7). This combination of factors points to reduced range and frequency of mobility, an increase in stockpiling of sites with raw materials from local sources as well as a reduced range of distant sources, and a discontinuation of artefact scavenging and recycling. The strategy is clearly what Kuhn (1995) calls "place provisioning" – one that is most suited to more regular movements within the landscape, where the types and frequency of subsistence opportunities can be predicted. In the context of greater predictability of use and lower residential mobility, the peak in occupational intensity also suggests an increase in people visiting the shelters, more frequent visitation, longer visitations, or an overall increase in population density such that all forms of site use are intensified. This period of intensive and predictable occupation of shelters coincides with increased rainfall and reduced interannual variability. These are exactly the sorts of conditions under which we should expect population growth and technological strategies to emerge that take advantage of higher resource abundance and more predictable availability of food and raw materials.

After 5,000 cal. yr. B.P., there is a change in technology back toward the higher levels of reduction that existed in the initial period of occupation. However, the nature of technological strategies employed after 5,000 cal. yr. B.P. appears to differ from that of those employed earlier on. Standardized retouched implements begin to make their appearance from 5,000 to 6,000 cal. yr. B.P., including bifacial points and late reduction stage scrapers (identified as burrens), and raw material richness and the proportion of high-quality exotic stone increases again. Cores begin to drop out of the record and the size of artefacts begins to decrease markedly.

The rate of change intensified at 3,000 cal. yr. B.P., including a marked increase in technological diversity, and a big increase in the use of what is likely situational gear. Reduction intensity and the extension of reduction potential also peaks between 2,000 and 3,000 cal. yr. B.P., with the most reduced stages of many retouched implement forms (i.e., bifacial points, tulas, and burins) and the end points in core reduction (i.e., bipolar cores and flakes) peaking at this time, and then declining soon after. Edge-ground axes, arguably the most extendable and most costly implements to produce, also make their first appearance at this time. The rise in diversity represents a far greater investment in technology in terms of time and labor that can only have been recouped through the extension of artefact use-lives. The greater attention to design and standardization of form at this time was no doubt targeted at increasing the efficiency of tools in performing particular tasks and may also have aided in reducing the risk of subsistence failure by increasing capture rates for mobile prey (as in the case of points) and reducing handling times (as in grindstones and tulas), while also building in an element of flexibility through the transformation and recycling of tool-bits to guard against potential technological shortfall (in the case of burination and the reworking of broken artefacts).

Frequent hafting was almost certainly a key element in technological change after 5,000 cal. yr. B.P., as seen in the diminution of implement forms and an increased concern for standardizing the dimensions of implements (see Clarkson 2004, 2007). Standardization and the use of invasive retouching and bifacial reduction over this period also likely improved the maintainability of tools, by allowing the use of interchangeable forms within costly, predesigned hafts, and by ensuring that problems in implement geometry (such as steep edge angles and the accumulation of step and hinge terminations) could be overcome through careful invasive flaking across the surfaces of implements (Macgregor 2005).

The nature of technological change over the period from 5,000 to 1,500 cal. yr. B.P. can be characterized as a shift from place provisioning toward an extreme form of individual provisioning, where very little besides small, standardized, and highly retouched implements was manufactured. Rates of diverse patch visitation were high, as was the long-distance import of raw materials, implying that mobility had greatly increased over this period. The increase in toolkit diversity, on the other hand, could point to higher logistical rather than residential mobility. This implies that resources may have become more mobile/clumped after 5,000 cal. yr. B.P. (Smith 1983), and that longer, dedicated foraging trips under increasingly time-limited circumstances were required after this time. The rise in risk reduction strategies after 5,000 cal. yr. B.P., such as increased maintainability of toolkits, use of higher-quality raw materials, and increased diversity and increased effectiveness of tools, points to a period of increased subsistence risk at this time. Climatic data indicate that interannual variability peaked between 3,500 and 2,000 cal. yr. B.P. The change in technology toward pronounced individual provisioning points to the use of mechanisms that evolved to cope with decreased certainty of access to resources such as food, water, and stone and increased logistical mobility to reconcile the differences between the location of people and fluctuating resources. Interestingly, Fitzhugh (2003) predicts that foragers facing less than minimum subsistence returns are more likely to invest in technologies that enhance capture rate of larger, high-ranked prey, but as these are driven to decline, the focus should shift to hardier, and more reproductively stable r-selected species. The appearance of points after 5,000 cal. yr. B.P., and around the time of intensified climatic variation, may represent an instance in which foragers attempted first to improve success rates in hunting larger, higher-ranked game such as macropods, but were also led to improve handling times for more reliable, lower-ranking resources such as seeds (as represented by a later rise in the frequency of grindstones) once high-ranked game became depleted.

The final phase of technological change takes place after 1,500 cal. yr. B.P., at which point a second peak in stone artefact discard occurred. This last period also witnessed a decline, though not a total disappearance, in the frequency of some reduction sequences producing highly retouched implement forms. Raw material richness and the proportion of exotic materials decreased, and therefore so presumably did the level of logistical mobility. The fact that most technologies persisted throughout this last period, however, suggests that changes back toward a system of lower mobility and increased abundance and predictability of resource availability after this time were likely to be minor in comparison with the complete system change that took place around 7,000 cal. yr. B.P. This is understandable given that interannual variation in rainfall continued to the present day, and that despite a reduction in overall amplitude, oscillations are still capable of producing regular floods and droughts. Subsistence risk therefore likely remained high right up until the arrival of Europeans, and many of the technological and social strategies set in place after 5,000 cal. yr. B.P. for coping with risk, unpredictable resource abundance, and increased mobility appear to have continued to some degree until historical times.

New technologies and implement forms, such as large Leilira blades and serrated pressure retouching, also appear in the past 1,000 years (Figure 13.7). Leilira blades are ethnographically known to have been traded over large areas and are dated in Wardaman Country to the past 330 years. The appearance of serrated pressure retouch in the past 1,000 years may also indicate interregional contacts with the Kimberley region at this time, as this technique is common (although in undated contexts) in that region, but appears always to have been rare in Wardaman Country. Kimberley points are also a well-documented exchange item, traded over many hundreds of kilometers in the recent past. The emergence of both of these new systems of manufacture and exchange may signify ongoing, albeit altered, social networks for the purpose of (among other things) social storage and ensuring access to resources in bad times (Cashdan 1985; Myers 1982).

CHANGING WORLD VIEWS AND SYMBOLIC ENGAGEMENT WITH PLACE

Given these changing levels of occupational intensity and therefore visitation frequency and familiarity with rockshelters as domestic places, it is significant that ochre deposition peaks in these sites at times of least occupational intensity (Figure 13.9). This suggests that places were more intensively decorated and maintained when visited less frequently, perhaps indicating that they acquired greater significance in ritual activities that required privacy to perform restricted activities and store powerful ritual items without fear of loss or exposure to unsuitable eyes.

Major changes in rock art styles have also been documented in Wardaman Country that coincide with the changes in land use and occupational intensity documented above. These include an early dominance of engraved rock art prior to 3,000 years ago (David et al. 1992; Mulvaney 1969; Watchman et al. 2000), followed by a change to the use of large figurative art panels (perhaps around 3,000 years ago) (Watchman et al. 2000) coincident with infrequent and highly mobile use of rockshelters, and finally the creation of large striped anthropomorphs and other changes in rock art styles in the last 300 years (Attenbrow et al. 1995; David 2002), coincident with the appearance of Leilira trade blades and Kimberley points.

It is not surprising that major readjustments in land use and engagement with landscape should be marked by equivalent changes in other spheres of human life, such as world views, ritual, and art. David (2002) has recently argued, for instance, that ontology – or the system of meaning and preunderstanding with which people interpret the world and their own place in it – is fundamentally shaped by our relationship to and experience of landscape, material objects, and other people, such that a change in any one of these variables will likely also result in a change in systems of belief and preunderstandings about the world. Such changes in worldview are likely signaled by major alterations in land use and the experience of places based on frequency and nature of use, and are expressed in Wardaman Country, among other things, through changing rock art styles and frequency of artistic activity.

SOCIAL STORAGE AND THE SPREAD OF NEW TECHNOLOGIES

In concluding this analysis, it is tempting to try to explain why many of the new retouched technologies only became common in Wardaman Country in the mid to late Holocene, despite their appearance in other parts of Australia much earlier. One possibility is that increasing subsistence risk after 5,000 cal. yr. B.P. (with the onset of heightened ENSO-driven variability) led people to begin to establish forms of social storage through risk reduction reciprocity with neighboring groups. Such social networks may have brought the inhabitants of Wardaman Country into contact with new technologies developed in many parts of Australia that were successful in reducing risk in particular ways in particular regions. Their early appearance in Wardaman

Country in low numbers around 5,000 cal. yr. B.P. might therefore represent a slow trickle of technological information across kinship and linguistic boundaries, but culminating in an efflorescence in use of these new technologies once economic risk intensified around 3,000 years ago.

CONCLUSIONS

This study has worked hard to incorporate state-of-the-art measures of retouch and reduction intensity into broad-based reconstruction of changing Holocene land use and society in Wardaman Country. Although the study draws heavily on hunter–gatherer theory developed in processual and behavioral archaeologies, as well as evolutionary ecology as a means of relating technological changes to optimal subsistence and mobility strategies (each of which arguably requires further testing and clarification in its own right), the data nevertheless stand on their own as a compelling record of major technological change that points to human responsiveness to changing social and environmental conditions, however they are interpreted. This study hopefully demonstrates the potential to move beyond stale debates about the potential for reduction to dramatically alter implement form, to interpretations of reduction data that are behaviorally and culturally meaningful. Such studies provide an opportunity to illustrate ways to reconnect lithics with social and economic theory after many decades of unproductive culture-historical research, and perhaps even position lithics as an important or even primary evidential source for much of human evolution. Studies of Australian lithic reduction and land use are particularly relevant to global lithic studies because they provide an opportunity to examine one component of hunter–gatherer diversity among fully modern populations that have been relatively independent of technological and economic developments in Eurasia. This is of great importance for understanding the independent evolution of technology and the diversity of hunter–gatherer responses over vast time-scales.

REFERENCES CITED

Andrefsky, W. 2006. Experimental and Archaeological Verification of an Index of Retouch for Hafted Bifaces. *American Antiqtuity* 71:743–758.

2007. Cobble Tool or Cobble Core: Exploring Alternative Typologies. In *Tools versus Cores: Alternative Approaches to Stone Tool Analysis*, edited by S. P. McPherron, pp. 253–66. Cambridge Scholars Publishing, Cambridge.

Attenbrow, V., B. David, and J. Flood. 1995. Mennge-ya and the Origins of Points: New Insights into the Appearance of Points in the Semi-arid Zone of the Northern Territory. *Archaeology in Oceania* 30:105–19.

Binford, L. R. 1979. Organizational and Formation Processes: Looking at Curated Technologies. *Journal of Anthropological Research* 35:255–73.

——— 1980. Willow Smoke and Dog's Tails: Hunter–Gatherer Settlement Systems and Archaeological Site Formation. *American Antiquity* 45:4–20.

Blades, B. 1999. Aurignacian Settlement Patterns in the Vezere Valley. *Current Anthropology* 40:712–23.

Bleed, P. 1986. The Optimal Design of Hunting Weapons: Maintainability or Reliability. *American Antiquity* 51:737–47.

Brantingham, P. J. 2003. A Neutral Model of Stone Raw Material Procurement. *American Antiquity* 68:487–509.

Cashdan, E. 1985. Coping with Risk Reciprocity among the Basara of Northern Botswana. *Man* 20:454–74.

Clarkson, C. 2002. An Index of Invasiveness for the Measurement of Unifacial and Bifacial Retouch: A Theoretical, Experimental and Archaeological Verification. *Journal of Archaeological Science* 1:65–75.

——— 2004. Technological Provisioning and Assemblage Variation in the Eastern Victoria River Region, Northern Australia: A Darwinian Perspective. Ph.D. thesis, Australian National University.

——— 2005. Tenuous Types: 'Scraper' Reduction Continuums in Wardaman Country, Northern Australia. In *Lithics "Down Under": Australian Perspectives on Stone Artefact Reduction, Use and Classification*, edited by C. Clarkson and L. Lamb, pp. 21–34. BAR International Series S1408. Archaeopress, Oxford.

——— 2006. Explaining Point Variability in the Eastern Victoria River Region, Northern Territory. *Archaeology in Oceania* 41:97–106.

——— 2007. *Lithics in the Land of the Lightning Brothers: The Archaeology of Wardaman Country, Northern Territory*. Terra Australis. ANU E Press, Canberra.

Clarkson, C., and B. David. 1995. The Antiquity of Blades and Points Revisited: Investigating the Emergence of Systematic Blade Production South-West of Arnhem Land, Northern Australia. *The Artefact* 18:22–44.

Clarkson, C., and S. O'Connor. 2005. An Introduction to Stone Artefact Analysis. In *Archaeology in Practice: A Student Guide to Archaeological Analyses*, edited by J. Balme and A. Patterson. Blackwell, New York.

Clarkson, C., and L. A. Wallis. 2003. The Search for El Nino/Southern Oscillation in Archaeological Sites: Recent Phytolith Analysis at Jugali-ya Rockshelter, Wardaman Country, Australia. In *Phytolith and Starch Research in the Australian–Pacific–Asian Regions: The State of the Art*, edited by D. M. Hart and L. A. Wallis, pp. 137–152. Terra Australis 19. Pandanus Books, Canberra.

David, B. 1991. Archaeological Excavations at Yiwarlarlay 1: Site Report. *Memoirs of the Queensland Museum* 30:373–80.

— 2002. *Landscapes, Rock-Art and the Dreaming: An Archaeology of Preunderstanding*. Leicester University Press, London.

David, B., D. Chant, and J. Flood. 1992. Jalijbang 2 and the Distribution of Pecked Faces in Australia. *Memoirs of the Queensland Museum* 32:61–77.

David, B., J. Collins, B. Barker, J. Flood, and R. Gunn. 1995. Archaeological Research in Wardaman Country, Northern Territory: The Lightning Brothers Project 1990–91 Field Seasons. *Australian Archaeology* 41:1–10.

David, B., M. David, J. Flood, and R. Frost. 1990. Rock Paintings of the Yingalarri Region: Preliminary Results and Implications for an Archaeology of Inter-regional Relations in Northern Australia. *Memoirs of the Queensland Museum* 28:443–62.

David, B., I. McNiven, V. Attenbrow, and J. Flood. 1994. Of Lightning Brothers and White Cockatoos: Dating the Antiquity of Signifying Systems in the Northern Territory, Australia. *Antiquity* 68:241–51.

Dewar, R. 2003. Rainfall Variability and Subsistence Systems in Southeast Asia and the Western Pacific. *Current Anthropology* 44:369–88.

Dibble, H. 1987. Reduction Sequences in the Manufacture of Mousterian Implements of France. In *The Pleistocene Old World Regional Perspectives*, edited by O. Soffer, pp. 33–45. New York: Plenum Press, New York.

— 1988. Typological Aspects of Reduction and Intensity of Utilization of Lithic Resources in the French Mousterian. In *Upper Pleistocene Prehistory of Western Eurasia*, edited by H. Dibble and A. White, pp. 181–98. University of Pennsylvania, Philadelphia.

— 1995. Middle Paleolithic Scraper Reduction: Background, Clarification, and Review of Evidence to Date. *Journal of Archaeological Method and Theory* 2:299–368.

Fitzhugh, Ben. 2003. The Evolution of Complex Hunter-Gatherers. Interdisciplinary Contributions to Archaeology. Kluwer Academic/Plenum Publishers, New York.

Gagan, M. K., A. R. Chivas, and P. J. Isdale. 1994. High-Resolution Isotopic Records of the Mid-Holocene Tropical Western Pacific. *Earth and Planetary Sciences* 121:549–58.

Gagan, M., E. J. Hendy, S. G. Haberle, and W. S. Hantoro. 2004. Postglacial Evolution of the Indo–Pacific Warm Pool and El Niño–Southern Oscillation. *Quaternary International* 118–19:127–43.

Goodyear, A. C. 1989. A Hypothesis for the Use of Crypto-crystalline Raw Materials among Paleoindian Groups of North America. In *Eastern Paleoindian Lithic Resource Use*, edited by C. G. Ellis and J. C. Lothrop, pp. 1–9. Westview Press, Boulder.

Gordon, D. 1993. Mousterian Tool Selection, Reduction, and Discard at Ghar, Israel. *Journal of Field Archaeology* 20:205–18.

Gould, R. A., and S. Saggers. 1985. Lithic Procurement in Central Australia: A Closer Look at Binford's Idea of Embeddedness in Archaeology. *American Antiquity* 50:117–36.

Haberle, S. G., and B. David. 2004. Climates of Change: Human Dimensions of Holocene Environmental Change in Low Latitudes of the PEPII Transect. *Quaternary International* 118–19:165–79.

——— 2007. Looking the Other Way: A Materialist/Technological Approach to Classifying Tools and Implements, Cores and Retouched Flakes, with Examples from Australia. In *Tools versus Cores: Alternative Approaches to Stone Tool Analysis*, edited by S. P. McPherron, pp. 198–222. Cambridge Scholars Publishing, Cambridge.

Hiscock, P., and V. Attenbrow. 2002. Early Australian Implement Variation: A Reduction Model. *Journal of Archaeological Science* 30:239–49.

——— 2003. Morphological and Reduction Continuums in Eastern Australia: Measurement and Implications at Capertee 3. *Tempus* 7:167–74.

——— 2005. *Australia's Eastern Regional Sequence Revisited: Technology and Change at Capertee 3*. BAR S1397. Archaeopress, Oxford.

Hiscock, P., and C. Clarkson. 2005a. Experimental Evaluation of Kuhn's Geometric Index of Reduction and the Flat-Flake Problem. *Journal of Archaeological Science* 32:1015–22.

——— 2005b. Measuring Artefact Reduction – An Examination of Kuhn's Geometric Index of Reduction. In *Lithics "Down Under": Australian Perspectives on Stone Artefact Reduction, Use and Classification*, edited by C. Clarkson and L. Lamb, pp. 7–20. BAR International Series S1408. Archaeopress, Oxford.

Holdaway, S. 1991. Resharpening Reduction and Lithic Assemblage Variability Across the Middle to Upper Paleolithic Transition. Ph.D. dissertation, University of Pennsylvania, Philadelphia.

Horn, H. S. 1968. The Adaptive Significance of Colonial Nesting in the Brewer's Blackbird (*Euphagus cyanocephalus*). *Ecology* 49:682–94.

Kershaw, A. P. 1995. Environmental Change in Greater Australia. *Antiquity* 69:656–76.

Koutavas, A., J. Lynch-Steiglitz, T. M. J. Marchitto, and J. P. Sachs. 2002. El Niño–Like Pattern in Ice Age Tropical Pacific Sea Surface Temperature. *Science* 297:226–31.

Kuhn, S.L. 1990. A Geometric Index of Reduction for Unifacial Stone Tools. *Journal of Archaeological Science* 17:585–93.

1995. *Mousterian Lithic Technology.* Princeton University Press, Princeton.

Macgregor, O. 2005. Abrupt Terminations and Stone Artefact Reduction Potential. In *Lithics "Down Under": Australian Perspectives on Stone Artefact Reduction, Use and Classification*, edited by C. Clarkson and L. Lamb, pp. 57–66. BAR International Series S1408. Archaeopress, Oxford.

McGlone, M. S., A. P. Kershaw, and V. Markgraf. 1992. El Niño/Southern Oscillation Climatic Variability in Australasian and South American Paleoenvironmental Records. In *El Niño: Historical and Paleclimatic Aspects of the Southern Oscillation*, edited by H. F. Diaz and V. Markgraf, pp. 435–62. Cambridge University Press, Cambridge.

McNiven, I. 1992. Delamere 3: Further Excavations at Yiwarlarlay (Lightning Brothers Site), Northern Territory. *Australian Aboriginal Studies* 1992:67–73.

Mulvaney, D. J. 1969. *The Prehistory of Australia.* Thames and Hudson, London.

Myers, F. 1982. Always Ask: Resource Use and Land Ownership among Pintupi Aborigines. In *Resource Managers*, edited by N. Williams and E. S. Hunn, pp. 173–96. Australian Institute of Aboriginal Studies, Canberra.

Nelson, M. C. 1991. The Study of Technological Organization. *Archaeological Method and Theory* 3:57–100.

Nott, J., and D. Price. 1999. Waterfalls, Floods and Climate Change: Evidence from Tropical Australia. *Earth and Planetary Science Letters* 171:267–76.

Schulmeister, J., and B. Lees. 1995. Pollen Evidence from Tropical Australia for the Onset of ENSO-Dominated Climate at c. 4000 BP. *The Holocene* 5:10–18.

Shott, M. J. 1986. Technological Organization and Settlement Mobility: An Ethnographic Examination. *Journal of Anthropological Research* 42:15–51.

Smith, E. A. 1983. Anthropological Applications of Optimal Foraging Theory: A Critical Review. *Current Anthropology* 24:625–51.

Torrence, R. 1986. *Production and Exchange of Stone Tools: Prehistoric Obsidian in the Aegean.* Cambridge University Press, Cambridge.

Ugan, A., J. Bright, and A. Rogers. 2003. When Is Technology Worth the Trouble? *Journal of Archaeological Science* 30:1315–29.

Watchman, A. L., B. David, I. McNiven, and J. Flood. 2000. Microarchaeology of Engraved and Painted Rock Surface Crusts at Yiwarlarlay (the Lightning Brothers Site), Northern Territory, Australia. *Journal of Archaeological Science* 27:315–25.

14 NATHAN B. GOODALE, IAN KUIJT, SHANE J.
MACFARLAN, CURTIS OSTERHOUDT, AND BILL
FINLAYSON

LITHIC CORE REDUCTION TECHNIQUES: MODELING EXPECTED DIVERSITY

Abstract

We define diversity in core reduction systems as the degree of deviation from the most efficient means to proceed from the start to the end product exhibited in a given core reduction system. Because lithic core reduction systems are often characterized along a continuum of high or low degree of diversity, some archaeologists have suggested that assemblage diversity is linked to raw material availability and quality. In this paper we provide a model that predicts when humans would favor less systematic core reduction techniques as opposed to those that are more systematic. The model incorporates three factors influencing diversity in core reduction techniques: raw material availability, raw material quality, and the ratio of producers to consumers. We provide the model and then estimate where several case examples from different archaeological contexts fit within the expectations. This allows us to generate hypotheses about the relationship of producers and consumers who manufactured the assemblages.

We extend our thanks to William Andrefsky, Jr., for inviting us to be a part of his organized symposium at the 71st Annual Meeting of the Society of American Archaeology in San Juan, Puerto Rico. Additionally, his editorial comments have significantly improved the quality of the paper. We also thank the discussants Margaret Nelson and Michael Shott and several anonymous reviewers for their helpful comments and critiques regarding our arguments presented here. We also thank Diane Curewitz, who provided very helpful technical edits to the final draft of this manuscript. As always, any omissions or flaws in logic are completely our responsibility.

INTRODUCTION

The process of lithic core reduction is often described as systematic (nearly uniform) or unsystematic (highly variable) (Bleed 2001; Brantingham et al. 2000; Root 1997). For example, some core reduction systems represent human interaction with raw materials that are much more prone to knapping error and failure rates, whereas others appear to follow very specific chains (for an example of each, see Bleed 1996, 101–2). Core reduction systems that are highly uniform usually have less/little sign of rejuvenation due to knapping error, whereas other systems are almost cyclical in nature, indicated by a series of rejuvenation events and techniques that compensate for knapping error and/or raw material failure.

Some core reduction systems are described on a continuum (Shott 1996), ranging from nearly uniform (low diversification) at one end of the axis to unsystematic (high diversification) at the other. Diversity represented within a particular reduction system is likely the result of interaction between human behavior (e.g., social organization or knapping skill) and raw material quality and availability. Subsequently, we equate the *diversity* in reduction techniques to the degree of deviation from uniformity.

Some goals have a potential single most efficient solution. For example, there is always the opportunity to maximize the usefulness of lithic raw material by constraining the reduction sequence to a small degree of diversity around the optimal operation chain. Many times, however, goals can be achieved with less efficient strategies that could produce a high degree of deviation from the optimal operation chain. In light of this, we define *diversity* with respect to core reduction systems as the degree of deviation from the most efficient means to proceed from the start (such as the selection of a cobble to the setup of the core) to the end product (tool blank production) exhibited in a given core reduction sequence. Efficiency is quantifiable with time, energy, and raw material use in relation to the production sequence (Costin 1991: 37). For the purposes of this paper, we are only concerned with the end product of core reduction (tool blanks), not the subsequent negotiations of tool production and maintenance.

Debitage assemblages demonstrate how diversity in core reduction systems is a byproduct of human decision-making processes. Some extraneous factors, such as raw material availability and quality, condition human decision-making with regard to core reduction strategies. Although previous models indicate that the relationship between core reduction techniques and raw material quality and quantity is important, often the relationship between these two variables does not anticipate or fit the diversity that is present in the archaeological assemblage (Andrefsky 1994; Brantingham et al. 2000). This suggests that other variables are influencing the system. One additional variable that can help to explain these situations is the ratio of producers to consumers in the given society.

Drawing upon optimality theory (Foley 1985), we develop a predictive model of core reduction systems that focuses on three aspects influencing the diversity represented in core reduction techniques: raw material availability, raw material quality, and the ratio of producers to consumers. After presenting the model, we turn to several case studies from different archaeological contexts. The case studies demonstrate the continuous relationship between the three variables of interest. This approach departs from previous analyses that use a discontinuous approach or hold several variables constant. This allows us to capture greater subtleties than would have been acknowledged through applying discontinuous or static approaches. The utility of the model is two-fold: (1) it explains the variance in lithic diversity measures not captured in previous analyses and (2) it provides a method for estimating the producer:consumer ratio in particular archaeological contexts.

OPTIMALITY THEORY AND LITHIC REDUCTION

Natural selection has the consequence of optimizing design features for individual gene propagation (Krebs and Davies 1997). Design features that optimize somatic interests (e.g., access to resources such as food and space) have the potential to be converted into individual reproductive success (Krebs and Davies 1997; Smith and Winterhalder 1992). Where resource access is highly competitive, and variation in strategies solving for a particular goal exists, selection should favor the strategy that can solve the problem with the least cost in relation to the other strategies present (Foley 1985). The rationale is that organisms have limited energetic budgets. Individuals that solve particular adaptive problems efficiently can divert energetic surpluses into reproductive or other somatic interests (Kaplan et al. 2000). This is not to say that humans (or other organisms) are optimally adapted to their environment; rather, natural selection tends toward the optimal solution given the range of available phenotypes present in the environment (Foley 1985; Smith and Winterhalder 1992) and contingent on their evolutionary history (Prentiss and Clarke, this volume).

Humans are a cognitively and behaviorally plastic organism (Flinn 2005), suggesting that selection pressures have favored a human phenotype that can adaptively respond to fluctuating social and ecological pressures (Flinn 1996). Additionally, humans are at times aware of diminishing returns that are the product of certain strategies. This allows individuals to adjust investment accordingly (Kaplan and Lancaster 2000). Thus, humans will generally pursue behavioral strategies (for specific goals) that tend to optimize opportunity costs within specific socioecological settings (Smith 2000).

The degree to which optimization is likely to occur is dependent upon the selection pressures surrounding a particular resource (Foley 1985). For resources characterized as having a large impact on fitness (i.e., resources associated with strong selection pressures), individuals can achieve greater fitness returns by selecting strategies that focus attention on the attainment of that resource (Hames 1992; Winterhalder 1983). As a result, optimization of strategies to attain that resource is a likely outcome. Conversely, when a resource has a limited effect on fitness (i.e., resources associated with low selection pressures), selection could tend toward optimization; however, due to the limited energetic budgets of individuals, selection should favor phenotypes that divert their time and energy to the acquisition of other resources that do have high fitness outcomes (Hames 1992; Winterhalder 1983). As a consequence, satisfactory solutions become viable and diversity in strategy sets becomes tolerated for resources that have limited effect on fitness. Winterhalder (1983) provides a graphical model that demonstrates the conditions favoring decisions to invest an additional unit of time and energy into a focal activity (conditions of limited energy) or to divert these scarce resources into other activities (conditions of limited time).

For human populations that rely on lithic resources for access to food or other somatic interests, the nature and access of lithic resources impacts survivorship. Lithic resources approximate a zero-sum game (when one individual accesses the lithic resource, it represents a loss for other individuals in the population). When the lithic resource is proportionally present at high density compared to a hypothetical population, the depletion of the lithic resource may seem inconsequential to individuals within the populace. Thus access to the lithic resource can be conceptualized as having low fitness consequences, as there is little competition. Alternatively, when a lithic resource exists at proportionally low density in comparison to a hypothetical population, its depletion is consequential. Therefore, it can be characterized as having high fitness consequences, as it is likely to be under intense competition.

Optimality reasoning would lead one to conclude that when use of a lithic resource is highly competitive, strategies for converting the lithic resource into a usable end product will be constrained, with the likely solution (or solutions) being the most economical given the range of possible solutions in the environment. A possible outcome is that only a few individuals might specialize in production from the resource, while other individuals consume the few types that are created. If a resource is quickly being depleted, individuals may better redirect their time and energy into other goals or somatic interests. The rationale is that not everyone can effectively engage in an economic enterprise where there are constraints on the resource.

Alternatively, for a lithic resource under low selection pressure, optimality reasoning indicates that strategies for converting the lithic resource into a usable end product will diversify. The rationale is that individuals can maximize opportunity costs by not investing heavily in the manufacture of the resource, but investing in some other arena where high selection pressures exist. Thus, satisfactory solutions are likely to emerge with the manufacture of lithic products. Because the cost of accessing and manufacturing the lithic resource is low, many individuals can access and manufacture its products with few negative repercussions. As a result, a greater proportion of individuals may act as both producers and consumers of the end products.

IMPORTANT PARAMETERS IN CORE REDUCTION DIVERSITY

Arguably, diversity is largely dependent on human decisions in relation to availability, quality, and the ratio of producers to consumers. We now provide our understanding of how this system operates and define the variables presented in our model.

Modeling Diversity and Raw Material Availability

A number of studies argue that there is a link between raw material availability and the constraints on technological design and conformity (Beck et al. 2002; Kuhn 1996). Raw material availability can be modeled as the kcal/hr expended to procure and transport the resource. This would equal the distance one has to travel to the source and the size of the package (Beck et al. 2002).

The simplest function between diversity and availability is a linear relationship, where diversity is zero when availability is zero. In this situation, when availability increases, diversity also increases at a constant rate. A slightly more realistic function shows diversity increasing as the square root of availability (a). In other words, the function shows a curve where diversity increases drastically with changes in low availability. The slope is less extreme as availability approaches the maximum, but is still increasing (Figure 14.1):

$$d(a) \propto \sqrt{a}. \qquad (14.1)$$

Modeling Diversity and Raw Material Quality

Researchers (Andrefsky 1994; Brantingham et al. 2000; Kuhn 1996) have argued that raw material quality affects the degree of diversity in reduction sequences and raw material breakage patterns (Amick and Mauldin 1997). Raw material quality is quantifiable along several dimensions: (1) percent crystallinity, (2) average crystal size, (3) range in crystal size, and (4) abundance of impurities (Brantingham et al. 2000: 257). All four aspects influence fracture mechanics. As noted by Brantingham et al. (2000: 257), "Regardless of quantity, poor quality rocks usually lead to informal technologies." This, however, is not always the case, and systematic reduction sequences have been found in association with poor-quality raw materials (Brantingham et al. 2000).

Raw material quality can also be shown on a continuum. The lowest-quality material would hypothetically be the lowest quality that could still be manipulated by a flintknapper. The highest quality would

LITHIC CORE REDUCTION TECHNIQUES

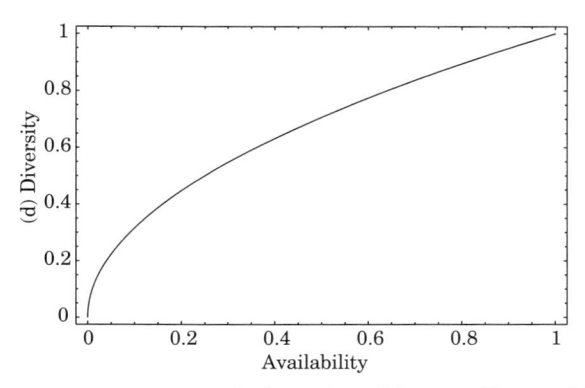

FIGURE 14.1. Functional relationship of diversity (d) to availability (a).

be comparable to a raw material with very low percent crystallinity, on average small crystal size, a small range in crystal size, and low abundance or zero impurities.

We hypothesize that the relationship between diversity and quality is more complex than a simple linear function. Although more data are needed to specifically model this relationship, especially given that it is highly contingent on specific sites and raw materials, the function we used is presented in Figure 14.2 and equation (14.2). With this equation, we propose that diversity scales as an exponentially decreasing function of quality. From this perspective, diversity is highest (or unity) at lowest quality ($q = 0$, the lowest-quality material that can actually still be knapped), and diversity decreases as q increases to the maximum ($q = 1$, the highest-quality material). It is further hypothesized that at low qualities, diversity falls rapidly as q increases, but at higher qualities (smaller grain size, smaller density of inclusions, etc.), diversity does not change nearly as rapidly. The simplest function that meets these criteria is a decaying exponential, where the parameter α controls the rate of the falloff and e is equal to the base of the natural logarithms ($e \approx 2.718$):

$$d(q) \propto e^{-\alpha q} \qquad (14.2)$$

In our model we utilize $\alpha = 3$ as an arbitrary starting point. With further detailed analysis of raw material quality from a given archaeological context, an explicit estimate of α could be obtained.

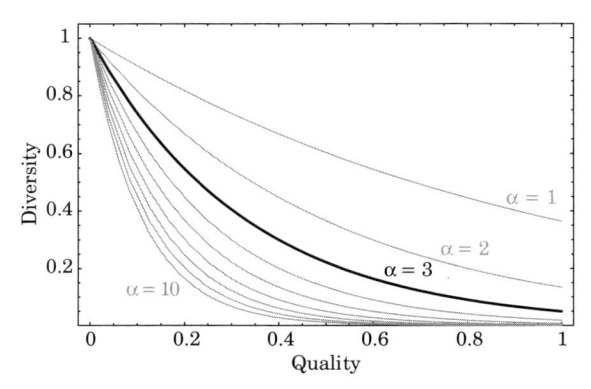

FIGURE 14.2. Functional relationship of diversity (d) to quality (q).

We chose this value for α because it provides an expectation that diversity will increase substantially with increases in poor-quality material but will also have a slope that is less steep with higher-quality material. We also assume that this curve will never reach zero diversity, because the model is built for a reductive technology (core reduction), and that human interaction with reductive technologies will always produce some degree of diversity.

Ratio of Producers to Consumers

The ratio of producers to consumers is a remarkably complicated variable to explain in mathematical terms. It is not clear how the relationship between diversity in core reduction systems and the producer:consumer ratio would actually pattern under specific conditions. Adopting a conservative approach, we have chosen the simplest linear model (Figure 14.3). We define μ to be the ratio of producers to consumers, $\mu = P/C$, where diversity increases at a constant rate as the ratio of producers to consumers increases. We recognize that this is largely based on parameters guiding knowledge transmission in different contexts. However, we believe that this allows a starting point that we and others can test to model human behavior and the diversity of core reduction techniques:

$$d(\mu) \propto P/C. \qquad (14.3)$$

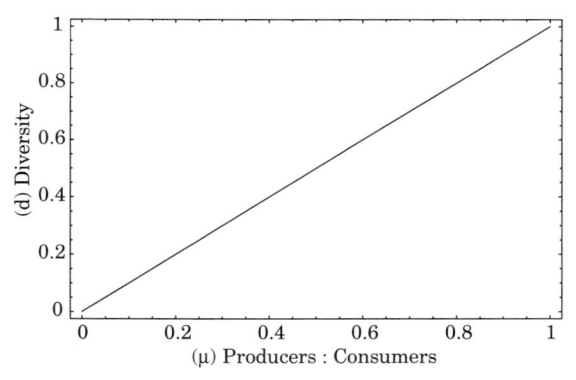

FIGURE 14.3. Functional relationship of diversity (d) to the ratio of producers to consumers (μ).

A MODEL OF CORE REDUCTION DIVERSITY (CRD)

The CRD model is based on the three parameters discussed above. In the following equation,

$$d(a, q, \mu) \propto \mu \sqrt{a} e^{-\alpha q}, \qquad (14.4)$$

diversity is proportional to the ratio of producers to consumers (μ), the square root of availability (a), and the base of the natural logarithms (e) to the negative power of α times quality (q). The equation is presented in Figure 14.4. In this plot, quality changes in increments of .1 in each graphic from 0 (the lowest-quality raw material) to 1 (the highest-quality material).

This model provides a technique that can estimate the ratio of producers to consumers (μ). Therefore, we can solve for μ by inverting the last expression (eq. (14.4)) and writing it as

$$\mu(a, v, q) \propto \frac{d}{\sqrt{a}} e^{\alpha q}. \qquad (14.5)$$

This equation is plotted in Figure 14.5, where availability changes in each plot by increments of .1 from very costly to attain ($a = 0.1$) to readily available ($a = 1.0$). As seen in Figures 14.4 and 14.5, case examples can be explicitly plotted on the graphs based on the quantifiable variables: raw material quality, raw material availability, and diversity in core reduction techniques. If the relationships between the variables are an accurate estimate of data sets, then one should be

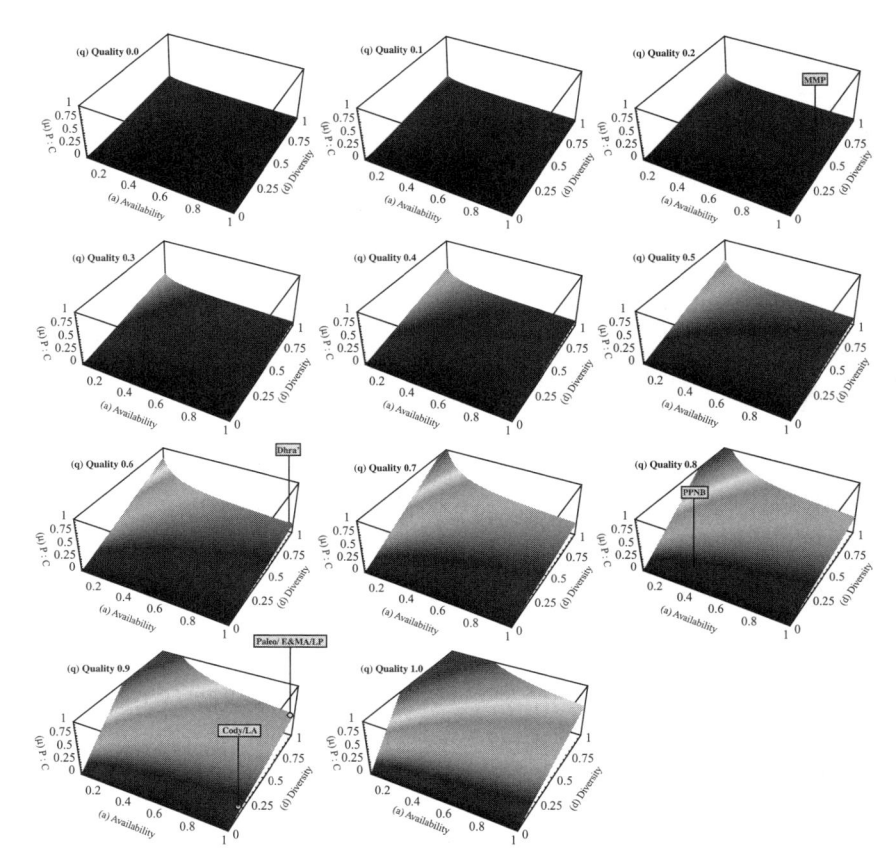

FIGURE 14.4. Plot of equation (14.4), where quality is decreasing by increments of .1 in each graph. Case examples discussed in text are labeled as (1) MMP = Mongolian Middle Paleolithic (Brantingham et al. 2000), (2) Dhra' = Dhra' Early Neolithic (Goodale et al. 2002), 3) PPNB = Middle Pre-Pottery Neolithic (Wilke and Quintero 1994), and (4) Paleo/E&M A/ LP = Paleoindian, Early and Middle Archaic, Late Prehistoric (Root 1997).

able to approximate the ratio of producers to consumers in a given community. We have plotted several cases in Figures 14.4 and Figures 14.5 where we would expect them to be a best fit in the model.

CASE EXAMPLES

To evaluate the potential utility of this model, we now explore several case studies from different archaeological contexts around the world

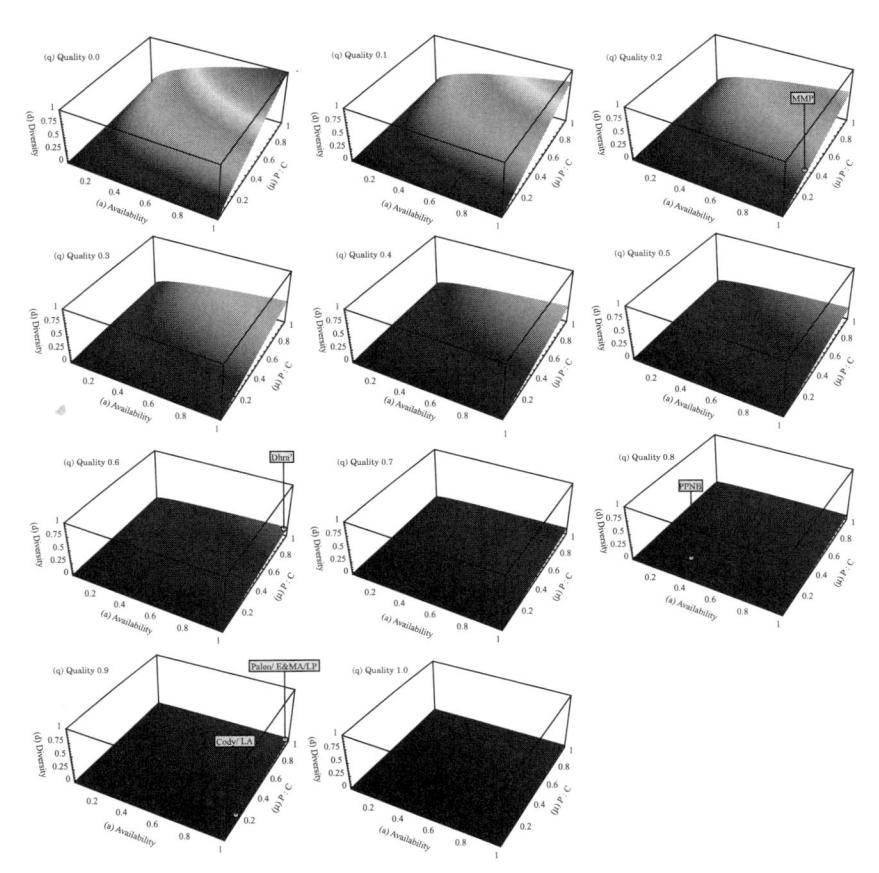

FIGURE 14.5. Plot of equation (14.5), where availability is decreasing by increments of .1 in each graph. Case examples discussed in text are labeled as (1) MMP = Mongolian Middle Paleolithic (Brantingham et al. 2000), (2) Dhra' = Dhra' Early Neolithic (Goodale et al. 2002), (3) PPNB = Middle Pre-Pottery Neolithic (Wilke and Quintero 1994), and (4) Paleo/E&M A/ LP = Paleoindian, Early and Middle Archaic, Late Prehistoric (Root 1997).

that reflect different occupational histories. Each case is plotted in Figures 14.4 and Figures 14.5 for reference. Cases act as working hypotheses about the ratios of producers to consumers reflected by the given assemblages. Each case provides the quality, availability, and diversity reflected in each assemblage, allowing an estimate of the producer:consumer ratio.

Near East Early Neolithic

The early Neolithic Site of Dhra', Jordan, exhibits a very large lithic assemblage composed of over one million pieces of debitage, tools, and cores (Finlayson et al. 2003; Goodale et al. 2002). The lithic assemblage is so large that a specific study of lithic core reduction techniques has been difficult. However, we have observed debitage elements that can provide the basic and most efficient means of how Pre-Pottery Neolithic A (PPNA) knappers produced the final product or tool blanks. We have also observed a number of diagnostic by-products that suggest that the knappers at Dhra' had to overcome a number of production errors and raw material failures.

The knappers at Dhra' primarily exploited one type of raw material (although there is some variability in the assemblage, the use of other nonlocal raw materials equates to less than 1%). The raw material, flint, is found in an outcrop approximately 50 m from the site (Goodale et al. 2002). It can be described as medium-quality, with small to medium crystallinity, but with frequent impurities and random planes subsequent to the formation processes.

In the case of Dhra', the raw material is readily available with low procurement and transport costs and is characteristic of medium quality. As shown in Figure 14.6, the debitage indicates that there were often circumstances where the knappers at Dhra' adjusted for knapping error and raw material failure. This likely facilitated a situation where it was not necessary for any knapper at Dhra' to be highly proficient and also allowed anyone in the community to participate as both producer and consumer. In this example, we see highly available raw material, a medium quality that we would approximate at .6 in our model, and a high degree of diversity in the core reduction system, where knappers often had to negotiate production errors or raw material failure. The hypothesis is that Dhra' is best characterized as reflecting a high ratio of producers to consumers.

Near East Middle Neolithic

During the Middle Pre-Pottery Neolithic, something quite different appears to happen in terms of uniformity in core reduction sequences. We see the advent of a highly systematic type of core

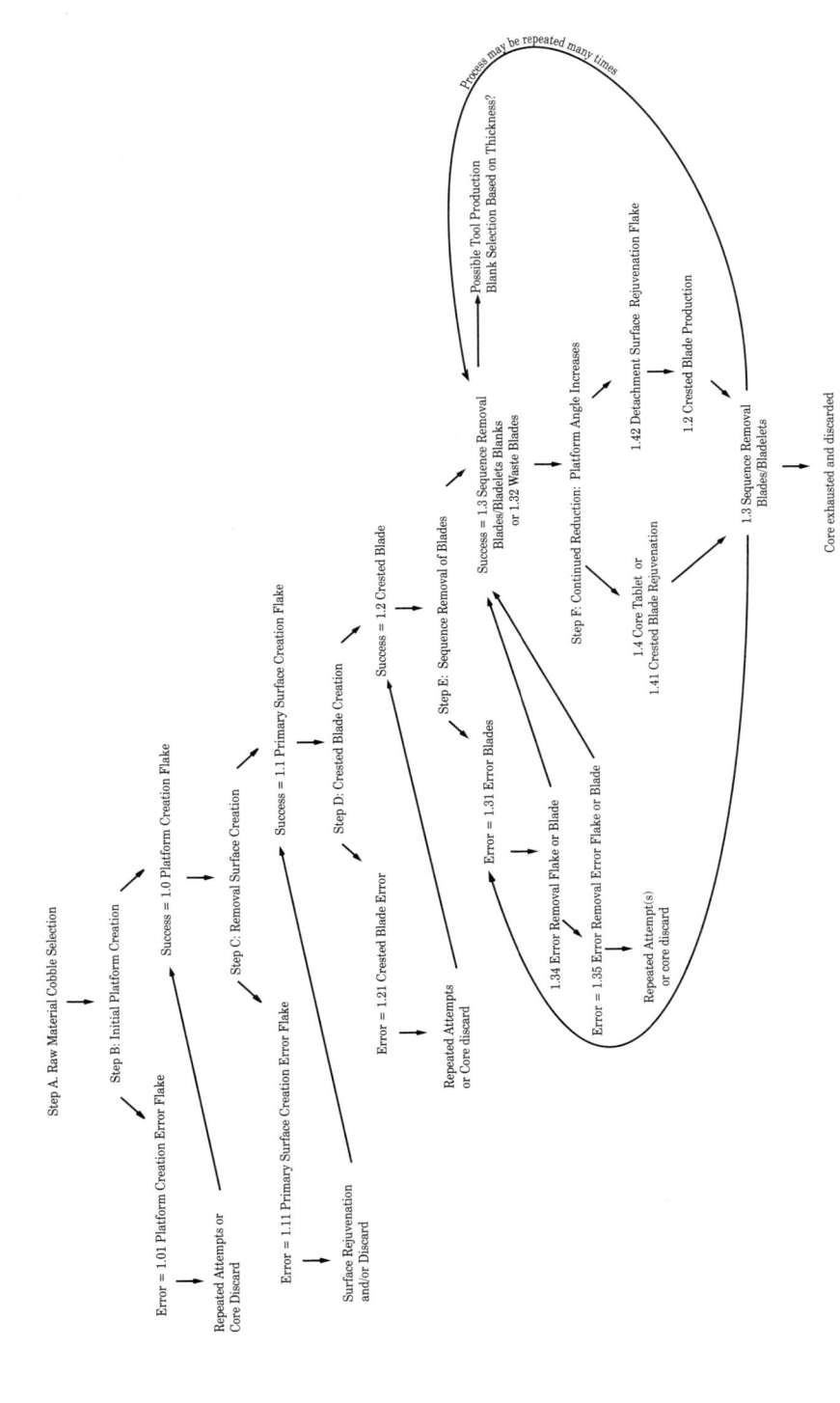

FIGURE 14.6. The highly variable reduction system exhibited in the Dhra' debitage and core assemblages.

reduction referred to as the naviform technique (Quintero and Wilke 1995). This type of core reduction has more specific operational chains (Wilke and Quintero 1994) that were hypothetically selected for under the social requirement for standardized long and straight blade tool blanks (Quintero and Wilke 1995). Naviform core technology utilized specific, high-quality raw material, which was not locally available (Quintero and Wilke 1995: 20). The naviform technique allows a higher degree of control over blade morphology than was previously possible with other core reduction technologies (such as that exhibited in the Dhra' assemblage). In comparison to the early Neolithic knappers at Dhra', who were producing highly variable products, middle Neolithic naviform producers were able to maximize the end product in the form of long and thin blades. Quintero and Wilke (1995) note the important manner in which knappers prepared their naviform cores with a consistent length of 12–15 cm and a width of 1.5–3.5 cm. They go on to suggest (1995: 26) that the socioeconomic conditions that accompanied the development of specialized blade-making flourished with demographic and economic growth. This would also hypothetically correlate with a greater degree of roles in the community, where select individuals were rewarded for flintknapping skills. Our hypothesis is that the process of naviform core reduction is characterized by expensive raw material acquisition, high quality, and a low degree of diversity, emphasizing a low producer:consumer ratio.

Mongolia Middle Paleolithic

Brantingham et al. (2000) provide a very interesting case of core reduction techniques from the Middle Paleolithic of East Asia. The raw material primarily exploited at the site is locally available and is on average of very poor quality. There are a few examples of core reduction that appear highly unsystematic, where the knappers negotiated the failures of the raw material, producing highly diversified core reduction techniques. However, they focus on another example of reduction technique that appears highly systematic and demonstrates that knappers focused on the most efficient chain that the raw material would allow. Brantingham et al. (2000) are unsure why this strategy was favored. Based on our model, we suggest that the highly uniform core reduction technique is representative of a low ratio of producers to consumers and that select individuals in the community paid the cost to learn how to negotiate the poor-quality material. Our hypothesis for the highly systematic core reduction technique is representative of poor quality and highly available raw material with a low degree of diversity, emphasizing a low ratio of producers to consumers.

North America Paleoindian to Late Prehistoric

Drawing on the Paleoindian to Late Prehistoric occupations of the Benz site in North Dakota, Root (1997) makes a compelling argument linking the ratio of producers to consumers to the efficiency of biface production. The site contained several "features" composed of clusters of lithic debitage that "likely mark the places where individual knappers made tools (Root 1997: 35)." The knappers at the Benz site exploited locally available and abundant high quality Knife River Flint. In his analysis, Root (1997: Table 7) provides estimates for the number of tools made in each feature by dated occupation. He concludes that the periods of highest efficiency are the Cody Complex and Late Archaic occupations. In opposition, the Paleoindian, Early and Middle Archaic, and Late Prehistoric occupations have the lowest scores for efficiency in biface reduction. This is an interesting pattern and we suggest that it may be linked with fluctuating social systems and changes in the ratio of producers to consumers through time. Root (1997: 42) also suggests that in the periods of highest efficiency, knappers were producing bifaces for exchange in the area, which was likely negotiated by shifts in social organization enabling an expansion of the number of community roles. In essence, Root's hypothesis (1997: 42) is similar to ours by suggesting that participation in production and consumption was no longer equal.

DISCUSSION

The case studies presented highlight the flexibility of human behavior negotiating the constraints of resources (or lack thereof) and the ability of humans to produce a range of diversity in reduction techniques. This range of diversity may be predicated on a number of factors, including how humans interact with their social and natural environments. Natural selection has favored a human phenotype that is behaviorally and cognitively flexible (Flinn 1996). Humans are aware of strategies that produce diminishing marginal returns on investment (Kaplan and Lancaster 2000). As a result of these propensities, humans can alternate strategies toward specific goals as social and environmental circumstances fluctuate (Kaplan and Lancaster 2000). The cost–benefit structure of engaging in any economic activity is shaped by the level of skill required for involvement and the competitiveness of the particular context (Kaplan and Lancaster 2000). This structure helps negotiate whether an individual engages in the production of a lithic core reduction technology or spends time and energy in other arenas. Linked to this is the availability of resources in the environment, the quality of the resources available, and the number of other individuals already engaged in the enterprise. The balancing of these three conditions affects the diversity (or lack thereof) in production techniques. If competition is high, costs will be high to engage in the economic activity, which leads to fewer individuals engaged in production. As a result, the diversity of lithic reduction techniques will be constrained. However, if competition is low, costs in engaging in the economic activity will be low, leading to more individuals engaging in production. As a result, diversity in reduction techniques should expand. Since researchers can estimate lithic availability, indices of lithic quality, and indices of diversity in reductive techniques, it is possible to extrapolate the producer:consumer ratio (at least in terms of our general model).

When lithic quality is low, availability of resources is low, and diversity in technique is low, one can expect a low ratio of producers to consumers. This is due to the fact that poor-quality resources require a greater degree of skill to manipulate in an efficient manner. To gain such a high degree of skill, one must go through a learning process. The time and energy required to learn such a technique would have been high. In an environment such as this, a tradeoff is present: (1) does one invest the time and energy in learning the lithic reduction craft; or (2) does one allocate energy into other arenas where time and energy produce greater returns from investment. In an environment of high stress, the strategy of learning lithic reductive techniques may be frequency-dependent. In other words, as the number of individuals learning and investing in lithic reduction techniques increases and the quantity of the resource decreases, the value of the time and energy expended on the craft decreases. Human behavior should be sensitive to this relationship, and people will hypothetically tend to allocate their time and energy into other arenas where they may receive a greater return on investment. Consequently, few producers will be favored in proportion to the number of consumers.

A high ratio of producers to consumers is consistent with conditions where lithic quality is high, availability is high, and diversity in reduction technique is high. This is due to the fact that the resource is relatively inexpensive (in terms of energy expended for access and in terms of investment required for learning how to manufacture the resource). With low costs, there is less incentive to invest heavily into learning skills associated with the lithic technology. As a result, more individuals are likely to be producers. Included in this expansion of the individuals in the production phase may be a younger age bracket, which also shapes the level of diversity witnessed in reduction techniques. As argued by Bock (2005), younger individuals have less motor control (which is a function of time involved in the production of the craft), resulting in greater degree of variability in production techniques within and between individuals.

CONCLUSIONS

Understanding the social, economic, and technical constraints for different chipped stone reduction pathways helps us examine differences in human behavior. The ability to estimate the producer:consumer ratio contributes toward this goal. It deals with a question that has been associated with studies of craft specialization throughout the study of anthropological archaeology (Costin 1991). The model and mathematical estimate focus on several independent, nonconstant parameters that scale along a continuum rather than holding several of them as static (for example, Beck et al. 2002).

Although we have not directly tested the model, we have presented case studies as hypotheses. By adding a third variable that is articulated with a well-supported principle in evolutionary analyses (optimality), it is possible to explain some of the diversity in the archaeological record. As an example, it explains the anomalous occurrence of low diversity despite low quality and high availability in the Middle Paleolithic of Mongolia. In future studies, if we can determine the relationship between population size and the producer:consumer ratio, we may be able to directly test this relationship.

REFERENCES CITED

Amick, Daniel S., and Raymond P. Mauldin. 1997. Effects of Raw Material on Flake Breakage Patterns. *Lithic Technology* 22:18–32.

Andrefsky, William A., Jr. 1994. Raw-Material Availability and the Organization of Technology. *American Antiquity* 59:21–34.

Beck, Charlotte, Amanda K. Taylor, George T. Jones, Cynthia M. Fadem, Caitlyn R. Cook, and Sara A. Millward. 2002. Rocks Are Heavy: Transport Costs and Paleoarchaic Quarry Behavior in the Great Basin. *Journal of Anthropological Archaeology* 21:481–507.

Bleed, Peter. 1996. Risk and Cost in Japanese Microblade Technology. *Lithic Technology* 21:95–107.

——— 2001. Trees or Chains, Links or Branches: Conceptual Alternatives for Consideration of Stone Tool Production and Other Sequential Activities. *Journal of Archaeological Method and Theory* 8(1):101–27.

Bock, John. 2005. What Makes a Competent Adult Forager. In *Hunter-Gatherer Childhoods*, edited by B. Hewlett and M. Lamb, pp. 109–28. Aldine Transaction, Somerset, NJ.

Brantingham, Jeffrey P., John W. Olsen, Jason A. Rech, and Andrei I. Krivoshapkin. 2000. Raw Material Quality and Prepared Core Technologies in Northeast Asia. *Journal of Archaeological Science*, 27:255–71.

Costin, Cathy Lynne. 1991. Craft Specialization: Issues in Defining, Documenting, and Explaining the Organization of Production. In *Archaeological Method and Theory V.3*, edited by Michael B. Shiffer, pp. 1–56. University of Arizona Press, Tucson.

Finlayson, Bill, Ian Kuijt, Trina Arpin, Meredith Chesson, Samantha Dennis, Nathan Goodale, Seji Kadowaki, Lisa Maher, Sam Smith, Mark Schurr, and Jode McKay. 2003. Dhra', Excavation Project, 2002 Interim Report. *Levant* 35:1–38.

Flinn, M. V. 1996. Culture and the Evolution of Social Learning. *Evolution and Human Behavior* 18:23–67.

——— 2005. Culture and Developmental Plasticity: Evolution of the Social Brian. In *Evolutionary Perspectives on Human Development*, edited by Robert L. Burgess and Kevin MacDonald, pp. 73–98. Sage Publications, Thousand Oaks, CA.

Foley, R. 1985. Optimality Theory in Anthropology. *Man* 20(2):222–42.

Goodale, Nathan B., Ian Kuijt, and Bill Finlayson. 2002. Results on the 2001 Excavation at Dhra', Jordan: Chipped Stone Technology, Typology, and Intra-assemblage Variability. *Paléorient* 28(1):125–40.

Hames, Raymond. 1992. Time Allocation. In *Evolutionary Ecology and Human Behavior*, edited by E. A. Smith and B. Winterhalder, pp. 203–35. Aldine de Gruyter, New York.

Kaplan, H., K. Hill, J. Lancaster, and A. M. Hurtado. 2000. A Theory of Human Life History Evolution: Diet, Intelligence, and Longevity. *Evolutionary Anthropology* 9(5):1–30.

Kaplan, H. S., and J. B. Lancaster. 2000. The Evolutionary Economics and Psychology of the Demographic Transition to Low Fertility. In *Adaptation and Human Behavior: An Anthropological Perspective*, edited by Lee Cronk, Napoleon Chagnon, and William Irons, pp. 283–322. Aldine de Gruyter, New York.

Krebs, J. R., and N. B. Davies, eds. 1997. *Behavioral Ecology: An Evolutionary Approach*. 4th ed. Blackwell Publishing, Oxford.

Kuhn, Steven L. 1996. Middle Paleolithic Responses to Raw Material Quality: Two Italian Cases. *Quaternaria Nova*, 6:261–77.

Quintero, Leslie, and Philip J. Wilke. 1995. Evolution and Economic Significance of Naviform Core-and-Blade Technology in the Southern Levant. *Paléorient* 21(1):17–33.

Root, Matthew J. 1997. Production for Exchange at the Knife River Flint Quaries, North Dakota. *Lithic Technology* 22:33–50.

Shott, Michael J. 1996. Stage versus Continuum in the Debris Assemblage from Production of a Fluted Biface. *Lithic Technology* 21:6–22.

Smith, Eric Alden. 2000. Three Styles in the Evolutionary Analysis of Human Behavior. In *Adaptation and Human Behavior*, edited by Lee Cronk, Napoleon Chagnon, and William Irons, pp. 27–46. Aldine de Gruyter, New York.

Smith, Eric Alden, and Bruce Winterhalder. 1992. Natural Selection and Decision Making. In *Evolutionary Ecology and Human Behavior: An Anthropological Perspective*, edited by Eric Alden Smith and Bruce Winterhalder, pp. 25–60. Aldine de Gruyter, New York.

Wilke, Philip J., and Leslie Quintero. 1994. Naviform Core-and-Blade Technology: Assemblage Character as Determined by Replicative Experiments. In *Neolithic Chipped Stone Industries of the Fertile Crescent: Proceedings of the First Workshop on PPN Chipped Lithic Industries, Berlin 1993*, edited by H. G. Gebel and S. K. Kozlowski, pp. 33–60. Free University of Berlin, Berlin.

Winterhalder, Bruce. 1983. Opportunity Cost Foraging Models for Stationary and Model Predators. *American Naturalist* 122:73–84.

INDEX

Ahler, S., 5, 10, 181, 196
Alaska, 259, 262, 265
allometric, 29, 30, 32, 36
allometry, 26, 138
Amick, D., 5, 10, 139
Ancestral Pueblo, 176, 178, 180, 185, 188
Andrefsky, W., 4, 14, 27, 33, 50, 86, 98, 138, 152, 188, 196, 206, 219, 230, 237, 258, 290, 319
Apache, 181
Archaic, 33, 175, 183, 222, 280
argillite, 141
arrises, 30
arrow, 12, 32, 175, 182, 188, 209
artifact
  curation, 25
  density, 36
  discard, 14
  form, 152, 170
  function, 10, 151, 195
  life cycle, 67
  life history, ix, 10, 150
  production, 10
  recycling, 13, 180
  retouch, 12
  transformation, ix, 261
Attenbrow, V., 6, 27, 119, 294, 311
Australia, 23, 30, 35, 107, 287, 312

Bamforth, D., 7, 9, 13, 24, 76, 217, 224, 250, 258
Bandelier, 180
Barton, M., 86
beads, 153, 162
Bergerac, 142
biface, 10, 27, 87, 89, 94, 139, 175, 208, 225, 261, 331
Binford, L., 7, 24, 76, 144, 217, 238, 258, 303
Birch Creek, 25, 196
Black Perigord, 107
Blades, B., 86, 138, 151, 206
blank
  size, 4, 50, 51
Bleed, P., 6, 24, 263, 318
Bobtail Wolf site, 219
Bordes, F., 24, 107, 117
Bradbury, A., 8, 26, 139, 225, 237
Brantingham, P., 6, 305, 322
Broken Mammoth, 262

Carr, P., 5, 139, 144, 237
Chalk Basin, 87
Charente, 142
Chestnut Ridge, 143
cladistics, 259
Clarke, D., 257

Clarkson, C., 6, 23, 55, 96, 106, 162, 286, 290, 308
Classic period, 176, 187
Coalition period, 177
Cobden, 242
cognitive, 107, 236, 287, 332
collector, 8, 144, 270
Columbia Plateau, 217
Combe Grenal, 106
Cooper, R., 233
core
   platform rotation, 35, 288
   reduction, 287, 289, 302, 317, 324
   size, 36
   technology, 330
cortex, 89, 123, 288
CRD, 325
curation
   artifact, 12
   assessing, 7, 94
   concept, 7, 24, 151, 238
   definition, 86
   index, 153, 165
   process, 8, 39

dart, 11, 32, 175, 209
Darwinian, 257
Davis, Z., 265
Dead Sea, 153
Denali, 261
Deneba's Canvas 8, 96
Dhra', 153, 328
Dibble, H., 27, 98, 107, 111, 122, 163, 239
distal end, 80, 110, 164
Diuktai, 261
diversity, 13, 36, 61, 107, 146, 275, 294, 302, 318, 321, 322
dorsal cortex, 89, 290
drilling, 11, 155

EKCI, 31, 153, 166
el-Khiam, 152, 162, 164
Eren, M., 6, 49, 170, 206
ERP, 50, 60

evolution, 260
evolutionary
   approaches, 13
   ecology, 236, 312
   framework, 280
   history, 278, 320
   theory, 258
expediency, 236
expedient
   technology, 180
   tools, 8, 145, 176, 188
experimental, 11, 29, 51, 88, 292
exterior platform angle, 240

Finlayson, B., 150, 317
flake
   blank, 4, 35
   tool, 12, 23
flintknapper, 5, 101, 322
Flint Ridge, 227, 242
Folsom, 31, 219
forager, 4, 144, 220, 299, 309
Fort Payne, 242
France, 50, 111
Fumel, 142

geometric index, 292
GIUR, 117, 126, 292
Goodale, N., 153, 280, 328
Great Basin, 33, 198

hafted biface, 172, 184, 199, 225
haft element, 27, 164, 183
hammer
   hard, 242
   soft, 240
   type, 11, 34
Harper, C., 12, 175
Hiscock, P., 6, 27, 68, 106, 113, 238, 288
Holdaway, S., 112, 287
HRI, 206, 225

II (index of invasiveness), 50, 55, 80, 291
impact damage, 199, 204
IR (index of reduction), 54, 60, 68, 76

Jicarilla, 181
Jordan, 153, 328

Kanawha, 33, 220
Kaolin, 242
Keatley Creek, 270
Kelly, R., 7, 24, 196, 237, 279
kiva, 187
Knife River Flint, 219, 331
Kuhn, S., 6, 26, 29, 50, 68, 87, 113, 162, 206, 234, 258
Kuijt, I., 150, 317

La Colombiere, 50
land-use, 4, 196, 229
Late Archaic, 33, 176, 221, 331
Late Woodland, 34, 221
leatherworking, 160
Levallois, 78
Levant, 150

MacDonald, D., 170, 216, 275
Macfarlan, S., 317
mass loss, 12, 50, 244
Mass Predictor Equation, 29, 51
maximum utility, 24, 151
Mellars, P., 107, 131
microblade, 259, 261
microwear, 152
Middle Atlantic, 220
Middle Paleolithic, 330
Mid-Fraser, 35
mobility, 76, 86, 113, 137, 217, 258
Mongolia, 10, 330
Mousterian, 106, 130

natural selection, 260
Navajo, 181
Neanderthal, 107
Near East, 150, 280
Nelson, M., 4, 24, 217
Neolithic, 61, 150, 328
New Mexico, 176
North America, 175, 219, 234, 259

objective piece, 5, 23, 89
obsidian, 25, 196, 243
Odell, G., 7, 76, 196, 234
Old Cordilleran, 261
Onion Portage, 262
optimality, 13, 319
Oregon, 25, 87, 196
organization
    technological, 24, 146, 151, 188, 217
    technology, 14, 234
organizational strategies, 8, 86, 211

Paleolithic, 10, 30, 50, 142, 240, 330
Parry, W., 7, 24, 180, 275
Paulina Lake, 204
Pelcin, A., 5, 29, 206, 239
Pennsylvania, 141, 223
phenotype, 320
platform, 5, 35, 89, 113, 138, 164, 239, 288
point sharpness, 155
Prendergast, M., 49
Prentiss, A., 13, 257
Prentiss, W., 260
Pre-Pottery Neolithic, 152, 328
production
    errors, 328
    event, 93
    life, 94
    phases, 5
    process, 5, 23
    rules, 133
    stage, 6, 87
    techniques, 332
projectile point, 4, 152, 175, 181, 195, 225, 266
provisioning, 11, 25, 112, 195, 287, 306
proximal flake, 89
Pueblo, 176

quarry, 87, 203
Quina, 107, 116
Quinn, 31, 150

raw material
  abundance, 9
  availability, 8, 113, 152, 172, 196, 217, 302, 322
  flaws, 100
  quality, 102, 217, 230, 239, 318, 322
  reduction, 34, 137, 144
  selection, 12, 197, 225, 287, 305
  type, 51
reduction
  concept, 5, 25, 50, 55
  core, 113, 288, 318, 324, 330
  event, 95, 170
  hypothesis, 108, 114
  indices, 29, 50, 51, 83, 121
  intensity, 123, 131, 137, 140, 287, 290, 294, 300
  measure, 120
  process, 23, 133, 241
  sequence, ix, 5, 78, 100, 126, 235, 296, 318, 322
  stage, 76, 113, 139, 167, 288
  tactics, 278
  thesis, 23, 27, 136
refitting, 235
resharpening, 4, 23, 50, 87, 96, 112, 182, 207, 238
retouch
  amount, 26
  degree of, 8, 86, 112, 164, 182, 200, 230, 238, 266, 274
  indices, 12, 51, 152
  intensity, 9, 13, 94, 107, 127, 132, 137, 139, 150, 206, 209, 287, 291, 300
  invasiveness, 82, 96

location, 119, 122
measures, 11
patterns, 30, 50
retouch-tion, 79, 81
ridge count, 96
ridges, 30, 96, 164
Rio Grande, 175
RMS, 260, 278
rockart, 311

scar count, 30, 96, 139
Senonian, 142
sharpness index, 157
Shea, J., 5, 32, 164, 240
Shott, M., 6, 12, 23, 24, 27, 32, 35, 39, 50, 51, 76, 86, 151, 237, 318
Siberia, 261
Simek, J., 6, 234, 258
Skink Rockshelter, 33, 216
Ste. Genevieve, 242

Torrence, R., 4, 25, 234, 258, 280
transport
  raw materials, 262, 266, 322
  tools, 7, 25, 141, 145, 258
transverse fracture, 161

uniface, 80, 263
Upper Mercer, 33, 223
use-life, 26, 76, 170, 209, 302

Wardaman, 286, 306
West Virginia, 25, 143, 216
Wilson, J., 29, 86

XRF, 11, 196